Mathematics and Its Applications

Volume 467

Topics in Number Theory

Topics in Number Theory

In Honor of B. Gordon and S. Chowla

edited by

Scott D. Ahlgren
George E. Andrews
and
Ken Ono

Department of Mathematics,
Pennsylvania State University,
University Park, PA, U.S.A.

KLUWER ACADEMIC PUBLISHERS
DORDRECHT / BOSTON / LONDON

A C.I.P. Catalogue record for this book is available from the Library of Congress.

ISBN 0-7923-5583-0

Published by Kluwer Academic Publishers,
P.O. Box 17, 3300 AA Dordrecht, The Netherlands.

Sold and distributed in North, Central and South America
by Kluwer Academic Publishers,
101 Philip Drive, Norwell, MA 02061, U.S.A.

In all other countries, sold and distributed
by Kluwer Academic Publishers,
P.O. Box 322, 3300 AH Dordrecht, The Netherlands.

Printed on acid-free paper

TABLE OF CONTENTS

Preface

From July 31 through August 3, 1997, the Pennsylvania State University hosted the Topics in Number Theory Conference. The conference was organized by Ken Ono and myself. By writing the preface, I am afforded the opportunity to express my gratitude to Ken for beng the inspiring and driving force behind the whole conference. Without his energy, enthusiasm and skill the entire event would never have occurred.

We are extremely grateful to the sponsors of the conference: The National Science Foundation, The Penn State Conference Center and the Penn State Department of Mathematics.

The object in this conference was to provide a variety of presentations giving a current picture of recent, significant work in number theory. There were eight plenary lectures:

H. Darmon (McGill University),
"Non-vanishing of *L*-functions and their derivatives modulo *p*."

A. Granville (University of Georgia),
"Mean values of multiplicative functions."

C. Pomerance (University of Georgia),
"Recent results in primality testing."

C. Skinner (Princeton University),
"Deformations of Galois representations."

R. Stanley (Massachusetts Institute of Technology),
"Some interesting hyperplane arrangements."

F. Rodriguez Villegas (Princeton University),
"Modular Mahler measures."

T. Wooley (University of Michigan),
"Diophantine problems in many variables: The role of additive number theory."

D. Zeilberger (Temple University),
"Reverse engineering in combinatorics and number theory."

The papers in this volume provide an accurate picture of many of the topics presented at the conference including contributions from four of the plenary lectures.

On Saturday evening, August 2, the conference banquet was held at the Nittany Lion Inn. The evening had a two-fold purpose. The first was to honor Basil Gordon at 65. Basil has been the Ph.D. advisor to many of the most active young number theorists of today. In addition, his work has been a powerful inspiration to many, many others, including me.

Besides honoring Basil Gordon, we also paid tribute to the late Sarvadaman Chowla. Chowla was a professor at Penn State from 1963 to 1976. He was a towering presence in number theory, throughout his tenure at Penn State. He brought a continuous stream of famous mathematical visitors. In addition he inspired and directed many graduate students and young assistant professors (me included) with his exciting and eccentric "Chowla Seminar." It was only fitting that evening that Penn State honor his memory by announcing the creation of the S. Chowla Assistant Professorship and by also recognizing Kevin James (one of the contributors to this volume) as the second S. Chowla Assistant Professor. The plaque honoring S. Chowla Assistant Professorship and a framed photo of Chowla were unveiled in the presence of his daughter Paromita Chowla, who recently retired from Penn State.

It is then with great pleasure that I commend to you these proceedings of the Topics in Number Theory Conference.

George E. Andrews
University Park, Pennsylvania
July 25, 1998.

MOTIVATING THE MULTIPLICATIVE SPECTRUM

ANDREW GRANVILLE AND K. SOUNDARARAJAN

Dedicated to the memory of S. D. Chowla

ABSTRACT. In this article, we describe and motivate some of the results and notions from our ongoing project [2]. The results stated here are substantially new (unless otherwise attributed) and detailed proofs will appear in [2].

1. DEFINITIONS AND PROPERTIES OF THE SPECTRUM

Let S be a subset of the unit disc U. By $\mathcal{F}(S)$ we denote the class of completely multiplicative functions f such that $f(p) \in S$ for all primes p. Our main concern is: What numbers arise as mean-values of functions in $\mathcal{F}(S)$?

Precisely, we define

$$\Gamma_N(S) = \left\{ \frac{1}{N} \sum_{n \le N} f(n) : f \in \mathcal{F}(S) \right\} \quad \text{and} \quad \Gamma(S) = \lim_{N \to \infty} \Gamma_N(S).$$

Here and henceforth, if we have a sequence of subsets J_N of the unit disc $U := \{|z| \le 1\}$, then by writing $\lim_{N\to\infty} J_N = J$ we mean that $z \in J$ if and only if there is a sequence of points $z_N \in J_N$ with $z_N \to z$ as $N \to \infty$. We call $\Gamma(S)$ the *spectrum* of the set S and our main object is to understand the spectrum. Although we can determine the spectrum explicitly only in a few cases (see Theorem 1 below for the most interesting of these cases), we are able to qualitatively describe it, and obtain a lot of its geometric structure. For example, we can always determine the boundary points of the spectrum (that is the elements of $\Gamma(S) \cap \mathbb{T}$ where $\mathbb{T}$ is the unit circle). Another property is that the spectrum is always connected. Qualitatively, the spectrum may be described in terms of Euler products and solutions to certain integral equations.

We begin with a few immediate consequences of our definition:

- $\Gamma(S)$ is a closed subset of the unit disc U.
- $\Gamma(S) = \Gamma(\overline{S})$ (where $\overline{S}$ denotes the closure of S). Henceforth, we shall assume that S is always closed.
- If $S_1 \subset S_2$ then $\Gamma(S_1) \subset \Gamma(S_2)$.
- $\Gamma(\{1\}) = \{1\}$.

One of the main results of [2], which formed the initial motivation to study the questions discussed herein, is a precise description of the spectrum of $[-1, 1]$.

The first author is a Presidential Faculty Fellow. He is also supported, in part, by the National Science Foundation. The second author is supported by an Alfred P. Sloan dissertation fellowship

S.D. Ahlgren et al. (eds.), *Topics in Number Theory*, 1–15.

Theorem 1. *The spectrum of the interval* $[-1,1]$ *is the interval* $\Gamma([-1,1]) = [\delta_1, 1]$ *where* $\delta_1 = 2\delta_0 - 1 = -0.656999\ldots$, *and*

$$\delta_0 = 1 - \frac{\pi^2}{6} - \log(1+\sqrt{e})\log\frac{e}{1+\sqrt{e}} + 2\sum_{n=1}^{\infty}\frac{1}{n^2}\frac{1}{(1+\sqrt{e})^n}$$
$$= 0.17150049\ldots.$$

Theorem 1 tells us that for any real-valued completely multiplicative function f with $|f(n)| \le 1$,

$$\sum_{n\le x} f(n) \ge (\delta_1 + o(1))x. \tag{1}$$

In 1994, Roger Heath-Brown conjectured that there is some constant $c > -1$ such that $\sum_{n\le x} f(n) \ge (c+o(1))x$. Richard Hall [5] proved this conjecture, and, in turn, conjectured (as did Hugh L. Montgomery independently) the stronger estimate (1). Both Hall and Montgomery noticed that the estimate (1) is best possible by taking

$$f(q) = \begin{cases} 1 & \text{for primes } q \le x^{1/(1+\sqrt{e})} \\ -1 & \text{for primes } x^{1/(1+\sqrt{e})} \le q \le x. \end{cases} \tag{2}$$

In this example, the reader can verify (or see [5]) that equality holds in (1). Our proof shows that this is essentially the only case when equality holds in (1). The proof of Theorem 1 is too complicated to be given here; however, in §3, we shall sketch its salient features and prove a weaker estimate with $-(2-\frac{2}{\sqrt{e}}) = -0.78693\ldots$ replacing δ_1 in (1).

By applying our Theorem to the completely multiplicative function $f(n) = (\frac{n}{p})$, for some prime p, we deduce that $\ge (\delta_0 + o(1))x$ integers below x are quadratic residues $\pmod p$. In fact, the constant δ_0 here is best possible. To see this, we choose p such that $(\frac{q}{p})$ is given as in the Hall-Montgomery example (2); that infinitely many such primes exist follows from quadratic reciprocity and Dirichlet's theorem on primes in arithmetic progressions.

For a given $f \in \mathcal{F}(S)$, the mean-value of f (that is, $\lim_{x\to\infty} x^{-1}\sum_{n\le x} f(n)$), if it exists, is obviously an element of the spectrum $\Gamma(S)$. We begin by trying to understand the subset of the spectrum consisting of such mean-values.

Suppose that $1 \in S$ and consider $f \in \mathcal{F}(S)$ satisfying $f(p) = 1$ for all large primes p. Writing $f(n) = \sum_{d|n} g(d)$ for some multiplicative function $g(d)$, an easy argument shows that

$$\lim_{x\to\infty}\frac{1}{x}\sum_{n\le x} f(n) = \sum_{d=1}^{\infty}\frac{g(d)}{d}$$
$$= \prod_p \left(1 + \frac{f(p)}{p} + \frac{f(p^2)}{p^2} + \ldots\right)\left(1 - \frac{1}{p}\right)$$
$$=: \Theta(f,\infty). \tag{3}$$

Thus the element $\Theta(f,\infty) \in \Gamma(S)$, and if we define $\Gamma_\Theta(S)$ to be the closure of the set of values $\Theta(f,\infty)$ we have $\Gamma_\Theta(S) \subset \Gamma(S)$. Here, we assumed that $1 \in S$ and if $1 \notin S$ we define $\Gamma_\Theta(S) = \{0\}$. We call $\Gamma_\Theta(S)$ the Euler product spectrum of S.

Proving an old conjecture of Erdős and Wintner, Wirsing [10] showed that every real multiplicative function with $|f(n)| \le 1$ has a mean-value. In fact, he proved that (3) always holds for such functions. Thus, when $S \subset [-1,1]$ Wirsing's Theorem gives that $\Gamma_\Theta(S)$ is precisely the set of mean-values of elements in $\mathcal{F}(S)$. The critical point in Wirsing's Theorem is to show that if f is real valued and $\sum_p (1-f(p))/p$ diverges, then $x^{-1}\sum_{n\le x} f(n) \to 0$.

The situation is more delicate for complex valued multiplicative functions. For example, the function $f(n) = n^{i\alpha}$ (α a non-zero real) does not have a mean-value; indeed $\sum_{n\le x} f(n) \sim x^{1+i\alpha}/(1+i\alpha)$. Again, note that here $\sum_p (1-\text{Re } p^{i\alpha})/p$ diverges but $x^{-1}\sum_{n\le x} n^{i\alpha}$ does not tend to 0. Halász [3] realised that the problem with this example is that the set $\{f(p)\}$ is everywhere dense on $\mathbb{T}$. He proved that if $\sum_p (1-\text{Re } f(p)p^{-i\alpha})/p$ diverges (which obviously does not hold for the troublesome example $n^{i\alpha}$) for all real α then $x^{-1}\sum_{n\le x} f(n) \to 0$; and he quantified how fast this tends to 0.

Over the years Halász' Theorem has been considerably refined, and recently Hall [4] found the following useful formulation: Let D be a convex subset of U containing 0. If $f \in \mathcal{F}(D)$ then

$$\sum_{n\le x} f(n) \ll x\exp\left(-\eta(D)\sum_{p\le x}\frac{1-\text{Re } f(p)}{p}\right), \tag{4}$$

where $\eta(D)$ is a constant defined in terms of the geometry of D, and $\eta(D) > 0$ when the perimeter length of the closure of D is $< 2\pi$. Using (4), it is easy to check that if $1 \notin S$ then $\Gamma(S) = \{0\}$ (recall that S is assumed to be closed). Thus we may assume henceforth that $1 \in S$.

But all this doesn't answer when (3) holds. We now give a criterion for the set S, such that for all $f \in \mathcal{F}(S)$, (3) is true, thus generalizing Wirsing's Theorem. To formulate this fluidly, and for subsequent results, we introduce the notion of the *angle* of a set. For any $V \subseteq U$, define

$$\text{Ang}(V) := \sup_{\substack{v\in V\\ v\ne 1}} |\arg(1-v)|. \tag{5}$$

Note that each such $1-v$ has positive real part, so $0 \le \text{Ang}(V) \le \pi/2$. We adopt the convention that $\text{Ang}(\{1\}) = \text{Ang}(\emptyset) = 0$. Sometimes we will speak of the angle of a point $z \in U$ ($z \ne 1$); by this we mean $\text{Ang}(z) = |\arg(1-z)|$.

Theorem 2. *Suppose $S \subset U$ and $\text{Ang}(S) < \pi/2$. Then (3) holds for every $f \in \mathcal{F}(S)$; that is, every $f \in \mathcal{F}(S)$ has a mean-value. Thus,*

$$\Gamma_\Theta(S) = \left\{\lim_{N\to\infty}\frac{1}{N}\sum_{n\le N} f(n) : \ f \in \mathcal{F}(S)\right\}$$
$$= \left\{\prod_p\left(\frac{p-1}{p-f(p)}\right) : \ f \in \mathcal{F}(S)\right\}.$$

If $S \subset [-1,1]$ then $\mathrm{Ang}(S) = 0$, and so Theorem 2 generalizes Wirsing's result. If $\alpha \neq 0$ is real then $\mathrm{Ang}(\{p^{i\alpha}\}) = \pi/2$, and thus Theorem 2 avoids the example $f(n) = n^{i\alpha}$. What happens when $\mathrm{Ang}(S) = \pi/2$? It will follow from results stated below that here $\Gamma(S) = \Gamma_\Theta(S) = U$. Thus the spectrum is quite easy to understand here.

In general, the spectrum contains more elements than simply the Euler products. For example, the spectrum of Euler products for $S = [-1,1]$ is simply the interval $[0,1]$. However, as Theorem 1 shows, the spectrum of S is more exotic. We now describe a family of integral equations whose solutions belong to the spectrum. In Theorem 3, we shall show that all points of the spectrum may be obtained by suitably combining an Euler product and a solution to one of these integral equations.

Recall that we assume S is closed and $1 \in S$. We define $\Lambda(S)$ to be the set of values $\sigma(u)$ obtained as follows. Let $\chi(t)$ be any measurable function with $\chi(t) = 1$ for $0 \le t \le 1$ and with $\chi(t)$ belonging to the convex hull of S, for all t. Next define $\sigma(u) = 1$ for $0 \le u \le 1$ and for $u > 1$ by the integral equation

(6) $$u\sigma(u) = \sigma * \chi = \int_0^u \sigma(u-t)\chi(t)dt.$$

Here, and throughout, $f * g$ denotes the convolution of the two functions f and g: that is, $f * g(x) = \int_0^x f(t)g(x-t)dt$. As we prove in [2] (see Theorem 7 below), there is a unique solution $\sigma(u)$ to (6). This solution is continuous and satisfies $|\sigma(u)| \le 1$ for all u.

That the integral equation (6) is relevant to the study of multiplicative functions was already observed by Wirsing [10]. This connection may be seen from the following Proposition.

Proposition 1. *Let f be a multiplicative function with $|f(n)| \le 1$ for all n and $f(n) = 1$ for $n \le y$. Let $\vartheta(x) = \sum_{p \le x} \log p$ and define*

$$\chi(u) = \chi_f(u) = \frac{1}{\vartheta(y^u)} \sum_{p \le y^u} f(p) \log p.$$

Then $\chi(t)$ is a measurable function taking values in the unit disc and with $\chi(t) = 1$ for $t \le 1$. Let $\sigma(u)$ be the corresponding unique solution to (6). Then

$$\frac{1}{y^u} \sum_{n \le y^u} f(n) = \sigma(u) + O\left(\frac{u}{\log y}\right).$$

The converse to Proposition 1 is also true:

Proposition 1 (Converse). *Let $S \subset U$ and suppose χ takes on values in the convex hull of S with $\chi(t) = 1$ for $t \le 1$. Given $\epsilon > 0$ and $u \ge 1$ there exist arbitrarily large y and $f \in \mathcal{F}(S)$ with $f(n) = 1$ for $n \le y$ and*

$$\left|\chi(t) - \frac{1}{\vartheta(y^u)} \sum_{p \le y^u} f(p) \log p\right| \le \epsilon \quad \textit{for almost all } 0 \le t \le u.$$

Consequently, if $\sigma(u)$ is the solution to (6) for this χ then

$$\sigma(t) = \frac{1}{y^t}\sum_{n\le y^t} f(n) + O(\epsilon\log(u+2)) + O\left(\frac{u}{\log y}\right) \quad \text{for all } t \le u.$$

If J and K are two subsets of the unit disc, we define $J \times K$ to be the set of elements $z = jk$ where $j \in J$ and $k \in K$.

Theorem 3. *For any $S \subset U$, $\Gamma(S) = \Gamma_\Theta(S) \times \Lambda(S)$.*

Researchers in the field have long "known" that results like Proposition 1 and Theorem 3 can used when needed (see [10] and [6], for instance). But this appears to be the first attempt to provide a complete proof of such a result in this generality. The idea of the proof of Theorem 3 is to decompose $f \in \mathcal{F}(S)$ into two parts: $f_s(p) = f(p)$ for $p \le y$ and $f_s(p) = 1$ for $p > y$, and $f_l(p) = 1$ for $p \le y$ and $f_l(p) = f(p)$ for $p > y$. For appropriately chosen y, the average of f until x is approximated by the product of the averages of f_s and f_l. If y is small enough compared with x, then the average of f_s is approximated by $\Theta(f_s, \infty) \in \Gamma_\Theta(S)$. Proposition 1 shows that if y is not too small, the average of f_l is approximated by the solution to an integral equation. Combining these, one gets that $\Gamma(S) \subset \Gamma_\Theta(S) \times \Lambda(S)$. The proof that $\Gamma_\Theta(S) \times \Lambda(S) \subset \Gamma(S)$ is similar, invoking the converse of Proposition 1.

As the case $S = [-1, 1]$ illustrates, $\Gamma_\Theta(S)$ represents the easy part of the spectrum while $\Lambda(S)$ is more mysterious. Here Theorems 1 and 3 tell us that $\Lambda(S) \subset [\delta_1, 1]$. That is, given any measurable function χ with $\chi(t) = 1$ for $t \le 1$, and $-1 \le \chi(t) \le 1$ always, then $\sigma(u) \ge \delta_1$ for all u (where σ is the corresponding solution to (6)). An important example is the function $\chi(t) = 1$ for $t \le 1$ and $\chi(t) = -1$ for $t > 1$. Denote by $\rho_1(u)$ the corresponding solution to (6). Then $\rho_1(u)$ satisfies a differential-difference equation very similar to that satisfied by the Dickman-de Bruijn function. Namely, $\rho_1(u) = 1$ for $u \le 1$ and for $u > 1$,

$$u\rho_1'(u) = -2\rho_1(u-1).$$

It is not hard to verify that $\rho_1(u)$ decreases for u in $[1, 1+\sqrt{e}]$ and increases for $u > 1 + \sqrt{e}$. The absolute minimum $\rho_1(1+\sqrt{e})$ is guaranteed by Theorem 1 to be $\ge \delta_1$ and in fact $\rho_1(1+\sqrt{e}) = \delta_1$. By continuity, $\rho_1(u)$ takes on all values in the interval $[\delta_1, 1]$ showing that $\Lambda(S) = [\delta_1, 1]$.

We now describe a lot of properties of $\Gamma_\Theta(S)$ which are also inherited by $\Gamma(S)$; in many cases, we get an explicit description of $\Gamma_\Theta(S)$. To state our results we introduce the set $\mathcal{E}(S)$ defined as follows. Let $\mathcal{E}(S) = \{0\}$ if $1 \notin \overline{S}$ and if $S \subset U$ is closed with $1 \in S$, we define

$$\mathcal{E}(S) = \{e^{-k(1-\alpha)} : \ k \ge 0, \ \ \alpha \text{ is in the convex hull of } S\}.$$

Thus $\mathcal{E}(S)$ consists of various 'spirals' connecting 1 to 0.

Theorem 4. *For all subsets S of U,*

$$\mathcal{E}(S) \times [0,1] \supset \Gamma_\Theta(S) = \Gamma_\Theta(S) \times \mathcal{E}(S) \supset \mathcal{E}(S).$$

If the convex hull of S contains a real point other than 1, *(in other words, if* $-1 \in S$ *or if* $\alpha, \beta \in S$ *with* $\mathrm{Im}(\alpha) > 0 > \mathrm{Im}(\beta)$*) then*

$$\Gamma_{\Theta}(S) = \mathcal{E}(S) = \mathcal{E}(S) \times [0,1],$$

and $\Gamma_{\Theta}(S)$ *is* starlike *(that is, if* $z \in \Gamma_{\Theta}(S)$ *then* $\Gamma_{\Theta}(S)$ *contains the line joining* 0 *and* z*).*

If $\mathrm{Ang}(S) = \pi/2$, then it is easy to see that $\mathcal{E}(S) = U$. Hence in this case $\Gamma(S) = \Gamma_{\Theta}(S) = U$, as we claimed earlier. Theorems 3 and 4 enable us to deduce some basic properties of the spectrum.

Corollary 1. *For all subsets S of U,* $\Gamma(S) = \Gamma(S) \times \mathcal{E}(S)$. *Consequently, the spectrum of S is connected. If the convex hull of S contains a real point other than* 1, *then the spectrum is starlike. If* 1, $e^{i\alpha}$ *and* $e^{i\beta}$ *are distinct elements of S, then the spectrum contains the disc centered at the origin with radius* $\exp(-2\pi/(|\cot(\alpha/2) - \cot(\beta/2)|))$. *In fact,* $\mathcal{E}(S)$ *contains this disc.*

The sets $\Gamma_{\Theta}(S)$ and $\Gamma(S)$ have the property that multiplying by $\mathcal{E}(S)$ leaves them unchanged. It turns out that $\Lambda(S)$ also has this property: $\Lambda(S) = \Lambda(S) \times \mathcal{E}(S)$. This gives us the following variant of Theorem 3 which reveals that $\Lambda(S)$ contains all the information about the spectrum.

Theorem 3′. *For all subsets S of U,*

$$\Lambda(S) \subset \Gamma(S) \subset \Lambda(S) \times [0,1].$$

If the convex hull of S contains a real point different from 1 *then* $\Gamma(S) = \Lambda(S)$.

Next we give a bound on the spectrum and determine $\Gamma(S) \cap \mathbb{T}$.

Theorem 5. *Suppose S is a subset of U with* $1 \in \overline{S}$. *The spectrum of S is U if and only if* $Ang(S) = \pi/2$. *If* $Ang(S) = \theta < \pi/2$, *then there exists a positive constant* $A(\theta)$, *depending only on* θ, *such that* $\Gamma(S)$ *is contained in a disc centered at* $A(\theta)$ *with radius* $1 - A(\theta)$. *In fact,* $A(\theta) = (28/411)\cos^2\theta$ *is permissible. Thus*

$$\Gamma(S) \cap \mathbb{T} = \begin{cases} \{1\} & \text{if } Ang(S) < \pi/2 \\ \mathbb{T} & \text{if } Ang(S) = \pi/2. \end{cases}$$

Applied to the set $S = [-1,1]$, Theorem 5 shows that there exists $c > -1$ such that $\Gamma(S) \subset [c,1]$. Thus Theorem 5 generalises Hall's result on Heath-Brown's conjecture.

By a simple calculation, we can show that $\mathcal{E}(S)$, $\Gamma_{\Theta}(S)$ and S all have the same angle: $\mathrm{Ang}(\Gamma_{\Theta}(S)) = \mathrm{Ang}(\mathcal{E}(S)) = \mathrm{Ang}(S)$. From Theorems 3 and 3′ we see that $\mathrm{Ang}(\Gamma(S)) = \mathrm{Ang}(\Lambda(S)) \geq \mathrm{Ang}(S)$.

Conjecture 1. *The angle of the set equals the angle of the spectrum. Thus*

$$Ang(\Gamma_{\Theta}(S)) = Ang(\mathcal{E}(S)) = Ang(\Lambda(S)) = Ang(\Gamma(S)) = Ang(S).$$

We support this conjecture by showing that $\mathrm{Ang}(S)$ and $\mathrm{Ang}(\Gamma(S))$ are comparable in the situations $\mathrm{Ang}(S) \to 0$ and $\mathrm{Ang}(S) \to \pi/2$.

Theorem 6. *Suppose $S \subset U$ and $Ang(S) = \theta = \pi/2 - \delta$. Then $Ang(S) \ll Ang(\Gamma(S))$ and*

$$\frac{\pi}{2} - \delta = Ang(S) \le Ang(\Gamma(S)) \le \frac{\pi}{2} - \frac{\sin\delta}{2}.$$

The first part of the Theorem says that Ang(S) and Ang($\Gamma(S)$) are comparable when Ang(S) is small. The second part of the Theorem is mainly interesting in the complementary case when Ang(S) is close to $\pi/2$. In fact, when δ is small we see that we are away from the truth only by a factor of 2 (as $\sin\delta \sim \delta$).

Example. Let $k \ge 3$ and S_k denote the set of k-th roots of unity. If $f \in \mathcal{F}(S_k)$ then $f(n) \in S_k$ for all n. Hence $\Gamma(S_k)$ is contained in the convex hull of S_k: that is, in the regular k-gon with vertices the k-th roots of unity. Notice that this implies Ang($\Gamma(S_k)$) $\le$ Ang(S_k), so that Ang(S_k) $=$ Ang($\Gamma(S_k)$) supporting Conjecture 1. Applying Corollary 1 with the two points $e^{\pm 2\pi i/k}$, we conclude that $\Gamma(S_k)$ is starlike and contains the disc centered at 0 with radius $\exp(-\pi\tan(\pi/k))$.

We define *the projection of* (a complex number) z *in the direction* $e^{i\alpha}$ to be Re $(e^{-i\alpha}z)$. Theorem 1 may be re-interpreted as stating that if $z \in \Gamma(\{\pm 1\})$ then the projection of z along -1 is $\le -\delta_1$. Evidently if $1 \in S$ then $1 \in \Gamma(S)$ so there is always a $z \in \Gamma(S)$ whose projection in the direction 1, is 1, and thus uninteresting to us. This motivates us to define the *maximal projection of the spectrum of a set* $S \subset \mathbb{T}$ as

$$\max_{1 \ne \zeta \in S}\ \max_{z \in \Gamma(S)} \text{Re } (\zeta^{-1} z).$$

Conjecture 2. *If $S \subset \mathbb{T}$ with $1 \in S$ and Ang $(S) = \theta$ then the maximal projection of $\Gamma(S)$ is*

$$\max_{1 \ne \zeta \in S}\ \max_{z \in \Gamma(S)} \text{Re } (\zeta^{-1} z) = 1 - (1 + \delta_1)\cos^2\theta.$$

One half of this conjecture is easy to establish: namely, the maximal projection is $\ge 1 - (1+\delta_1)\cos^2\theta$. To see this, let $z = x^{-1}\sum_{n \le x} f(n)$ where f is the completely multiplicative function defined by $f(p) = 1$ for all $p \le x^{1/(1+\sqrt{e})}$, and $f(p) = \zeta$ for $x^{1/(1+\sqrt{e})} \le p \le x$, where $\zeta \in \mathbb{T}$ and Ang(ζ) $= \theta$. Then, a simple calculation (analogous to the calculation in the Hall-Montgomery example (2)) gives that the projection of z along ζ is $1 - (1+\delta_1)\cos^2\theta + o(1)$.

Theorem 7. *Conjecture 2 is true for the sets $S = \{1, -1\}$ and $S = \{1, -1, i, -i\}$. If $S \subset \mathbb{T}$ with $1 \in S$ and Ang $(S) = \theta$ then the maximal projection of $\Gamma(S)$ is $\le 1 - (56/411)\cos^2\theta$.*

To facilitate comparison between Theorem 7 and Conjecture 2, we observe that $1 + \delta_1 = 0.3430\ldots$ whereas $56/411 = 0.1362\ldots$. Thus Theorem 6 is not too far away from the (conjectured) truth.

2. Solutions to Integral Equations

In §1, we described the Euler product spectrum and deduced information about the geometric part of the spectrum. We now outline some of our results on the integral equation (6). Our tool in analysing (6) is the Laplace transform, which, for a measurable function $f:\ [0,\infty)\to\mathbb{C}$ is given by

$$\mathcal{L}(f,s)=\int_0^\infty f(t)e^{-ts}dt$$

where s is some complex number. If f is integrable and grows sub-exponentially (that is, for every $\epsilon>0$, $|f(t)|\ll_\epsilon e^{\epsilon t}$ almost everywhere) then the Laplace transform is well defined for all complex numbers s with Re $(s)>0$. Laplace transforms occupy a role in the study of differential equations analogous to Dirichlet series in multiplicative number theory.

Theorem 8. *Suppose that $\chi(t)=1$ for $t\le 1$ and $|\chi(t)|\le 1$ for all t. There exists a unique solution $\sigma(u)$ to (6), which satisfies $\sigma(u)=1$ for $0\le u\le 1$ and $|\sigma(u)|\le 1$ for all u. In terms of χ the solution σ is given explicitly by*

$$\sigma(u)=1+\sum_{k=1}^{\infty}\frac{(-1)^k}{k!}I_k(u;\chi) \tag{7a}$$

where

$$I_k(u;\chi):=\int_{\substack{t_1,\ldots,t_k\ge 1\\ t_1+\cdots+t_k\le u}}\prod_{1\le i\le k}\left(\frac{1-\chi(t_i)}{t_i}\right)dt_1\ldots dt_k. \tag{7b}$$

In terms of Laplace transforms we have

$$\mathcal{L}(\sigma,s)=\frac{1}{s}\exp\left(-\mathcal{L}\left(\frac{1-\chi(t)}{t},s\right)\right).$$

The formula (7ab) has the appearance of an inclusion-exclusion identity, and indeed if we translate this result back in terms of multiplicative functions (via Proposition 1) then we find it can be so interpreted. Therefore if χ is real valued one can obtain upper and lower bounds by truncating the sum in (7a) at odd or even values of k, as in the combinatorial sieve. This is one of the main tools in our proof of Theorem 1.

Proposition 2. *Suppose χ is real valued so that $\chi(t)=1$ for $t\le 1$ and $-1\le\chi(t)\le 1$ for all t. Then the corresponding solution $\sigma(u)$ to (6) satisfies, for all $m\ge 0$,*

$$1+\sum_{k=1}^{2m+1}\frac{(-1)^k}{k!}I_k(u;\chi)\le\sigma(u)\le 1+\sum_{k=1}^{2m}\frac{(-1)^k}{k!}I_k(u;\chi). \tag{8}$$

Our other main tools in the proof of Theorem 1 (and many other results stated in §1) are the non-increasing property of the average $v^{-1}\int_0^v|\sigma(t)|dt$, and a convolution identity connecting $I_1(t,\chi)$ and $I_2(t,\chi)$, which we now describe.

Lemma 1. *Suppose χ is a complex valued measurable function with $\chi(t) = 1$ for $t \le 1$ and $|\chi(t)| \le 1$ for all t and let σ be the corresponding solution to (6). Then*

$$A(v) := \frac{1}{v}\int_0^v |\sigma(t)|dt$$

is a non-increasing function of v. Hence, for all $u \ge v$,

$$|\sigma(u)| \le A(v) = \frac{1}{v}\int_0^v |\sigma(t)|dt. \tag{9}$$

Proof. From (6), we have $|\sigma(u)| \le A(u)$ for all u. Hence (9) follows once we have shown that $A(v)$ is non-increasing. Differentiating the definition of $A(v)$, we have $A'(v) = |\sigma(v)|/v - A(v)/v \le 0$, so $A(v)$ is non-increasing and the result follows.

Lemma 2. *Let χ be as in Lemma 1. Then*

$$\int_0^u I_2(t,\chi)dt = \int_0^u I_1(t,\chi)I_1(u-t,\chi)dt.$$

Proof. For brevity, let $\chi_1(t) = (1-\chi(t))/t$. Then $I_1(t,\chi) = (1*\chi_1)(t)$ and $I_2(t,\chi) = (1*\chi_1*\chi_1)(t)$. The left side of our claimed identity is

$$\begin{aligned}(1*I_2)(u) &= (1*1*\chi_1*\chi_1)(u) = ((1*\chi_1)*(1*\chi_1))(u)\\ &= (I_1*I_1)(u),\end{aligned}$$

which is the right side of the claimed identity.

A fundamental question is: when does $\lim_{u\to\infty}\sigma(u)$ exist? If it exists, what does it equal? In view of Proposition 1 and its converse, this is equivalent to the question of the existence of the mean-value of a multiplicative function. Thus Theorem 2 has the following implication for integral equations.

Theorem 2′. *Suppose $S \subset U$ and $Ang(S) < \pi/2$. Let $\chi(t) = 1$ for $t \le 1$ and suppose $\chi(t)$ lies in the convex hull of S for all t. Then*

$$\lim_{u\to\infty}\sigma(u) = \exp\left(-\int_1^\infty \frac{1-\chi(t)}{t}dt\right) = \exp(-I_1(\infty,\chi)).$$

Translated in terms of integral equations, Hall's result (4) can be re-interpreted as follows: Suppose that D is a closed, convex subset of the unit disc containing 0 and suppose that χ is measurable with $\chi(t) = 1$ for $t \le 1$ and $\chi(t) \in D$ for $t > 1$. Then

$$\sigma(u) \ll \exp\left(-\eta(D)\int_1^u \frac{1-\operatorname{Re}\chi(t)}{t}dt\right). \tag{10}$$

In [2] we give a direct proof of this result using Laplace transforms. The advantages of our treatment is that the details are considerably simpler, and so we prove (10) giving explicit values to all of the implicit constants. Translating back to multiplicative functions, this allows us to give a completely explicit version of (4) (we are still refining the constants obtained, so do not state this result here). To whet the reader's appetite we now sketch a proof of the integral equations analogue of Montgomery's lemma (see [8] and [9]), which is the key to proving results such as (4) and (10).

For $t > 0$ we define $H(t) = \max_{y\in\mathbf{R}} |\mathcal{L}(\sigma, t+iy)|$ and

$$M(t) = \min_{y\in\mathbf{R}} \operatorname{Re} \mathcal{L}\left(\frac{1-\chi(v)e^{-ivy}}{v}, t\right).$$

Note that

$$\begin{aligned}\mathcal{L}(\sigma, t+iy) &= \frac{1}{t+iy}\exp\left(-\mathcal{L}\left(\frac{1-\chi(v)}{v}, t+iy\right)\right)\\ &= \frac{1}{t+iy}\exp\left(-\mathcal{L}\left(\frac{1-\chi(v)e^{-ivy}}{v}, t\right)\right.\\ &\qquad\qquad \left.+\mathcal{L}\left(\frac{1-e^{-ivy}}{v}, t\right)\right).\end{aligned}$$

The identity

$$\operatorname{Re} \mathcal{L}\left(\frac{1-e^{-ivy}}{v}, t\right) = \log|1+iy/t|$$

is easily proved by differentiating both sides with respect to y, so that

$$t|\mathcal{L}(\sigma, t+iy)| = \exp\left(-\operatorname{Re} \mathcal{L}\left(\frac{1-\chi(v)e^{-ivy}}{v}, t\right)\right)$$

combining the last two estimates. Therefore $tH(t) = \exp(-M(t))$.

Proposition 3. *For any $\alpha > 0$ and any $u \geq 1$, we have*

$$|\sigma(u)| \leq \frac{e^{2\alpha u}}{u}\int_\alpha^\infty \frac{H(t)}{t}dt = \frac{e^{2\alpha u}}{u}\int_\alpha^\infty \frac{\exp(-M(t))}{t^2}dt.$$

Note that $H(t) \leq \int_0^\infty |\sigma(v)|e^{-tv}dv \leq 1/t$ and so the integral in the Proposition converges.

Proof. By Lemma 1 we have

$$\begin{aligned}|\sigma(u)| &\leq \frac{1}{u}\int_0^u |\sigma(v)|dv \leq \frac{e^{2\alpha u}}{u}\int_0^\infty |\sigma(v)|e^{-2\alpha v}dv\\ &= \frac{e^{2\alpha u}}{u}\int_{v=0}^\infty 2v|\sigma(v)|\int_{t=\alpha}^\infty e^{-2tv}dtdv\\ &= \frac{e^{2\alpha u}}{u}\int_{t=\alpha}^\infty\left(\int_{v=0}^\infty 2v|\sigma(v)|e^{-2tv}dv\right)dt. \qquad (11)\end{aligned}$$

We shall prove that for all $t > 0$

$$2\int_0^\infty v|\sigma(v)|e^{-2tv}\,dv \le \frac{H(t)}{t}, \tag{12}$$

which when inserted in (11) furnishes the Proposition.

By Cauchy-Schwarz

$$\left(2\int_0^\infty v|\sigma(v)|e^{-2tv}dv\right)^2 \le \frac{2}{t}\int_0^\infty |v\sigma(v)|^2 e^{-2tv}dv.$$

By Plancharel's formula (since the Fourier transform is an isometry on L^2)

$$\int_0^\infty |v\sigma(v)|^2 e^{-2tv}dv = \frac{1}{2\pi}\int_{-\infty}^\infty |\mathcal{L}(v\sigma(v), t+iy)|^2 dy.$$

Now $v\sigma(v) = (\sigma * \chi)(v)$ and, since $\mathcal{L}(f*g, x) = \mathcal{L}(f,x)\mathcal{L}(g,x)$, this is

$$= \frac{1}{2\pi}\int_{-\infty}^\infty |\mathcal{L}(\sigma, t+iy)|^2 |\mathcal{L}(\chi, t+iy)|^2 dy$$
$$\le H(t)^2 \frac{1}{2\pi}\int_{-\infty}^\infty |\mathcal{L}(\chi, t+iy)|^2 dy.$$

Applying Plancharel again we get

$$\frac{1}{2\pi}\int_{-\infty}^\infty |\mathcal{L}(\chi, t+iy)|^2 dy = \int_0^\infty |\chi(v)|^2 e^{-2tv}dv$$
$$\le \int_0^\infty e^{-2tv}dv = \frac{1}{2t}.$$

Assembling the above estimates we obtain (12) and hence the Proposition.

Given χ, how does the solution $\sigma(u)$ vary with u? In terms of multiplicative functions we want to know how the average $x^{-1}\sum_{n\le x} f(n)$ varies with x. By considering the problem example $f(n) = n^{i\alpha}$ we see that these averages can fluctuate a lot:

$$\left|\frac{1}{x}\sum_{n\le x} n^{i\alpha} - \frac{w}{x}\sum_{n\le x/w} n^{i\alpha}\right| = \frac{1}{|1+i\alpha|}|1 - w^{-i\alpha}|.$$

However, Elliott [1] observed that the absolute values of these averages is slowly varying. He showed that for any multiplicative function f with $|f(n)| \le 1$ and all $2 \le w \le \sqrt{x}$

$$\frac{1}{x}\left|\sum_{n\le x} f(n)\right| - \frac{w}{x}\left|\sum_{n\le x/w} f(n)\right| \ll \left(\frac{\log w}{\log x}\right)^{1/19}.$$

Translated in terms of integral equations, this means that $|\sigma(u)|$ satisfies the strong Lipschitz estimate:

$$|\sigma(u)| - |\sigma(v)| \ll \left|\frac{u-v}{u+v}\right|^{1/19}.$$

Estimates like this are of interest because, as Hildebrand [7] observed, they allow one to (slightly) extend the range of validity of Burgess' character sum estimate. Our next Theorem improves the Lipschitz exponent 1/19 considerably.

Theorem 9. *Suppose that χ is a measurable function with $\chi(t) = 1$ for $t \le 1$ and $|\chi(t)| \le 1$ for all $t > 1$. Let $\sigma(u)$ denote the corresponding solution to the integral equation (6). Then for all u, v we have*

$$|\sigma(u)| - |\sigma(v)| \ll \left|\frac{u-v}{u+v}\right|^{\kappa}$$

where κ is any real number less than $2(1-2/\pi)/3 = 0.24225\ldots$ and the implied constant depends only on κ.

The constant $2(1-2/\pi)/3$ can be improved slightly by our methods. But we have been unable to attain the optimal exponent 1.

3. $\Gamma([-1,1]) \subset [-(2-2/\sqrt{e}), 1]$ AND A SKETCH OF THE PROOF OF THEOREM 1

Notice that $\Gamma([-1,1]) = \Lambda([-1,1])$ by Theorem 3′. Let χ be any real valued measurable function with $\chi(t) = 1$ for $t \le 1$ and $-1 \le \chi(t) \le 1$ for $t > 1$ and let $\sigma(u)$ be the corresponding solution to (6). Then we need to show that $\sigma(u) \ge -(2-2/\sqrt{e})$ for all u.

Let u_0 be such that $I_1(u_0, \chi) = 1$; if no such u_0 exists then $I_1(u, \chi) < 1$ for all u, and define $u_0 = \infty$. If $u \le u_0$ then, using the lower bound of (8) with $m = 0$, $\sigma(u) \ge 1 - I_1(u,\chi) \ge 1 - I_1(u_0, \chi) = 0$ which is stronger than our claimed bound.

Now suppose $u > u_0$ so that by (9), and since $\sigma(t) \ge 0$ for $t \le u_0$

$$|\sigma(u)| \le A(u_0) = \frac{1}{u_0}\int_0^{u_0} |\sigma(t)| dt = \frac{1}{u_0}\int_0^{u_0} \sigma(t) dt.$$

Using the upper bound of (8) with $m = 1$ we get

$$A(u_0) \le 1 - \frac{1}{u_0}\int_0^{u_0} I_1(t,\chi) dt + \frac{1}{2u_0}\int_0^{u_0} I_2(t,\chi) dt.$$

Using Lemma 2 here, we see with a little manipulation

$$A(u_0) \le \frac{1}{2} + \frac{1}{2u_0}\left((1-I_1) * (1-I_1)\right)(u_0). \tag{13}$$

Notice that for $0 \le t \le u_0$,

$$\begin{aligned} 1 - I_1(t,\chi) &= \int_t^{u_0} \frac{1-\chi(v)}{v} dv \le \min(1, 2\log(u_0/t)) \\ &= \begin{cases} 1 & \text{if } t \le u_0/\sqrt{e} \\ 2\log(u_0/t) & \text{if } u_0/\sqrt{e} < t \le u_0. \end{cases} \end{aligned}$$

Substituting this into (13) we deduce that

$$\begin{aligned} A(u_0) &\le \frac{1}{2} + \frac{1}{2u_0}\left(\int_0^{u_0(1-1/\sqrt{e})} 2\log\frac{u_0}{u_0-t} dt \right. \\ &\qquad \left. + \int_{u_0(1-1/\sqrt{e})}^{u_0/\sqrt{e}} dt + \int_{u_0/\sqrt{e}}^{u_0} 2\log\frac{u_0}{t} dt\right) \\ &= 2 - \frac{2}{\sqrt{e}}, \end{aligned}$$

which proves our desired bound.

We have shown more than just $\sigma(u) \geq -(2-2/\sqrt{e})$ above; in fact we have shown that either $\sigma(t)$ is non-negative for all $t \leq u$ or $|\sigma(u)| \leq (2-2/\sqrt{e})$. Analogously, in [2] we prove that either $\sigma(t)$ is non-negative for all $t \leq u$ or $|\sigma(u)| \leq |\delta_1|$.

We now give the barest outline of our proof that $\sigma(u) \geq \delta_1$. As above we may assume that $u > u_0$ and for simplicity we assume that $u > u_0(1+1/\sqrt{e})$ (the case $u_0 < u \leq u_0(1+1/\sqrt{e})$ succumbs to similar arguments). Here we know that

$$\begin{aligned} |\sigma(u)| &\leq A(u_0(1+1/\sqrt{e})) \\ &= \frac{1}{u_0(1+1/\sqrt{e})}\left(\int_0^{u_0} \sigma(t)dt + \int_{u_0}^{u_0(1+1/\sqrt{e})} |\sigma(t)|dt\right). \end{aligned} \tag{14}$$

We shall bound the above quantities in terms of the parameters

$$\lambda = I_1(u_0(1-1/\sqrt{e}), \chi), \qquad \text{and} \qquad \tau = I_1(u_0/\sqrt{e}, \chi),$$

which satisfy $0 \leq \lambda \leq \tau \leq 1$.

With these conditions, one can easily show (by techniques similar to those used above) that $I_1(tu_0, \chi)$ exceeds

$$\begin{cases} \max\left(0, \lambda + 2\log(t(1-1/\sqrt{e})^{-1})\right) & \text{if } 0 \leq t \leq 1-1/\sqrt{e} \\ \max(\lambda, \tau + 2\log(\sqrt{e}t)) & \text{if } 1-1/\sqrt{e} \leq t \leq 1/\sqrt{e} \\ \max(\tau, 1+2\log t) & \text{if } 1/\sqrt{e} \leq t \leq 1. \end{cases}$$

By inserting these bounds into (13), we deduce that

$$\frac{1}{u_0}\int_0^{u_0} \sigma(t)dt \leq 2 - \frac{2}{\sqrt{e}} - E_1(\lambda,\tau) = \frac{1}{\sqrt{e}}\int_0^{\sqrt{e}} \rho_1(t)dt - E_1(\lambda,\tau), \tag{15}$$

where $E_1(\lambda,\tau)$ is an explicit (but complicated) non-negative function of τ and λ.

For $u_0 \leq t \leq u_0(1+1/\sqrt{e})$, we use the inclusion exclusion inequalities (8) (with $m = 0$ or 1) to obtain inequalities of the form

$$|\sigma(t)| \leq |\rho_1(\sqrt{e}t/u_0)| + E_2(\lambda,\tau,t/u_0),$$

for some non-negative function $E_2(\lambda,\tau,t/u_0)$. The key is to obtain very precise estimates for $E_2(\lambda,\tau,t/u_0)$ such that

$$\frac{1}{u_0(1+1/\sqrt{e})}\int_{u_0}^{u_0(1+1/\sqrt{e})} E_2(\lambda,\tau,t/u_0)dt \leq \frac{E_1(\lambda,\tau)}{1+1/\sqrt{e}}.$$

In fact, equality above holds only when $\lambda = \tau = 0$. Combining this with (14) and (15), we obtain (after a suitable change of variables)

$$|\sigma(u)| \leq \frac{1}{1+\sqrt{e}}\int_0^{1+\sqrt{e}} |\rho_1(t)|dt.$$

Miraculously, the right side above equals $|\rho_1(1+\sqrt{e})| = |\delta_1|$ and everything is proved!

4. Open Problems

Our main objective is to explicitly determine the spectra of interesting sets S (as in Theorem 1). For example, what is the spectrum of S_k, the k-th roots of unity for $k \geq 3$? Failing this, one can ask for information on various geometric aspects of the spectrum, as in Conjectures 1 and 2.

What is the largest disc centered at the origin that can be contained in the spectrum? Corollary 1 gives a lower bound on this. In the other direction, one can ask for the largest positive $A(\theta)$ such that, for any S with $\text{Ang}(S) = \theta$, $\Gamma(S)$ is contained in the disc centered at $A(\theta)$ with radius $1 - A(\theta)$. Theorem 5 gives that $A(\theta) \geq (28/411)\cos^2\theta$, and using Corollary 1, we can check that $A(\theta) \leq (1 - \exp(-\pi\cot\theta))/2 \leq (38/75)\cos\theta$. Is it true that $A(\theta) \asymp \cos\theta$ as $\theta \to \pi/2$?

What is $\mu(S) := \min_{z\in\Gamma(S)} \text{Re}\,(z)$? When $S = [-1, 1]$, Theorem 1 shows that $\mu(S) = \delta_1$, and, as we saw in §1, this has the application that $\geq (1 + \delta_1 + o(1))x/2$ integers below x are quadratic residues $\pmod p$. It is especially interesting to determine $\mu(S_3)$ where S_3 denotes the cube roots of unity. It is easy to see that $\mu(S_3) \geq -1/2$ and we have shown that $\mu(S_3) > -1/2$. This demonstrates that a positive proportion of the integers below x are cubic residues $\bmod p$: suppose $p \equiv 1 \pmod 3$ and that χ is a cubic character $\pmod p$,

$$\begin{aligned}\#\{n \leq x : n \equiv m^3 \pmod p\} &= \frac{1}{3}\sum_{n\leq x}(1 + \chi(n) + \chi(n)^2)\\ &\geq \frac{1}{3}(1 + 2\mu(S_3) + o(1))x \gg x.\end{aligned}$$

We can determine explicit bounds here, but these are quite weak. In general, we know that $\mu(S) \geq -1 + 2A(\theta)$.

Let γ_m denote the largest real number such that if p is a prime then $\geq (\gamma_m + o(1))x$ integers below x are m-th power residues $\pmod p$. Theorem 1 gives $\gamma_2 = \delta_0$ and from the above paragraph we know $\gamma_3 > 0$. In fact, we have shown that $\gamma_m > 0$ for all m: that is, a positive proportion of the integers below x are m-th power residues $\pmod p$. It is not hard to show that $\gamma_m \leq \rho(m) = 1/m^{m+o(1)}$, where ρ denotes the Dickman-de Bruijn function. However, we have not been able to obtain good lower bounds for γ_m. So determining γ_m for $m \geq 3$ is a key open problem.

We may generalize the notion of spectrum by considering weighted averages of elements of $\mathcal{F}(S)$. Of particular interest is the *logarithmic spectrum*

$$\Gamma_0(S) = \lim_{N\to\infty}\left\{\frac{1}{\log N}\sum_{n\leq N}\frac{f(n)}{n} : \quad f \in \mathcal{F}(S)\right\}.$$

The logarithmic spectrum is contained in the convex hull of the spectrum, and its geometric properties are a lot easier to characterize. For example, while it is hard to determine the spectrum of $[-1, 1]$, it is an easy exercise that its logarithmic spectrum is $[0, 1]$. Further, the angle of the logarithmic spectrum equals the angle of the set; that is, the analogue of Conjecture 1 is true. In fact, if $\text{Ang}(S) = \theta = \pi/2 - \delta$ and $z \in \Gamma_0(S)$ with $|\arg(z)| = \varphi$ where $0 \leq \varphi \leq \pi$ then $|z| \leq (\cos\delta)^{\varphi/\delta}$. This allows us to bound $\Gamma_0(S)$ quite precisely. As for the spectrum, one can ask for the largest real number $A_0(\theta)$ such that, for all S with $\text{Ang}(S) = \theta = \pi/2 - \delta$, $\Gamma_0(S)$ is contained in the disc with center $A_0(\theta)$ and radius $1 - A_0(\theta)$. We can show that

$$\frac{\pi}{4}\delta + O(\delta^2) = \frac{1-(\cos\delta)^{\pi/\delta}}{2} \leq A_0(\theta)$$
$$\leq \frac{1-\exp(-\pi\cot\theta)}{2} = \frac{\pi}{2}\delta + O(\delta^2)$$

so that we have understood this up to a factor of 2.

Motivated by the logarithmic spectrum, we ask for the logarithmic density of m-th power residues (mod p). That is, what is the largest real number γ_m' such for all primes p

$$\sum_{\substack{n\leq x \\ n\equiv a^m \pmod p}} \frac{1}{n} \geq (\gamma_m' + o(1))\log x \ ?$$

By partial summation, it is clear that $\gamma_m' \geq \gamma_m$. Since $\Gamma_0([-1,1]) = [0,1]$, it is easy to see that $\gamma_2' = 1/2$. We do not know the precise value of γ_m' for any other value of m. However, by a combinatorial argument, we have shown that $\gamma_m' \geq 1/2^{m-1}$; and by an easy construction $\gamma_m' \leq \exp(-m/e + o(m))$.

References

[1] P.D.T.A. Elliott, *Extrapolating the mean-values of multiplicative functions*, Indag. Math **51** (1989), 409-420.

[2] A. Granville and K. Soundararajan, *The Spectrum of Multiplicative Functions*, In preparation.

[3] G. Halász, *On the distribution of additive and mean-values of multiplicative functions*, Stud. Sci. Math. Hungar **6** (1971), 211-233.

[4] R.R. Hall, *A sharp inequality of Halász type for the mean value of a multiplicative arithmetic function*, Mathematika **42** (1995), 144-157.

[5] R.R. Hall, *Proof of a conjecture of Heath-Brown concerning quadratic residues*, Proc. Edinburgh Math. Soc. **39** (1996), 581-588.

[6] A. Hildebrand, *Fonctions multiplicatives et équations intégrales*, Séminaire de Théorie des Nombres de Paris, 1982-83 (M.-J. Bertin, ed.), Birkhäuser, 1984, pp. 115-124.

[7] A. Hildebrand, *A note on Burgess' character sum estimate*, C.R. Acad. Sci. Roy. Soc. Canada **8** (1986), 35-37.

[8] H. L. Montgomery, *A note on the mean values of multiplicative functions*, Inst. Mittag Leffler **17** (1978).

[9] G. Tenenbaum, *Introduction to analytic and probabilistic number theory*, Cambridge University Press, 1995.

[10] E. Wirsing, *Das asymptotische Verhalten von Summen über multiplikative Funktionen II*, Acta Math. Acad. Sci. Hung **18** (1967), 411-467.

Department of Mathematics, University of Georgia, Athens, Georgia 30602, USA
E-mail address: andrew@sophie.math.uga.edu

Department of Mathematics, Princeton University, Princeton, New Jersey 08544, USA
E-mail address: skannan@math.princeton.edu

MODULAR MAHLER MEASURES I

F. RODRIGUEZ VILLEGAS

Department of Mathematics
University of Texas at Austin
Austin, TX 78712

ABSTRACT.
We relate Boyd's numerical examples, linking the Mahler measure $m(P_k)$ of certain polynomials P_k to special values of L-series of elliptic curves, to the Bloch-Beilinson conjectures. We study $m(P_k)$ as a function of the parameter k and find a relation to modular forms and certain formal similarities with the expansions of mirror symmetry of physics.

CONTENTS

Support for this work was provided in part by a grant of the NSF

S.D. Ahlgren et al. (eds.), Topics in Number Theory, 17–48.

Introduction The *logarithmic Mahler measure* of a Laurent polynomial $P \in \mathbb{C}[x_1^{\pm 1}, \dots, x_n^{\pm 1}]$ is defined as

$$m(P) = \int_0^1 \cdots \int_0^1 \log \left| P(e^{2\pi i\theta_1}, \dots, e^{2\pi i\theta_n}) \right| d\theta_1 \cdots d\theta_n$$

and its *Mahler measure* as $M(P) = e^{m(P)}$, the geometric mean of $|P|$ on the torus

$$T^n = \{(z_1, \dots, z_n) \in \mathbb{C}^n \,|\, |z_1| = \ldots = |z_n| = 1\}.$$

If α is an algebraic number and $P \in \mathbb{Z}[x]$ with $P(\alpha) = 0$ has relatively prime coefficients and minimal degree then Jensen's formula of complex analysis gives

$$M(P) = \text{height of } \alpha.$$

For example, if $\alpha = n/m \in \mathbb{Q}^*$ with m, n relatively prime integers then

$$M(mx - n) = \max\{|m|, |n|\}.$$

In the early 80's Smyth [Sm] discovered the following remarkable identity

$$m(1 + x + y) = L'(\chi, -1),$$

where $L(\chi, s)$ is the Dirichlet series associated to the quadratic character χ of conductor 3. A handful of similar formulas were known at the time, [Bo1], [Ra], in apparent isolation from the general (though mostly conjectural) world of special values of L-series. Recently Deninger [De] bridged the gap by showing how to interpret $m(P)$, when P does not vanish on T^n, as a Deligne period of a mixed motive. This combined with Beilinson's conjectures makes identities such as Smyth's somewhat less of a mystery, though certainly not less beautiful.

Deninger's work also prompted Boyd [Bo1] to search numerically for other identities of the form

$$m(P) = c \cdot L'(0),$$

where $P \in \mathbb{Z}[x^{\pm 1}, y^{\pm 1}]$, $c \in \mathbb{Q}^*$, and L is an L-series somehow associated to P. Perhaps we should clarify right away what we mean by 'numerically' here. Once P and $L(s)$ are chosen, both quantities $m(P)$ and $L'(0)$ can be computed as complex numbers to any desired degree of precision (though in general not without a certain amount of ingenuity). The ratio of such approximations is of course a rational number, but what one looks for is a situation where the ratio is very close to a rational number c of relatively low height (e.g. $c = 1$).

Boyd found many such 'numerically verified' identities. For example, let

$$P_k(x,y) = x + \frac{1}{x} + y + \frac{1}{y} + k, \qquad k \in \mathbb{C},$$

and let

$$B_k = \frac{L_k'(0)}{m(P_k)}, \qquad k \in \mathbb{N},$$

where $L_k(s)$ is the L-function of the elliptic curve E_k determined by $P_k = 0$ if $k \neq 4$ and $L(\chi, s-1)$ with χ the quadratic Dirichlet character of conductor 4 if $k = 4$. Then the numerical data suggests that

$$(?_1) \qquad B_k \in \mathbb{Q}^*, \qquad k \in \mathbb{N}.$$

(Here though not strictly necessary we assume E_k is modular, an essentially academic assumption in light of the work of Wiles.)

What is known is the following (see n° 15). The case $k = 4$, for which $P_k = 0$ is of genus 0, is very similar to that of Smyth's and indeed it can be proved [Bo1], by direct manipulations of the integral defining $m(P_4)$, that $B_4 = 1/2$. For $k = 1$ (despite P_1 vanishing on the torus T^2) Deninger showed that the conjecture of Bloch–Beilinson implies that $B_1 \in \mathbb{Q}^*$.

It turns out that a suitable modification of $(?_1)$ allows values of k with $k^2 \in \mathbb{N}$ and we find that for $k = 4\sqrt{2}$, where the corresponding elliptic curve has complex multiplication, we can prove that $B_k = 1$ and for $k = 3\sqrt{2}$, where E_k is isomorphic to the modular curve $X_0(24)$, a theorem of Beilinson's guarantees that $B_k \in \mathbb{Q}^*$. Apart from a few other similar values of k the truth of $(?_1)$ remains, however, unproven.

Boyd produced many other examples of families of Laurent polynomials which appear to satisfy the analogue of $(?_1)$, finding, with uncanny intuition, the right set of conditions that these polynomials should meet.

The purpose of this expository paper is to give, on one hand, a gentle introduction to the conjecture of Bloch–Beilinson (chapter II) and on the other, to clarify the relation of Boyd's examples to this conjecture (chapter III). As we will see, Boyd's construction gives a new way of producing elements in $K_2(E)$ for certain elliptic curves E defined over $\mathbb{Q}$.

In addition (chapter IV), we study $m(P_k)$, where P_k is a family of polynomials of a certain type, as a function of $k \in \mathbb{C}$. We discover a connection with modular forms. In some cases, $m(P_k)$ can be expressed in terms of an Eisenstein–Kronecker series (this was also proved by Deninger [De] for $k = 1$ in the family considered above) and $(?_1)$ is then related to a special case (due to Bloch) of the conjecture of Bloch–Beilinson. We also discuss certain formal similarities with expansions appearing in the "mirror symmetry" of physics.

We have strived to present the ideas with a minimum of cohomological machinery in an attempt to make them as widely accessible as possible. We give however few complete proofs and refer the reader to [RV] (a more technical and in depth presentation) for what is lacking here.

Acknowledgments Above all I would like to thank P. Sarnak for bringing Boyd's examples to my attention and for his unfailing encouragement and support over the years. I would like to thank D. Boyd and B. Poonen for several helpful conversations; D. Boyd and J. Tate for their comments on earlier versions of this paper; as well as D. Boyd, for his hospitality at UBC, where part of this work was done and B. Poonen for producing the polygons of fig. 1. Finally, I would like to thank the many people with whom I discussed the material of this paper.

I. – Generalities

1. Mahler's measure

When $n = 1$ Jensen's formula gives the identity

$$M(P) = |a_0| \prod_{|\alpha_\nu|>1} |\alpha_\nu| ,$$

where $P(x) = a_0 \prod_{\nu=1}^{d}(x - \alpha_\nu)$, and it is in fact in this guise that this quantity first seems to have arisen in an article by D. H. Lehmer in the early 1930's [Le]. Lehmer's primary interest was the discovery of new large primes and for this end he considered, after Pierce [Pi], the factors of the integers $\Delta_n = \prod_{\nu=1}^{d}(\alpha_\nu^n - 1)$ associated to a monic polynomial $P \in \mathbf{Z}[x]$. Here Pierce and Lehmer were following the old tradition of restricting the search for primes to numbers of a special kind; indeed, for $P = x - 2$ we have $\Delta_n = 2^n - 1$, the numbers considered by Mersenne.

The prime factors of Δ_n must satisfy certain congruences modulo n and are quite restricted. Factoring Δ_n is therefore much easier than factoring a random number of the same size. Pierce gives the following example.

$$\frac{\Delta_{61}}{\Delta_1} = 4459734401, \qquad P = x^3 + x + 1,$$

and we may test it is prime by means of 119 trial divisions as opposed to the 6655 we would need if we did not know its special form. In the same vein, Lehmer shows that

$$\frac{\Delta_{107}}{\Delta_1} = 323351425103273, \qquad P = x^3 - x - 1,$$

is prime. Incidentally, it is interesting to reflect on what was considered a large prime number early in the century.

To maximize the savings we would need to choose polynomials such that Δ_n remains relatively small for large n. It is easy to check, as Lehmer does, that Δ_n grows with n roughly like $M(P)^n$. Naturally, Lehmer then searched for monic polynomials $P \in \mathbb{Z}[x]$ with a small value of $M(P)$ (other than the minimum possible $M(P) = 1$ attained by products of cyclotomic polynomials, which give rise to uninteresting sequences Δ_n).

In what must have been a fantastic computational tour de force Lehmer found a polynomial of degree 10 with $M(P) = 1.17628\ldots$, and writes "All efforts put to find a better equation of degree 12 and 14 have been unsuccessful." Remarkably, this value of $M(P)$ still stands today as the smallest known despite many further intensive searches. In fact, it is not even known if there is a minimum value for $M(P) > 1$. We refer the reader to [Bo1-3] for further details and bibliography.

Mahler introduced his measure in the early 60's [Ma] in order to prove inequalities between various quantities measuring the size of a polynomial and its factors. These inequalities are useful in the theory of transcendental numbers (see for example [Ma], [Ph], and [So1]), where $M(P)$ enters as one of several possible height functions.

Mahler's measure also appears as the topological entropy of certain dynamical systems [LSW] and in connection with the spectra of certain discrete differential operators [Sa].

2. Special values of L-functions

The prototypical example of what we would like to discuss is given by Dirichlet's class number formula. To simplify the exposition we will only consider the case of a real quadratic field $F \subset \mathbb{R}$, where the group of units $\mathcal{O}_F^\star$ is of rank one. First, let us recall that, in general for any number field K, the zeta function of K, $\zeta_K(s)$, has an analytic continuation to all s (except for a simple pole at $s = 1$), satisfies a functional equation as $s \mapsto 1 - s$, and has a zero at $s = 0$ of order the rank of the finitely generated group $\mathcal{O}_K^\star$.

Dirichlet's class number formula implies the following

$$(*) \qquad \zeta_F'(0) \sim_{\mathbb{Q}^\star} \log(|\epsilon|), \qquad \epsilon \in \mathcal{O}_F^\star, \quad \epsilon \neq \pm 1$$

where for $a, b \in \mathbb{C}^\star$

$$a \sim_{\mathbb{Q}^\star} b \quad \Longleftrightarrow \quad a = rb \text{ for some } r \in \mathbb{Q}^\star.$$

The ingredients in this formula are:

(1) an arithmetic/geometric object $\mathcal{O}$ (the ring of integers $\mathcal{O}_F$ of the real quadratic field F),

(2) an associated finitely generated abelian group of rank one $K(\mathcal{O})$ (the group of units $\mathcal{O}_F^\star$),

(3) an associated L-function $L(\mathcal{O},s)$ with a simple zero at $s=0$ (the zeta function $\zeta_F(s)$), and

(4) a non-zero homomorphism $r: K(\mathcal{O}) \longrightarrow \mathbb{R}$ (the function $\log(|\cdot|)$).

With this setup (*) becomes

$$(**) \qquad L'(\mathcal{O},0) \sim_{\mathbb{Q}^*} r(\alpha), \qquad \alpha \in K(\mathcal{O}) \setminus K(\mathcal{O})_{\mathrm{tor}},$$

where $K(C)_{\mathrm{tor}}$ is the torsion subgroup of $K(C)$. (Note that in general, r necessarily vanishes on torsion elements as the additive group $\mathbb{R}$ has no torsion.)

We expect other lists of objects (1)-(4) to also satisfy (**). First let us notice some compatibilities in (*). The zeta function ζ_F is defined by an Euler product

$$\zeta_F(s) = \prod_p F_p(s)^{-1}, \qquad \mathrm{Re}\,(s) > 1,$$

where

$$F_p(s) = \prod_{\mathcal{P}|p} (1 - N\mathcal{P}^{-s})$$

is the Euler factor at p. For primes p unramified in $F/\mathbb{Q}$ we have

$$F_p(s)|_{s=1} = \frac{\#(\mathcal{O}_F/p\mathcal{O}_F)^*}{p^2}.$$

Note that $s=1$ corresponds to $s=0$ under the functional equation and that we may interpret $(\mathcal{O}_F/p\mathcal{O}_F)^*$ as $K(\mathcal{O} \bmod p)$, that is, as the group K of the reduction of $\mathcal{O}$ modulo p.

The other example we will discuss in this paper is that of an elliptic curve E defined over $\mathbb{Q}$. We will take a Néron model $\mathcal{E}$ of E over $\mathbb{Z}$ as $\mathcal{O}$, $L(E,s)$ as its corresponding L-function $L(\mathcal{O},s)$, and $K_2(\mathcal{E})$, a K-theory group to be described below, as $K(\mathcal{O})$. In this case (**) is then the conjecture of Bloch and Beilinson for elliptic curves.

II. – The conjecture of Bloch–Beilinson

In this chapter we give a short, low-brow introduction to the conjectures of Bloch and Beilinson for elliptic curves. We have benefited greatly (and borrowed freely) from a number of sources including [Be], [CCGLS], [De], [So2], [MS], [Ram], [Sch] to which we refer the reader for further details.

3. K_2 of fields In order to describe $K_2(\mathcal{E})$ we will proceed in stages. First, we define $K_2(F)$ for a field F. Our main reference will be Milnor's book [Mi].

In its original definition $K_2(F)$ is, loosely speaking, the group of non-trivial relations satisfied by elementary matrices of arbitrary size with entries in F [Mi §5]. By a theorem of Matsumoto [Mi §11], however, it can also be described as the universal target of symbols on F. More precisely, a ***symbol*** on F is a bimultiplicative map

$$c : F^\star \times F^\star \longrightarrow A$$

to an abelian group A (here written multiplicatively) such that

$$c(x, 1-x) = 1, \qquad x \neq 0, 1.$$

By Matsumoto, there is a (universal) symbol

$$\begin{array}{ccc} F^\star \times F^\star & \longrightarrow & K_2(F) \\ (x, y) & \mapsto & \{x, y\} \end{array}$$

such that composition gives a bijection

$$\left\{ \begin{array}{c} \text{homomorphism} \\ K_2(F) \longrightarrow A. \end{array} \right\} \longleftrightarrow \left\{ \begin{array}{c} \text{symbol on } F \\ \text{with values in } A \end{array} \right\}.$$

In other words, any symbol c on F with values in A is of the form $c(x, y) = \phi(\{x, y\})$ for a unique homomorphism $\phi : K_2(F) \longrightarrow A$.

Examples. 1) If $F = \mathbb{Q}_p$ then the ***Hilbert symbol***, with values in ± 1, is defined as

$$(a, b)_p = \begin{cases} 1 & \text{if } x^2 - ay^2 - bz^2 = 0 \text{ has a solution in } \mathbb{Q}_p \\ -1 & \text{otherwise} \end{cases}$$

2) Given a discrete valuation v on F with maximal ideal $\mathcal{M}$ and residue field k, Tate defined the ***tame symbol*** at v

$$(x, y)_v \equiv (-1)^{v(x)v(y)} \frac{x^{v(y)}}{y^{v(x)}} \mod \mathcal{M}.$$

Note that in particular $(x, y)_v = 1$ if $v(x) = v(y) = 0$.

We will let the corresponding map be

$$\lambda_v : K_2(F) \longrightarrow k^\star.$$

3) The analogue of 1) for $F = \mathbb{R}$ is the symbol

$$(x, y)_\infty = \begin{cases} -1 & \text{if } x, y < 0 \\ +1 & \text{otherwise.} \end{cases}$$

4) If F is a finite field then $K_2(F) = 1$; i.e. all symbols on F are trivial.

4. The regulator We will eventually only consider elliptic curves defined over $\mathbb{Q}$ but for now let C be an arbitrary compact Riemann surface and let $\mathbb{C}(C)$ be its field of rational functions. Let $f \in \mathbb{C}(C)^\star$ and let S be the set of its zeros and poles. Then, locally in $C \setminus S$, we can define a branch of $\log f$, which gives rise to a global, closed differential 1-form on $C \setminus S$

$$d\log f = \frac{df}{f} = d\log|f| + i\,d\arg f.$$

Given two non-zero functions $f, g \in \mathbb{C}(C)^\star$ we define

$$\eta(f,g) = \log|f|\, d\arg g - \log|g|\, d\arg f, \qquad f, g \in \mathbb{C}(C)^\star,$$

a real C^∞ differential 1-form on $C \setminus S$, where $S \subset C$ is the set of zeros and poles of f and g. The following properties of $\eta(f,g)$ are easy to check

$$\eta(g,f) = -\eta(g,f) \qquad f, g \in \mathbb{C}(C)^\star$$
$$\eta(f_1 f_2, g) = \eta(f_1, g) + \eta(f_2, g), \qquad f_1, f_2, g \in \mathbb{C}(C)^\star$$

Also, η is closed since

$$0 = \operatorname{Im}\left(\frac{df}{f} \wedge \frac{dg}{g}\right) = d\eta.$$

Given a finite set $S \subset C$ consider the first homology group $H_1(C \setminus S, \mathbb{Z})$ and let $H^1(C \setminus S, \mathbb{R})$ be the real vector space of linear forms $\xi : H_1(C \setminus S, \mathbb{Z}) \longrightarrow \mathbb{R}$. Given two finite sets $S \subset S' \subset C$ there is a natural injection $H^1(C\setminus S, \mathbb{R}) \longrightarrow H^1(C\setminus S', \mathbb{R})$, which we will regard as an inclusion. For a closed loop γ in $C \setminus S$ we let $[\gamma]$ be its homology class.

Since $\eta(f,g)$ is closed, we may associate to it an element of $H^1(C \setminus S, \mathbb{R})$, where S is any finite set containing the zeros and poles of f and g, by integration. More precisely, for γ a closed path in $C \setminus S$ the map

$$\gamma \mapsto \int_\gamma \eta(f,g)$$

only depends on the homology class of γ in $H_1(C \setminus S, \mathbb{Z})$ and hence determines an element $\bar{r}(f,g) \in H^1(C \setminus S, \mathbb{R})$.

A point $w \in C$ defines a valuation v on $\mathbb{C}(C)$. To simplify the notation we will let $(\cdot,\cdot)_w$ (and λ_w) denote the tame symbol given by v. The point w also determines a linear form $\operatorname{Res}_w : H^1(C \setminus \{w\}, \mathbb{R}) \longrightarrow \mathbb{R}$, the residue map, as follows. Let $\phi : U \longrightarrow \mathbb{C}$ be a local chart in C with $w \in U$ and $\phi(w) = 0$ and let γ be the preimage by ϕ of a small circle in $\mathbb{C}$ centered at zero and oriented counterclockwise. Then

$$\begin{array}{rccc} \mathrm{Res}_w : & H^1(C \setminus \{w\}, \mathbb{R}) & \longrightarrow & \mathbb{R} \\ & \xi & \mapsto & \frac{1}{2\pi}\,\xi([\gamma]) \end{array}.$$

Since the construction is local, Res_w actually gives a linear form $H^1(C\setminus S, \mathbb{R}) \longrightarrow \mathbb{R}$ for any finite set S, which is identically zero if w is not in S.

The differential form $\eta(f,g)$ form is related to the tame symbol as follows.

Lemma. *Let $w \in C$, $f, g \in \mathbb{C}(C)^\star$, and $S \subset C$ a finite set containing the zeros and poles of f and g. Then*

$$\mathrm{Res}_w\, \eta(f,g) = \log|(f,g)_w|.$$

Proof sketch. Both sides are bimultiplicative and skew-symmetric hence it is enough to check the cases: $(v(f), v(g)) = (0,0), (0,1), (1,1)$. For example, in the second case after taking coordinates we must verify that

$$\log|f(0)| = \frac{1}{2\pi i}\int_C \log|f|\,\frac{dz}{z} - \frac{1}{2\pi}\int_C \log|z|\,d\arg f,$$

where C is a circle around the origin with small radius oriented counterclockwise. By letting the radius of C go to zero we see that the second integral vanishes. The first integral equals the left hand side by Jensen's formula. □

Now comes the crucial point.

Corollary. *Let $f \in \mathbb{C}(C)$ with $f \neq 0, 1$ and $S \subset C$ the finite set of zeros and poles of f and $1-f$. Then*

$$\eta(f, 1-f) = 0 \quad \textit{in } H^1(C \setminus S, \mathbb{R}).$$

Proof. First assume that $C = \mathbb{P}^1$ and $f = z$ with $\mathbb{C}(C) = \mathbb{C}(z)$. Then the result follows from the lemma and the fact that $(\cdot,\cdot)_w$ is a symbol.

In the general case the claim follows by using the map $f : C\setminus S \longrightarrow \mathbb{P}^1 \setminus \{0,1,\infty\}$ to reduce to the previous one. □

Given a finite set $S \subset C$ we define

$$K_{2,S}(C) = \bigcap_{w \notin S} \ker \lambda_w \subset K_2(\mathbb{C}(C)).$$

(This notation is not standard.) By the preceding lemma and corollary, $(f,g) \mapsto \bar{r}(f,g)$ extends linearly to a map

$$\bar{r} : K_{2,S}(C) \longrightarrow H^1(C \setminus S, \mathbb{R}).$$

We may summarize neatly what we have proved in the following.

Proposition. *The diagram*

$$\begin{array}{ccc} K_{2,S}(C) & \xrightarrow{\bar{r}} & H^1(C \setminus S, \mathbb{R}) \\ \lambda_w \downarrow & & \downarrow \mathrm{Res}_w \\ \mathbb{C}^* & \xrightarrow{\log|\cdot|} & \mathbb{R} \end{array}$$

is commutative for every $w \in S$.

Note that

$$\bigcup_{S \subset C} K_{2,S}(C) = K_2(\mathbb{C}(C))$$

since $(f,g)_w = 1$ for all but finitely many $w \in C$. Hence, we could take direct limits over all $S \subset C$ and obtain a true symbol on $\mathbb{C}(C)$ (i.e. a map defined on all of $K_2(\mathbb{C}(C))$), but we will not need this fact.

Finally, for E an elliptic curve defined over $\mathbb{R}$, let $c_0 \in H_1(E, \mathbb{Z})$ be the class of the cycle determined by the connected component of its real points (with some orientation) and define

$$\begin{array}{rccc} r: & K_{2,\emptyset}(C) & \longrightarrow & \mathbb{R} \\ & \alpha & \mapsto & \frac{1}{2\pi}\,\bar{r}(\alpha)(c_0). \end{array}$$

5. An example To illustrate the previous discussion, let us consider $C = \mathbb{P}^1, f = P(z)$ and $g = z$, where $P \in \mathbb{C}[z]$ is a polynomial and z is a generator of $\mathbb{C}(C)$. Let S be the set of zeros of P in $\mathbb{C}^*$ and $S' = S \cup \{0, \infty\}$. Then, $\{P, z\}$ is an element of $K_{2,S'}(C)$ and $\bar{r}(P,z)$ an element of $H^1(C \setminus S', \mathbb{R})$. Note that $\{P,z\} \in K_{2,S}(C)$ if and only if the coefficients of the highest and lowest power of x in $P(-x)$ are 1.

On the other hand, let T be the circle $|z| = 1$ in C oriented counterclockwise. If P does not vanish on T then $[T]$ gives a class in $H_1(C \setminus S', \mathbb{Z})$ and we have

$$m(P) = \frac{1}{2\pi}\,\bar{r}(P,z)([T]),$$

since $\log|z|$ vanishes identically on T.

Let us now make a preliminary connection with values of L-series. Consider the following family of polynomials

$$P_k(x) = x^2 - kx + 1,$$

depending on a parameter $k \in \mathbb{C}$. If k is real and $|k| > 2$ then the roots of P_k, $S_k = \{\epsilon_k, \epsilon_k^{-1}\}$ with say $|\epsilon_k| > 1$, are real, $\{P_k, z\} \in K_{2,S_k}(C)$, and $\bar{r}(P_k, z) \in$

$H^1(C \setminus S_k, \mathbb{R})$. In this range of k we also have that P_k does not vanish on the unit circle T. Hence

$$\log|\epsilon_k| = m(P_k) = \frac{1}{2\pi}\,\bar{r}(P_k, z)([T]), \qquad k \in \mathbb{R}, \quad |k| > 2.$$

If k is an integer with $|k| > 2$ then $F_k = \mathbb{Q}(\epsilon_k)$ is a real quadratic field and by Dirichlet's class number formula (see n° 2)

$$\zeta'_{F_k}(0) \sim_{\mathbb{Q}^*} m(P_k), \qquad k \in \mathbb{Z}, \quad |k| > 2.$$

A similar analysis applies to the family $x^2 - kx - 1$.

As we shall see, Boyd's numerical examples are, in a sense, a generalization of this fact to two variable polynomials P_k determining elliptic curves.

6. $K_2(E)$ and $K_2(\mathcal{E})$

Let now E be an elliptic curve defined over $\mathbb{Q}$ and let $\mathcal{E}$ be a Néron model of E over $\mathbb{Z}$. (We will not really use any of the technical aspects of Néron models; the reader may just want to think of $\mathcal{E}$ as a good integral version of E, somewhat like the ring of integers in the case of a number field. For details on Néron models see [BLR]; for the arithmetic of elliptic curves see [Si].) Let us fix an algebraic closure $\overline{\mathbb{Q}} \subset \mathbb{C}$ of $\mathbb{Q}$.

Defining the group $K_2(E)$ itself would take us too far afield; for our purposes it will be enough to know that there is a surjective map

$$K_2(E) \longrightarrow \bigcap_w \ker \lambda_w \subset K_2(\mathbb{Q}(E)),$$

whose kernel is torsion; here $\mathbb{Q}(E)$ is the field of rational functions on E and w runs through all points in $E(\overline{\mathbb{Q}})$. Since non-algebraic points of $E(\mathbb{C})$ induce trivial valuations on $\mathbb{Q}(E)$ we obtain a map

$$K_2(E) \longrightarrow K_{2,\emptyset}(E),$$

(where on the right we view E as a Riemann surface, see n° 4).

Similarly, $K_2(\mathcal{E})$ may be considered, up to torsion, as the subgroup of $K_2(E)$ determined by finitely many extra conditions, one condition for each prime of bad reduction of E (only primes where E has split-multiplicative reduction matter, see [BG]).

Finally, combining these various maps with the regulator we obtain a map, also denoted by r,

$$r : K_2(\mathcal{E}) \longrightarrow \mathbb{R}.$$

7. The conjecture of Bloch–Beilinson

Conjecture. *(Bloch–Beilinson) Let E be a modular elliptic curve defined over $\mathbb{Q}$ and $\mathcal{E}$ a Néron model of E. Then $K_2(\mathcal{E})$ is of rank 1 and*

$$(?_2) \qquad L'(E,0) \sim_{\mathbb{Q}^*} r(\alpha), \qquad \alpha \in K_2(\mathcal{E}) \setminus K_2(\mathcal{E})_{\mathrm{tor}}.$$

(The assumption of modularity, an essentially academic one in light of the work of Wiles, is not really necessary as we could just as well formulate the conjecture in terms of $L(E,2)$ by means of the functional equation; as usual, the version at $s=0$ is cleaner.)

As in the case of Dirichlet's class number formula there are internal compatibilities in the conjecture. The L-function of E is given by

$$L(E,s) = \prod_p F_p(s)^{-1}, \qquad \mathrm{Re}\,(s) > 3/2,$$

where for primes p of good reduction

$$F_p(s) = (1-\alpha_p p^{-s})(1-\beta_p p^{-s}),$$
$$\alpha_p\beta_p = p, \quad \alpha_p + \beta_p = a_p = p+1-\#E(\mathbb{Z}/p\mathbb{Z}).$$

As a consequence of a theorem of Bass and Tate [Ta] the kernel of all tame symbols in $K_2(F)$, where F is the function field of the reduction of E modulo p, is finite of order $(1-\alpha_p p)(1-\beta_p p)$. We may formulate this result as follows: for p of good reduction

$$F_p(s)|_{s=2} = \frac{\#K_{2,\emptyset}(E \bmod p)}{p^3}.$$

Note that $s=0$ corresponds to $s=2$ under the functional equation satisfied by $L(E,s)$. (Compare this discussion with that of n° 2.)

III. – Boyd's examples

In writing this chapter, the paper [CCGLS], to which we were referred by Boyd, was of critical importance.

8. Construction Given a Laurent polynomial $P \in \mathbb{C}[x^{\pm 1}, y^{\pm 1}]$, with x, y indeterminates, we let $\Delta(P)$ be its Newton polygon; i.e. $\Delta(P) \subset \mathbb{R}^2$ is the convex hull of the finitely many points $(m,n) \in \mathbb{Z}^2$ for which the coefficient of $x^m y^n$ in P is non-zero. Notice that $\Delta(P)$ is by construction a convex polygon with vertices in $\mathbb{Z}^2$; we will call such a set a *convex lattice polygon* or *clp* for short. By *lattice points* we will mean points in $\mathbb{Z}^2$. Throughout the rest of the paper we will denote by $\mathbb{T} \subset \mathbb{A}^2$ the special affine open set $\{(x,y) \in \mathbb{C}^2 | xy \neq 0\}$.

Given a clp Δ, it is known [DK] that for a generic Laurent polynomial $P \in \mathbb{C}[x^{\pm 1}, y^{\pm 1}]$ with $\Delta(P) = \Delta$ the equation $P = 0$ defines a smooth affine curve in $\mathbb{T}$ of genus equal to the number of interior lattice points of Δ. For example, a generic polynomial $P \in \mathbb{C}[x,y]$ of degree $d > 0$ has Newton polygon the triangle of vertices $(0,0), (d,0), (0,d)$ and $P = 0$ has genus $\frac{1}{2}(d-1)(d-2)$, as is well known. Similarly, a hyperelliptic equation $y^2 = f(x)$ with $f \in \mathbb{C}[x]$ of degree $d > 0$ has generically genus $[\frac{1}{2}(d-1)]$, the number of lattice points inside the triangle of vertices $(0,0), (0,2), (d,0)$.

We will use τ to denote a side of a clp Δ and write $\tau < \Delta$. We will always parameterize a side clockwise around Δ and in such a way that $\tau(0), \tau(1), \cdots$ are the consecutive lattice points in τ. To every side we associate a one-variable polynomial

$$P_\tau(t) = \sum_{k=0}^{\infty} c_{\tau(k)} t^k \in \mathbb{C}[t], \qquad \tau < \Delta,$$

where

$$P = \sum_{(m,n) \in \mathbb{Z}^2} c_{(m,n)} x^m y^n \in \mathbb{C}[x^{\pm 1}, y^{\pm 1}],$$

(both sums are, naturally, finite). Note that all roots of P_τ are non-zero.

Let $P \in \mathbb{C}[x^{\pm 1}, y^{\pm 1}]$ be irreducible. The equation $P(x,y) = 0$ defines an affine curve C' in $\mathbb{T}$. Let C be the normalization of the projective closure of C'; C is a compact Riemann surface with function field isomorphic to the fraction field of $\mathbb{C}[x^{\pm 1}, y^{\pm 1}]/(P)$. We view x, y as rational functions on C and let $S \subset C$ be the set of zeros and poles of x and y. The following result is basically contained in [CCGLS §3].

Proposition. *Let $P(x,y), x, y, C$ and S be as above, $\Delta = \Delta(P)$ the Newton polygon of P. Let*

$$\mathcal{T} = \{t \in \mathbb{C} \mid \prod_{\tau < \Delta} P_\tau(t) = 0\}.$$

Then for every $w \in S$ there is a $t \in \mathcal{T}$ and a nonzero integer n such that

$$(x,y)_w = \pm t^n$$

and the map $S \longrightarrow \mathcal{T}$ defined by $w \mapsto t$ is surjective.

(The proof uses Puiseux expansions to associate valuations to the sides of Δ, see [Lef] and [CCGLS].)

It will be convenient to define one more concept. Let $P \in \mathbb{C}[x^{\pm1}, y^{\pm1}]$ and let $\Delta = \Delta(P)$ be its Newton polygon. We will say that P is *tempered* if the set of roots of $\prod_{\tau<\Delta} P_\tau$ consists of roots of unity only. (If the P_τ's are monic and have integer coefficients then this is equivalent, by a theorem of Kronecker, to requiring that $m(P_\tau) = 0$ for all $\tau < \Delta$.)

Corollary. *With the notation of the proposition we have*

$$\{x, y\}^N \in K_{2,\emptyset}(C) \quad \textit{for some } N \in \mathbf{N} \qquad \Leftrightarrow \qquad P \textit{ is tempered.}$$

(This fact was also proved independently by H. Bornhorn.)

We now turn to Boyd's construction. The corollary gives us a way to produce elements in $K_{2,\emptyset}(E)$, where E is an elliptic curve defined over $\mathbb{Q}$. Indeed, what we need is a tempered Laurent polynomial $P \in \mathbb{Q}[x^{\pm1}, y^{\pm1}]$ such that the affine curve $P = 0$ in $\mathbb{T}$ is birational to E. The simplest way to do this is to consider a clp with only one interior lattice point and assign numbers to the lattice points on its boundary so that the resulting polynomial is tempered. This leaves the coefficient corresponding to the unique interior point free and hence, in fact, we obtain a *family* of tempered polynomials P_k, depending on a parameter k, all with the same Newton polygon. For all but finitely many choices of $k \in \mathbb{Q}$ the equation $P_k(x, y) = 0$ will yield a curve $C_k/\mathbb{Q}$ of genus 1 together with an element $\{x, y\}^N \in K_{2,\emptyset}(C_k)$ (for some $N \in \mathbf{N}$ independent of k). (We say "curves of genus 1" instead of "elliptic curves" as the curves do not necessarily have rational points.) We describe these families of curves in more detail in the next n°.

We should emphasize that though there is not a single instance of an elliptic curve $E/\mathbb{Q}$ for which we know that $K_2(\mathcal{E}) \otimes \mathbb{Q}$ is one-dimensional (or even finite-dimensional!) it is actually quite hard to construct elements in this group. Most previous examples relied on the original method of Bloch involving functions with divisors supported on torsion points of E [BG], [Be]. Other constructions were made by Ross [Ro], Cohen and Zagier, Goncharov and Levin [GL], and Rolhausen [Rol].

9. Tempered families Let us say that two clp's are *equivalent* if they differ by a translation and/or a change of basis of $\mathbb{Z}^2$. Correspondingly, let us say that two Laurent polynomials in $\mathbb{C}[x^{\pm1}, y^{\pm1}]$ are *equivalent* if they differ by a factor $x^m y^n$ with $m, n \in \mathbb{Z}$ and/or an invertible change of variables $x \mapsto x^a y^b, y \mapsto x^c y^d$ of $\mathbb{T}$ with $a, b, c, d \in \mathbb{Z}$. Clearly, equivalent polynomials have equivalent Newton polygons and define curves in $\mathbb{T}$ which are birational over their field of definition. Also, two equivalent polynomials have the same Mahler measure. We may hence restrict ourselves to polynomials up to equivalence.

Our first task is to describe up to equivalence all clp with only one interior lattice point. We will call such a clp *reflexive* (after [Ba]). (In higher dimensions the notion of reflexive polyhedron requires more than just having only one interior lattice point, see [Ba].) It is known [Sc] that there are only finitely many. (More generally, for fixed $m, n \in \mathbf{N}$ there are only finitely many non-equivalent convex lattice polytopes in $\mathbb{R}^n$ with m interior lattice points [He].) In fact, there are only 16 classes and it is not hard to write down representatives (see fig. 1).

FIGURE 1. Equivalence classes of reflexive clp's

Tempered polynomials with relatively prime integer coefficients and fixed Newton polygon have finitely many possibilities for their P_τ's. If the Newton polygon is reflexive then it is easy to check that they should be among the following (up to changes of the form $\pm P_\tau(\pm t)$)

0	1	2	3	4
1	0	−2	0	1
1	0	−1	0	0
1	0	−1	0	1
1	0	0	0	−1
1	0	0	0	1
1	0	0	1	0
1	0	1	0	0
1	0	1	0	1
1	0	2	0	1
1	1	−1	−1	0
1	1	0	−1	−1
1	1	0	0	0
1	1	0	1	1

0	1	2	3	4
1	1	1	0	0
1	1	1	1	0
1	1	1	1	1
1	1	2	1	1
1	2	0	−2	−1
1	2	1	0	0
1	2	2	1	0
1	2	2	2	1
1	2	3	2	1
1	3	3	1	0
1	3	4	3	1
1	4	6	4	1

TABLE 1

(to clarify the notation: the first line in the second column corresponds to $P_\tau(t) = t^2 + t + 1$).

Examples (a) Consider the Hesse family of elliptic curves

$$x^3 + y^3 + z^3 - kxyz.$$

Upon setting $z = 1$ we obtain a tempered family whose Newton polygon is the triangle of vertices $(0,0), (3,0), (0,3)$ (after translation by $(1,1)$).

(b) The family

$$x + \frac{1}{x} + y + \frac{1}{y} - k$$

is tempered with Newton polygon the square of vertices

$$(0,1), \quad (0,-1), \quad (-1,0), \quad (1,0).$$

(c) Upon setting $z = 1$ in

$$(x-y)(x-z)(y-z) - kxyz$$

we obtain a tempered family with Newton polygon the hexagon of vertices

$$(1,0), \quad (2,0), \quad (2,1), \quad (1,2), \quad (0,2), \quad (0,1)$$

(after translation by $(1,1)$).

We fix the following notation: given a family of polynomials P_k as above and a value of $k \in \mathbb{Q}$ such that the affine curve $P_k = 0$ is non-singular in $\mathbb{T}$ we let $C_k/\mathbb{Q}$ be the smooth projective curve of genus 1 that it determines (after projective closure and normalization) and let E_k be its Jacobian, an elliptic curve defined over $\mathbb{Q}$,

Remark By analogy we may say that $P \in \mathbb{C}[x, 1/x]$ is tempered if the coefficients of its largest and smallest power of x are roots of unity. Then, for example, a polynomial $P \in \mathbb{Z}[x]$ is tempered if and only if its roots are units in $\overline{\mathbb{Q}}$. Any reflexive clp in one dimension is equivalent to the segment $[-1, 1]$ and hence, up to equivalence, the only tempered families of polynomials (defined over $\mathbb{Q}$) in one variable with a reflexive Newton polygon are (up to scaling) those of n° 5.

10. What fits the facts We may now give the general form of question $(?_1)$ of the introduction. Its formulation is due to Boyd and was based largely on the analysis of numerical data.

Conjecture. *With the above notation*

$$(?_3) \qquad B_k = \frac{L'(E_k, 0)}{m(P_k)} \in \mathbb{Q}^* \qquad \textit{for all sufficiently large } k \in \mathbb{Z}.$$

(The requirement that k be sufficiently large guarantees both that $P_k = 0$ is non-singular in $\mathbb{T}$ and that P_k does not vanish on T^2, see n° 11 below.)

In order to relate this question to the Bloch–Beilinson conjecture we need one more result. Let $P \in \mathbb{Q}[x^{\pm 1}, y^{\pm 1}]$ be a tempered Laurent polynomial, irreducible over $\overline{\mathbb{Q}}$ and not vanishing on the torus T^2. As in n° 8, let $C/\mathbb{Q}$ be the normalization of the projective closure of the curve in $\mathbb{T}$ determined by $P = 0$ and regard x, y as functions in $\mathbb{Q}(C)$.

Theorem. *With the above notation*

$$m(P) = \tfrac{1}{2\pi}\, \bar{r}(\{x, y\})([\gamma])$$

for some non-trivial cycle $\gamma \in H_1(C, \mathbb{Z})$ fixed by complex conjugation.

The homology class $[\gamma]$ is related to the class of the torus T in $H_2(\mathbb{P}^2 \setminus C, \mathbb{Z})$ by the tube homomorphism (see [Gr]). The identity is a consequence of the dimension 2 analogue of the lemma of n° 4 (see [RV] for more details).

Note that if C is an elliptic curve then $[\gamma]$ is a non-zero integer multiple of c_0 (the class of the connected component of the real points of C as in n° 4). Therefore, $m(P)$ is up to an non-zero integer factor precisely the quantity associated to $\{x, y\}$ appearing in the Bloch–Beilinson conjecture!

Hence all that remains to check is that some power of $\{x, y\}$ actually lies in $K_2(\mathcal{E}_k)$, where $\mathcal{E}_k$ is a Néron model of E_k. This we have not been able to confirm, but given the numerical evidence it seems safe to suggest that

$$(?_4) \qquad k \in \mathbf{Z} \quad \Rightarrow \quad \{x, y\}^M \in K_2(\mathcal{E}_k), \quad \text{for some } M \in \mathbf{N}.$$

Granting $(?_4)$ then, $(?_3)$ would follow from the Bloch–Beilinson conjecture.

IV. – Modular Mahler Measures

11. Expansions Let P_k be one of the tempered families of n° 9. By translating if necessary, we may assume that the unique lattice point of $\Delta(P_k)$ is the origin. We then have

$$P_k(x, y) = k - P(x, y),$$

for a Laurent polynomial $P \in \mathbb{C}[x^{\pm 1}, y^{\pm 1}]$ with no constant term. We will study $m(P_k)$ as a function of the complex parameter k.

Let $\mathcal{K} \subset \mathbb{C}$ be the image of T^2 under $(x, y) \mapsto P(x, y)$. Since T^2 is compact and the map continuous we see that $\mathcal{K}$ is compact. Clearly

$$P_k \text{ vanishes on } T^2 \quad \Leftrightarrow \quad k \in \mathcal{K}.$$

In what follows it will be convenient to use $\lambda = 1/k$ as the parameter rather than k. It is not hard to see that $\mathcal{K}$ cannot reduce to the origin. Define $R > 0$ as

$$\frac{1}{R} = \max_{(x,y) \in T^2} |P(x, y)|.$$

Then

$$1 - \lambda\, P(x, y) \text{ does not vanish on } T^2 \text{ for } |\lambda| < R.$$

We define

$$\tilde{m}(\lambda) = -\log \lambda - \sum_{n=1}^{\infty} \frac{a_n}{n} \lambda^n, \qquad |\lambda| < R, \quad \lambda \notin (-\infty, 0],$$

where $\log\lambda$ is the usual branch of logarithm and

$$a_n = \frac{1}{(2\pi i)^2}\int_{T^2} P(x,y)^n \, \frac{dx}{x}\frac{dy}{y}, \qquad n = 0,1,2,\ldots$$

$$= [P(x,y)^n]_0,$$

where for a Laurent polynomial $Q \in \mathbb{C}[x^{\pm1}, y^{\pm1}]$ we let $[Q]_0$ denote its the constant term. Note that $a_0 = 1$ and $a_1 = 0$.

Clearly $\tilde{m}$ is holomorphic and

$$m(P_k) = \operatorname{Re}(\tilde{m}(\lambda)), \qquad k = \frac{1}{\lambda}, \qquad |\lambda| < R.$$

Let

$$u_0(\lambda) = \frac{1}{(2\pi i)^2}\int_{T^2} \frac{1}{1-\lambda P(x,y)}\frac{dx}{x}\frac{dy}{y}$$
$$= \sum_{n=0}^{\infty} a_n\lambda^n, \qquad |\lambda| < R.$$

Then

$$\tilde{m}(\lambda) = -\log\lambda - \int_0^\lambda (u_0(t) - 1)\frac{dt}{t}.$$

12. Computing the coefficients Here is a way of computing the coefficients a_n in the expansion of $\tilde{m}$ and u_0 which is often useful in practice. Write

$$P(x,y) = \sum_{j=1}^{N} c_j\, x^{r_j} y^{s_j}, \qquad c_j \neq 0, \qquad j = 1, \ldots, N$$

and let

$$M = \begin{pmatrix} r_1 & r_2 & \ldots & r_N \\ s_1 & s_2 & \ldots & s_N \end{pmatrix} \in \mathbb{Z}^{2\times N}.$$

Then by the multinomial theorem

$$a_n = \sum_{\substack{w=(w_1,\ldots,w_N)\in\mathbb{Z}_{\geq 0}^N \\ w_1+\cdots+w_N=n \\ Mw=0}} \frac{n!}{w_1!\cdot\cdots\cdot w_N!}\cdot c_1^{w_1}\cdot\cdots\cdot c_N^{w_N}.$$

In particular, a_n only depends on the kernel $\ker M$ of M and on the constants $(c_1, \ldots, c_N)$. Note that in the situation of n° 9 (the only one we will consider) M always has rank 2 and hence $\ker M$ has rank $N-2$, where N is the number of non-zero monomials in P.

Example. The simplest case of the formula for a_n is when P has exactly 3 non-zero monomials (the minimum possible) and hence $\ker M$ has rank 1. In this case the Newton polygon of the family is a reflexive triangle, of which there are 5 up to equivalence (see fig. 1). The corresponding functions u_0 are hypergeometric. We have the following three possibilities (unlisted coefficients a_n are zero):

$\ker M$	u_0	coefficients
$\mathbf{Z}(1,1,1)$	$F(\frac{1}{3},\frac{2}{3};1;\ (-1)^\nu 3^3\lambda^3)$	$a_{3n}=(-1)^{\nu n}\frac{(3n)!}{n!^3}$
$\mathbf{Z}(1,1,2)$	$F(\frac{1}{4},\frac{3}{4};1;\ (-1)^\nu 2^6\lambda^4)$	$a_{4n}=(-1)^{\nu n}\frac{(4n)!}{n!^2\,(2n)!}$
$\mathbf{Z}(1,2,3)$	$F(\frac{1}{6},\frac{5}{6};1;\ (-1)^\nu 2^4 3^3\lambda^6)$	$a_{6n}=(-1)^{\nu n}\frac{(6n)!}{n!\,(2n)!\,(3n)!}$,

TABLE 2

where F is the standard hypergeometric function and $\nu = 0$ or 1.

These cases arise as follows (it is a general fact about reflexive simplices in any dimension [Ba]). If $w = (w_1, w_2, w_3)$ with $w_j \in \mathbf{Z}_{\geq 0}$ generates $\ker M$ let $d = w_1 + w_2 + w_3$ and $d_j = d/w_j$ then

$$\frac{1}{d_1} + \frac{1}{d_2} + \frac{1}{d_3} = 1, \qquad d_j \in \mathbf{Z}_{\geq 0}, \quad j = 1, 2, 3.$$

The solutions to this equation in positive integers are, respectively

$$(3,3,3), \qquad (2,4,4), \qquad (2,3,6).$$

(The degenerate case $(2,2,\infty)$ could be thought to correspond to the polynomials $x \pm 1/x - k$ of n° 5.)

Let us mention that examples with other Newton polygons may also yield coefficients a_n given as products of factorials (and hence u_0 hypergeometric). For example, for n° 9 (b) and (c) $a_n = 0$ for n odd and

$$a_{2n} = \binom{2n}{n}^2, \qquad \sum_{j=0}^{n} (-1)^j \binom{2n}{j}^3 = (-1)^n \frac{(3n)!}{(n!)^3},$$

respectively (the last equality is due to Dixon).

13. Modular expansions In the above examples u_0 turned out be a hypergeometric function, which satisfies a linear second order differential equation. In fact, this is true in all cases; u_0 satisfies a differential equation with only regular singularities, the Picard–Fuchs differential equation, of the form

$$\text{(PF)} \qquad A\frac{d^2u_0}{d\lambda^2} + B\frac{du_0}{d\lambda} + Cu_0 = 0,$$

where A, B, C are polynomials in λ. The reason is that the integral defining $u_0(\lambda)$ implies (see [Gr]) that it is a period of a holomorphic differential on the curve defined by $1 - \lambda P(x, y) = 0$. In our case is not hard to see that there must be a second solution u_1 of the form

$$u_1(\lambda) = u_0(\lambda)\log\lambda + v_1(\lambda), \qquad |\lambda| < R,$$

with v_1 holomorphic and $v_1(0) = 0$.

We define

$$\tau = \frac{1}{2\pi i}\frac{u_1}{u_0}, \qquad q = e^{2\pi i\tau} = \lambda + \cdots.$$

A loop around $\lambda = 0$ takes τ to $\tau + 1$ and hence fixes q. It follows that we may locally invert q and write $\lambda(\tau) = q + \cdots$ as a power series in q. We let

$$c(\tau) = u_0(\lambda(\tau)) = 1 + \cdots$$

$$e(\tau) = c\frac{q d\lambda/dq}{\lambda} = 1 + \sum_{n=1}^{\infty} e_n q^n.$$

The functions λ, c, e behave like modular forms (with singularities), under the action of the monodromy group of (**PF**), of weights $0, 1, 3$ respectively. Often we may relate them to usual modular forms (we give some examples below).

Finally, we obtain an expression for $\tilde{m}$ as a function of τ. Note that $\lambda = 0$ corresponds to $\tau = i\infty$. The change of variables $t = \lambda(\tau)$ yields the following.

Theorem. *With the above notation we have, locally around* $\tau = i\infty$,

$$\tilde{m}(\lambda(\tau)) = -2\pi i\tau - \sum_{n=1}^{\infty}\frac{e_n}{n}q^n, \qquad q = e^{2\pi i\tau}.$$

14. Example 1. Let us consider the Hesse family of elliptic curves (see n° 9 (a)). The corresponding Newton polygon is a reflexive triangle corresponding to the first row of table 2. For this example it will be more convenient to formulate things in terms of the parameter $\mu = \lambda^3$ instead of λ. (A simple change of variables shows that $m(x^3 + y^3 + 1 - kxy)$ only depends on k^3.) We find

$$u_0(\mu) = \sum_{n=0}^{\infty}\frac{(3n)!}{(n!)^3}\mu^n$$
$$= F(\tfrac{1}{3}, \tfrac{2}{3}; 1; 3^3\mu),$$

which satisfies the differential equation

$$\mu(27\mu-1)\frac{d^2u_0}{d\mu^2}+(54\mu-1)\frac{du_0}{d\mu}+6u_0=0;$$

a second solution around $\mu=0$ is

$$u_1(\mu)=u_0(\mu)\log\mu+15\mu+513/2\mu^2+5018\mu^3+\cdots.$$

In this case μ, c, e are modular of level 3:

$$c=1+6\sum_{n=1}^{\infty}\sum_{d|n}\chi(d)\,q^n,$$
$$e=1-9\sum_{n=1}^{\infty}\sum_{d|n}\chi(d)d^2\,q^n,$$
$$\mu=\tfrac{1}{27}(1-\frac{e}{c^3})=q-15q^2+171q^3-1679q^4+\cdots,$$

where $\chi=\left(\frac{d}{3}\right)$. In fact, μ is a Hauptmodul for the group $\Gamma_0(3)\subset SL_2(\mathbb{Z})$ with fundamental domain $\mathcal{F}$ formed by the geodesic triangle in the upper-half plane of vertices $i\infty, 0, (1+i/\sqrt{3})/2$ and its reflection along the imaginary axis. The value of $k^3=1/\mu$ at these vertices is respectively $\infty, 27, 0$.

The theorem of n° 13 yields

$$\tilde{m}(\mu)=\tfrac{1}{3}[\,-2\pi i\tau+9\sum_{n=1}^{\infty}\sum_{d|n}\chi(d)d^2\,\frac{q^n}{n}\,],$$

and provides an analytic continuation for $\tilde{m}$ to a cut μ-plane. The compact region $\mathcal{K}$ in this case consists of an hypocycloid with vertices at the roots of k^3-27. For every $k\in\mathbb{C}\backslash\mathcal{K}$ we have $\mathrm{Re}\,[\tilde{m}(1/k^3)]=m(x^3+y^3+1-kxy)$ since both sides are harmonic and agree on a neighborhood of $k=\infty$. In the interior of $\mathcal{K}$, however, the functions do not agree. In fact, at $k=0$, as we will see shortly, $\tilde{m}=0$, whereas by the formula of Smyth quoted in the introduction, $m(x^3+y^3+1)=m(x+y+1)=L'(\chi,-1)\neq 0$. On the other hand, the functions do agree, by continuity, on the boundary of $\mathcal{K}$. In particular, if we let τ approach the cusp 0 within the fundamental domain (so that k approaches 3) we find that

$$\begin{aligned} m(x^3+y^3+1-3xy) &= 3\lim_{y\to 0}\left[\sum_{n=1}^{\infty}\sum_{d|n}\chi(d)\,d^2\,\frac{e^{-2\pi ny}}{n}\right]\\ &= 3[\zeta(s)L(\chi,s-2)]\big|_{s=1}\\ &= 3L'(\chi,-1). \end{aligned}$$

(The second equality follows for example by shifting the line of integration in

$$\int_{\mathrm{Re}\, w=\eta_0} t^{-w}\Gamma(w)\zeta(s+w)L(\chi,s+w-2)\,dw\,, \qquad \eta_0 \gg 0.)$$

The above identity can also be proven directly as was pointed out to us by Boyd. Over $\mathbb{Q}(w)$, with w a primitive cubic root of unity,

$$x^3+y^3+1-3xy=(x+y+1)(x+w^2y+w)(x+wy+w^2)$$

and each term on the right hand side has measure $L'(\chi,-1)$ by Smyth's result.

Since e is an Eisenstein series we find, after some calculation, that

$$(\dagger)\qquad \begin{aligned} m(P_k) &= \tfrac{1}{3}\mathrm{Re}\,[\,-2\pi i\tau + 9\sum_{n=1}^{\infty}\sum_{d|n}\chi(d)d^2\,\frac{e^{2\pi i n\tau}}{n}\,] \\ &= \mathrm{Re}\,[\,\frac{3^3\sqrt{3}y}{4\pi^2}\sum_{m,n\in\mathbf{Z}}{}' \chi(n)\frac{1}{(m3\tau+n)^2(m3\overline{\tau}+n)}\,]\,, \end{aligned}$$

where $\tau = x+iy$ and, as before, $k^3 = 1/\mu(\tau)$ with $\tau \in \mathcal{F}$. (This shows that here $(?_3)$ actually reduces to a case of Bloch's original conjecture involving the regulator of functions with divisors supported on torsion points.)

The above calculation also allows us to prove $(?_3)$ for $k=-6$. Indeed, in this case $P_k=0$ determines an elliptic curve $E/\mathbb{Q}$ with complex multiplication by the ring of integers $\mathcal{O}_K$ of $K=\mathbb{Q}(w)$ (w a primitive 3rd root of unity) and $\tau=(1+\sqrt{-3})/2$; E has conductor 27 and minimal model $y^2+y=x^3$. (It is not hard to see that

$$y^2=x^3-27k^2x^2+216k(k^3-27)x-432(k^3-27)^2$$

is a Weierstrass equation for the curve $P_k=0$ over $\mathbb{Q}(k)$.)

By Deuring the L-function of E is $L(\overline{\psi},s)$, where ψ is a Hecke character of K; we find that $\psi((\alpha))=\alpha$ for $\alpha\in\mathcal{O}_K$ with $\alpha\equiv 1 \bmod 3\mathcal{O}_K$. Therefore, $2L(E,2)$ is visibly given by the last sum in $(\dagger)$ with $\tau=(1+\sqrt{-3})/2$ and we obtain

$$L'(E,0)=\tfrac{27}{4\pi^2}L(E,2)=\tfrac{1}{3}m(P_{-6}).$$

Finally, for $k=0$ we find $\tau=(1+i/\sqrt{-3})/2$ and by $(\dagger)$

$$\begin{aligned} \widetilde{m}(P_0) &= c\sum_{\alpha\in\mathcal{O}_K}\left(\frac{\alpha}{\sqrt{-3}}\right)\frac{1}{\alpha^2\overline{\alpha}}\,, \qquad c>0 \\ &= 0. \end{aligned}$$

(To see that the sum vanishes change α to $w\alpha$.)

It turns out that in this case ($?_3$) actually seems to hold for all sufficiently large k such that $k^3 \in \mathbf{Z}$ (and not just $k \in \mathbf{Z}$); the corresponding elliptic curve has Weierstrass model

$$y^2 = x^3 - 27m^2x^2 + 216m^3(m-27)x - 432m^4(m-27)^2,$$

where $m = k^3$.

We should add that the modular forms and hypergeometric functions in this example also appear in a related context in the recent work of Hosono, Saito and Stienstra [HSS] and of Zagier [Z].

15. Example 2 Let us consider the family $P_k = x + 1/x + y + 1/y - k$ of the introduction (also n°9 (b)); the situation is very similar to that of the previous example and we will hence be brief. We let $\mu = \lambda^2$ then

$$\begin{aligned} u_0(\mu) &= \sum_{n=0}^{\infty} \binom{2n}{n}^2 \mu^n \\ &= F(\tfrac{1}{2}, \tfrac{1}{2}; 1; 2^4\mu), \end{aligned}$$

which satisfies the differential equation

$$\mu(16\mu - 1)\frac{d^2u_0}{d\mu^2} + (32\mu - 1)\frac{du_0}{d\mu} + 4u_0 = 0;$$

a second solution at $\mu = 0$ is

$$u_1(\mu) = u_0(\mu)\log\mu + 8\mu + 84\mu^2 + 2960/3\mu^3 + \cdots.$$

In this case μ, c, e are modular of level 4:

$$\begin{aligned} c &= 1 + 4\sum_{n=1}^{\infty}\sum_{d|n} \chi(d)\, q^n, \\ e &= 1 - 4\sum_{n=1}^{\infty}\sum_{d|n} \chi(d) d^2\, q^n, \\ \mu &= \frac{\phi}{c^2} = q - 8q^2 + 44q^3 - 192q^4 + \cdots, \end{aligned}$$

where $\chi(n) = \left(\frac{n}{4}\right)$ and

$$\phi = \sum_{\substack{n \ge 1 \\ n \text{ odd}}} \sum_{d|n} d\, q^n = q + 4q^3 + 6q^5 + \cdots.$$

In fact, μ is a Hauptmodul for the subgroup of $SL_2(\mathbb{Z})$ with fundamental domain $\mathcal{F}$ formed by the geodesic triangle in the upper-half plane of vertices $i\infty, 0, 1/2$ and its reflection along the imaginary axis. The value of $k^2 = 1/\mu$ at these vertices is respectively $\infty, 16, 0$.

Although not a fact we will need, let us remark that by an identity of Jacobi (see e.g. [WW])

$$\theta^2 = F(\tfrac{1}{2}, \tfrac{1}{2}; 1; 16\mu), \qquad \text{where} \qquad \theta = \sum_{n\in\mathbb{Z}} q^{n^2},$$

and therefore $c = \theta^2$.

As in example 1, e is an Eisenstein series (and hence we are also in the original situation of Bloch involving functions supported on torsion points) and, as mentioned in the introduction, ($?_3$) appears to hold for all sufficiently large k such that $k^2 \in \mathbb{N}$ (the corresponding elliptic curve is A_m described below with $k^2 = 16m$).

The region $\mathcal{K}$ now consists of the interval $[-4, 4]$ and hence by continuity we find that

$$\begin{aligned} m(x + 1/x + y + 1/y - k) &= \tfrac{1}{2}\mathrm{Re}\,[\,-2\pi i\tau + 4\sum_{n=1}^{\infty}\sum_{d|n}\chi(d)d^2\,\frac{q^n}{n}\,] \\ &= \mathrm{Re}\,[\,\frac{16y}{\pi^2}\sum_{m,n\in\mathbb{Z}}{}'\,\chi(n)\frac{1}{(m4\tau+n)^2(m4\bar{\tau}+n)}\,] \end{aligned}$$

for *all* $k \in \mathbb{C}$ where $k^2 = 1/\mu(\tau)$ and $\tau \in \mathcal{F}$. (This identity was previously proved for $k = 1$ by Deninger [De].)

We also find, by taking limits as before, that $m(x+1/x+y+1/y-4) = 2L'(\chi, -1)$, which again may be established directly, and that we can prove the identities

$$\begin{aligned} &m(x + 1/x + y + 1/y - 4\sqrt{2}) = L'(A, 0), \\ &A : y^2 = x^3 - 44x + 112 \end{aligned}$$

$$\begin{aligned} &m(x + 1/x + y + 1/y - \frac{4}{\sqrt{2}}) = L'(B, 0) \\ &B : y^2 = x^3 + 4x. \end{aligned}$$

since A, B are CM elliptic curves.

Finally, for $k = 3\sqrt{2}$ we obtain the modular curve $X_0(24)$ and then a theorem of Beilinson applies and we can prove ($?_1$) in this case (see [MS] and [RV] for more details).

In the table 4 at the end of the paper we give some numerical data for this family. The table is organized as follows. For $m \in \mathbb{Z}$ let

$$A_m : \qquad y^2 = x^3 + 2m(2m-1)x^2 + m^2x$$

an elliptic curve of discriminant $2^8m(m-1)$; (A_m is one of the curves over $\mathbb{Q}(k^2)$ isomorphic to E_k where $16m = k^2$). The first column of the table is k^2; the second N, the conductor of A_m; the third ϵ, the sign of in the functional equation of the corresponding L-function; the fourth the coefficients of the minimal of A_m (in standard notation); and the fifth the (numerical) value of $B_k = L'(A_m, 0)/m(P_k)$. The rows are ordered according to the size of N.

16. Example 3. Finally, we consider an example involving the regulator of functions not supported on torsion points, i.e. where e is not an Eisenstein series. Let $P_k(x,y) = x^2/y - y/x - 1/xy - k$ corresponding to the third row of table 2. With $\mu = \lambda^6$ we have

$$\begin{aligned} u_0(\mu) &= \sum_{n=0}^{\infty} \frac{(6n)!}{(3n)!(2n)!n!}\mu^n \\ &= F(\tfrac{1}{6}, \tfrac{5}{6}; 1; 2^4 3^3 \mu), \end{aligned}$$

which satisfies the differential equation

$$\mu(432\mu - 1)\frac{d^2u_0}{d\mu^2} + (864\mu - 1)\frac{du_0}{d\mu} + 60u_0 = 0;$$

a second solution at $\mu = 0$ is

$$u_1(\mu) = u_0(\mu)\log\mu + 312\mu + 77652\mu^2 + 23485136\mu^3 \cdots .$$

We find that we can express μ, c, e in terms of the standard modular forms

$$E_4 = 1 + 240\sum_{n=1}^{\infty}\sum_{d|n} d^3\, q^n, \qquad E_6 = 1 - 504\sum_{n=1}^{\infty}\sum_{d|n} d^5\, q^n$$

as follows

$$\begin{aligned} c &= E_4^{1/4} = 1 + 60q - 4860q^2 + 660480q^3 - 105063420q^4 + \cdots \\ e &= \tfrac{1}{2}(E_4^{3/4} + \frac{E_6}{E_4^{3/4}}) = 1 - 252q + 53244q^2 - 11278368q^3 + \cdots \\ \mu &= \tfrac{1}{864}(1 - \frac{E_6}{E_4^{3/2}}) = q - 312q^2 + 87084q^3 - 23067968q^4 + \cdots . \end{aligned}$$

It is not hard to see that these q-series are indeed in $\mathbb{Z}[[q]]$ as suggested by their first few terms. Notice that in contrast with example 1, the functions c, e, μ now have singularities (at $\tau = (1+\sqrt{-3})/2$ where E_4 vanishes).

Let $\mathcal{F}$ be the domain in the upper-half plane formed the geodesic triangle of vertices $i\infty, 0, (1+\sqrt{-3})/2$ and its reflection along the imaginary axis. The function

μ gives a conformal mapping between $\mathcal{F}$ and some domain in the μ-plane. The values of $1/\mu$ at $i\infty, 0, (1+\sqrt{-3})/2$ are respectively $\infty, 432, 0$. As before the theorem of n° 13 provides an analytic continuation of $\tilde{m}$ to a cut μ-plane.

As a numerical example, consider $\tau = i \in \mathcal{F}$ at which μ takes the value 864. Then

$$m(x^3 - y^2 - 1 - \sqrt[6]{864}\, xy) = \tfrac{1}{6}[\, 2\pi - \sum_{n=1}^{\infty} \frac{e_n}{n} e^{-2\pi n}\,] = 1.11330531\cdots,$$

where as before $e = 1 + \sum_{n=1}^{\infty} e_n q^n$ (again ($?_3$) should hold for all sufficiently large k such that $k^6 \in \mathbb{N}$). This number seems to equal $\frac{1}{3}L'(E,0)$, where E is the elliptic curve with CM by $\mathbb{Z}[i]$ of minimal model $y^2 = x^3 - 27x$ and conductor 576. Despite the fact that E is a curve with CM we see no way of actually proving this.

Although e is not an Eisenstein series, by analogy (and inspired by the "mirror symmetry" of physics, which actually informed much of this chapter) we may still consider the numbers r_n such that

$$e_n = \sum_{d|n} r_d\, d^2, \qquad n = 1, 2, 3, \ldots .$$

(These identities define the numbers uniquely as one may see using the Möbius inversion formula

$$r_n = \frac{1}{n^2} \sum_{d|n} \mu(n/d)\, e_d,$$

where (here only) μ is the standard Möbius function.) A priori we can only assert that the r_n's are rational. Remarkably, it appears that they are actually integers for all n. Here is a short table.

n	e_n	r_n
1	−252	−252
2	53244	13374
3	−11278368	−1253124
4	2431713276	151978752
5	−531387193752	−21255487740
6	117213742465056	3255937602498
7	−26029619407786176	−531216722607876
8	5809495541986361340	90773367805541376
9	−1301648449222030804956	−16069733941012586748
10	292541140545091693518744	2925411405456230806590

TABLE 3

The power of d in the definition of r_n (namely 2) is somewhat natural since e is formally of weight 3; also, the analogue of the r_n's defined with a higher power of d are not integers. Notice that whether $r_n \in \mathbb{Z}$ for all $n \in \mathrm{N}$ may be formulated purely in terms of classical modular forms, independently of everything else we have done.

In order to obtain the expansions of this chapter it was not really necessary for the family of polynomials to be tempered. We may ask: (how) is the tempered condition reflected in the expansions? The examples we computed seem to suggest the following

$$(?_5) \qquad \begin{gathered} P_k \textit{ is tempered} \\ \Updownarrow \\ \textit{there exists an } N \in \mathrm{N} \textit{ such} \\ \textit{that } r_n N^n \in \mathbb{Z} \textit{ for all } n \in \mathrm{N}\,. \end{gathered}$$

(Here r_n are the analogue of the numbers defined above.)

16. Final remarks and speculations

(1) The conjectures of Beilinson are more general than those outlined here; as with Dirichlet's class number formula, if the rank in question is $r > 1$ one compares an L-value with the determinant of an $r \times r$ matrix constructed using the regulator map.

(2) Recent work of Bloch and Kato [BK] makes the conjectures more precise by attributing a meaning to the ratio of the L-value and the regulator (again in analogy to what happens with Dirichlet's class number formula).

(3) Given the formal similarities between the behavior of the coefficients e_n in the tempered case and those appearing in the expansions of "mirror symmetry", it is tempting to ask whether there is a deeper connection. For example, do the r_n's have a natural interpretation as the number of objects of some sort? Or perhaps they are involved in a theta-lift of the kind considered by Borcherds [Bor]?

(4) Much of chapters III and IV can be extended to higher dimensions (see [RV]). A particularly nice case is that of families of K3-surfaces, for example, those determined by $x^4 + y^4 + z^4 + w^4 - kxyzw = 0$.

References

[Ba] V. Batyrev, *Dual polyhedra and mirror symmetry for Calabi–Yau hypersurfaces in toric varieties*, J. Algebraic Geom. **3** (1994), 493–535.

[Be] A. Beilinson, *Higher regulators of modular curves*, Applications of algebraic K-theory to algebraic geometry and number theory, Part I, II (Boulder, Colo., 1983), Contemp. Math., vol. **55**, Amer. Math. Soc., Providence, R.I., 1986, pp. 1–34.

[Bo1] D. W. Boyd, *Mahler's measure and special values of L-functions*, Experiment. Math. **7** (1998), 37–82.

[Bo2] ———, *Speculations concerning the range of Mahler's measure*, Canad. Math. Bull. **24** (1981), 453–469.

[Bo3] ———, *Kronecker's theorem and Lehmer's problem for polynomials in several variables*, J. Number Theory **13** (1981), 116–121.

[Bor] R. Borcherds, *Automorphic forms on $O_{s+2,2}(\mathbb{R})$ and infinite products*, Invent. Math. **120** (1995), 161–213.

[BLR] S. Bosch, W. Lütkebohmert and M. Raynaud, *Néron models*, Ergebnisse der Mathematik und ihrer Grenzgebiete (3), Springer-Verlag, Berlin, 1990.

[BG] S. Bloch & D. Grayson, *K_2 and $L-$functions of elliptic curves: Computer Calculations*, Contemp. Math. **55** (1986), 79–88.

[BK] S. Bloch & K. Kato, *L-Functions and Tamagawa Numbers of Motives*, The Grothendieck Festschrift (P. Cartier et al, eds.), vol. I, Birkhäuser, Boston; Progress in Mathematics, vol. 86, 1990, pp. 333–400.

[CCGLS] D. Cooper, M. Culler, H. Gillet, D. D. Long, and P. B. Shalen, *Plane curves associated to character varieties of 3-manifolds*, Invent. Math. **118** (1994), 47–84.

[DK] V. I. Danilov and A. G. Khovansky, *Newton polyhedra and an algorithm for computing Deligne–Hodge numbers*, Math. USSR–Izv **29** (1987), 279–298.

[De] C. Deninger, *Deligne periods of mixed motives, K-theory and the entropy of certain $\mathbf{Z}^n$-actions*, J. Amer. Math. Soc. **10** (1997), 259–281.

[Gr] P. Griffiths, *On the periods of certain rational integrals: I, II*, Ann. of Math., 461–541.

[GL] A. B. Goncharov and A. M. Levin, *Zagier's conjecture on $L(E,2)$*, preprint, 1997.

[He] D. Hensley, *Lattice vertex polytopes with interior lattice points*, Pacific J. Math. **105** (1983), 183–191.

[Z] Hosono, Saito, and Stienstra, *On the mirror symmetry conjecture for Schoen's Calabi-Yau threefolds*, preprint 1998.

[Le] D. H. Lehmer, *Factorization of certain cyclotomic functions*, Ann. of Math. (2) **34** (1933), 461–479.

[Lef] S. Lefschetz, *Algebraic Geometry*, Princeton Univ. Press, Princeton, NJ, 1953.

[LSW] D. Lind, K. Schmidt, and T. Ward, *Mahler measure and entropy for commuting automorphisms of compact groups*, Invent. Math. **101** (1990), 593–629.

[Ma] K. Mahler, *On some inequalities for polynomials in several variables*, J. London Math. Soc. (2) **37** (1962), 341–344.

[MS] J. F. Mestre and N. Schappacher, *Séries de Kronecker et fonctions L de puisances symmétriques de courbes elliptiques sur* $\mathbb{Q}$, Arithmetic Algebraic Geometry (G. van der Geer, F. Oort, and J. Steenbrink, eds.), Progress in Mathematics **89**, Birkhäuser, 1991, pp. 209–245.

[Mi] J. Milnor, *Introduction to algebraic K-theory*, Annals of Mathematical Studies, vol. 72, Princeton Univ. Press, Princeton, N.J., 1971.

[Ph] P. Philippon, *Critères pour l'indépendence algébrique*, Publ. IHES **64** (1986), 5–52.

[Pi] T. Pierce, *The numerical factors of the arithmetic functions* $\prod_{i=1}^{n}(1 \pm \alpha_i^m)$, Ann. of Math. **18** (1916–17).

[Ram] D. Ramakrishnan, *Regulators, Algebraic Cycles, and Values of L-functions*, Algebraic K-theory and Algebraic Number Theory (M. R. Stein and R. K. Dennis, eds.), Contemporary Mathematics, vol. **83**, Amer. Math. Soc., Providence, R.I., 1989, pp. 183–310.

[Ra] G. A. Ray, *Relations between Mahler's measure and values of L-series*, Canad. J. Math. **39** (1987), 649–732.

[RV] F. Rodriguez Villegas, *Modular Mahler measures II*, in preparation.

[Ro] R. Ross, Ph. D. thesis, Rutgers University, 1989.

[Rol] K. Rolhausen, *Eléments explicites dans K_2 d'une courbe elliptique*, Institut de Recherche Mathématique Avancée, Strasbourg, 1996.

[Sa] P. Sarnak, *Spectral behavior of quasi-periodic potentials*, Comm. Math. Phys. **84** (1982), 377–401.

[Sch] N. Schappacher, *Les conjectures de Beilinson pour les courbes elliptiques*, Journées Arithmétiques, 1989 (Luminy, 1989), Astérisque, No. **198–200**, 1992, pp. 305–317.

[Sc] P. R. Scott, *On convex lattice polygons*, Bull. Austral. Math. Soc. **15** (1976), 395–399.

[Si] J. H. Silverman, *The arithmetic of elliptic curves*, Springer Verlag, Berlin and New York, 1986.

[So1] C. Soulé, *Geometrie d'Arakelov et théorie des nombres trascendants*, Journées Arithmetiques de Luminy (G. Lachaud, eds.), Astérisque, No. **198–200**, 1991, pp. 355–371.

[So2] ———, *Regulateurs*, Seminar Bourbaki, Vol. 1984/85, Astérisque, No. **133–134**, 1986, pp. 237–253.

[Sm] C. J. Smyth, *On measures of polynomials in several variables*, Bull. Austral. Math. Soc. **23** (1981), 49–63.

[WW] E. T. Whittaker and G. N. Watson, *A Course of Modern Analysis*, Cambridge University Press, Cambridge, 1984.

[Z] D. Zagier, *A modular identity arising from mirror symmetry*, preprint 1998.

Department of Mathematics
University of Texas at Austin

Preprint
E-mail address: `villegas@math.utexas.edu`

k^2	N	ϵ	a_1	a_2	a_3	a_4	a_6	B_k
−9	15	+	1	1	1	−5	2	1/5
1	15	+	1	1	1	0	0	1
25	15	+	1	1	1	−5	2	1/6
256	15	+	1	1	1	−80	242	1/11
−1	17	+	1	−1	1	−1	0	1/2
9	21	+	1	0	0	1	0	1/2
−2	24	+	0	−1	0	−4	4	2/3
4	24	+	0	−1	0	1	0	1
18	24	+	0	−1	0	−4	4	2/5
64	24	+	0	−1	0	−64	220	1/4
−16	32	+	0	0	0	−11	14	1/2
8	32	+	0	0	0	4	0	1
−4	40	+	0	0	0	−2	1	1
1024	42	+	1	1	1	−1344	18405	1/3
144	48	+	0	1	0	−384	2772	1/2
2	56	+	0	0	0	1	2	4
−768	63	+	1	−1	0	−7056	229905	4/7
32	64	+	0	0	0	−44	112	1
12	72	+	0	0	0	6	−7	2
−64	80	+	0	0	0	−107	426	1
−8	96	+	0	−1	0	−17	33	2
27	99	+	1	−1	1	−59	186	4/3
3	117	−	1	−1	1	4	6	4
36	120	+	0	1	0	−15	18	2
−48	144	+	0	0	0	−579	5362	2
7	147	+	1	1	1	48	48	4
−3	171	+	1	−1	1	−14	20	4
−32	192	−	0	−1	0	−129	609	2
−128	192	+	0	−1	0	−1537	23713	2
81	195	+	1	0	0	−110	435	2
20	200	+	0	0	0	−50	125	4
−25	205	−	1	−1	1	−22	44	2
15	225	+	1	−1	1	−5	−628	4
49	231	+	1	1	1	−34	62	2
400	240	+	0	−1	0	−3200	70752	2
1296	240	+	0	1	0	−34560	2461428	2
−256	272	−	0	0	0	−1451	21274	2

TABLE 4

k^2	N	ϵ	a_1	a_2	a_3	a_4	a_6	B_k
5	275	−	1	−1	1	20	22	8
24	288	−	0	0	0	−156	736	4
48	288	+	0	0	0	−291	1910	4
17	289	−	1	−1	1	−199	510	4
−81	291	+	1	1	1	−164	740	2
−36	312	+	0	−1	0	−39	108	4
784	336	−	0	−1	0	−12544	544960	2
6	360	+	0	0	0	33	34	16
−11	363	+	1	1	1	−789	8130	4
−27	387	+	1	−1	1	−221	1316	4
14	392	−	0	0	0	49	−686	8
80	400	−	0	0	0	−2675	53250	4
−18	408	+	0	1	0	−52	128	8
2304	429	−	1	0	0	−6864	218313	2
−484	440	+	0	0	0	−5042	137801	4
128	448	−	0	0	0	−1196	15920	4
−49	455	−	1	−1	1	−67	226	4
−200	480	+	0	−1	0	−3601	84385	4
−144	480	−	0	−1	0	−480	4212	4
−12	504	−	0	0	0	−66	205	8
13	507	−	1	1	1	81	−564	8
4096	510	+	1	1	1	−21760	1226417	4
−5	525	−	1	1	1	−63	156	8
−20	600	−	0	−1	0	−383	3012	8
10	600	+	0	−1	0	92	−188	16
11	605	−	1	−1	1	98	−316	12
169	663	−	1	1	1	−539	4592	4
72	672	+	0	−1	0	−337	2497	8
50	680	−	0	0	0	−143	658	8
9216	690	+	1	0	0	−110400	14109732	6
−112	784	+	0	0	0	−14651	682570	8
−6	792	−	0	0	0	−111	434	16
4624	816	−	0	−1	0	−443904	113984640	4
100	840	−	0	−1	0	−175	952	8
196	840	+	0	−1	0	−735	7920	8
576	840	+	0	1	0	−6720	209808	8
−192	936	−	0	0	0	−7491	249550	8

TABLE 4 (cont'd)

DIOPHANTINE PROBLEMS IN MANY VARIABLES: THE ROLE OF ADDITIVE NUMBER THEORY

TREVOR D. WOOLEY*

ABSTRACT. We provide an account of the current state of knowledge concerning diophantine problems in many variables, paying attention in particular to the fundamental role played by additive number theory in establishing a large part of this body of knowledge. We describe recent explicit versions of the theorems of Brauer and Birch concerning the solubility of systems of forms in many variables, and establish an explicit version of Birch's Theorem in algebraic extensions of $\mathbb{Q}$. Finally, we consider the implications of recent progress on explicit versions of Brauer's Theorem for problems concerning the solubility of systems of forms in solvable extensions, such as Hilbert's resolvant problem.

1. INTRODUCTION

The purpose of this paper is to provide an overview of the current state of knowledge concerning diophantine problems in many variables, and in particular to describe the fundamental role played by additive number theory in establishing a great part of this body of knowledge. Diophantine problems in few variables have attracted the enthusiastic attention of number theorists for millenia, and indeed the recent work of Wiles [81] concerning Fermat's Last Theorem has even attracted the attention of the mass media. Exercising considerable literary hyperbole, one might describe the current state of knowledge concerning diophantine problems as resembling the European view of the world towards the end of the 16th Century. Thus, while a scientific renaissance flourished within Europe itself, knowledge concerning much of the globe consisted of little more than wild speculation based on the exotic tales brought back by adventurous explorers. In a similar fashion, the past half century has delivered a remarkable level of understanding of the arithmetic of curves, and this in turn has provided reasonably satisfactory knowledge concerning the solubility of diophantine equations in 2 or 3 variables. In contrast, the solubility of diophantine equations in many variables is a wild frontier with, for the most part, only sketchy knowledge and speculative conjectures. Hopefully, rather than be deterred by the relative lack of knowledge in the latter area, readers will

1991 *Mathematics Subject Classification.* 11D72, 11G25, 11E76, (11E95, 14G20, 20D10).

Key words and phrases. Diophantine problems, local solubility, diophantine equations, forms in many variables, p-adic fields, solvable extensions, Hilbert's resolvant problem.

*Packard Fellow and supported in part by NSF grant DMS-9622773. This paper was completed while the author was enjoying the hospitality of the Department of Mathematics at Princeton University.

S.D. Ahlgren et al. (eds.), Topics in Number Theory, 49–83.

be tempted by the ripping yarns recounted herein to themselves become explorers of this vast untamed territory.

Before proceeding further we pause to more carefully describe the type of diophantine problems central to this paper. Usually we will be interested in the solubility of a system of polynomial equations or inequalities over the rational integers $\mathbb{Z}$ or field of rational numbers $\mathbb{Q}$, but sometimes we will consider such problems over more general fields. As intimated above, one may loosely divide such diophantine problems into two types.

(i) Diophantine problems in few variables. Consider a homogeneous polynomial $p(\mathbf{x}) \in \mathbb{Z}[x_1, x_2, x_3]$ of degree d. When d is large, one might expect there to be few, if any, primitive solutions of the equation $p(\mathbf{x}) = 0$ with $\mathbf{x} \in \mathbb{Z}^3 \setminus \{\mathbf{0}\}$, for the values taken by the polynomial $p(\mathbf{x})$ are naively expected to be sparse amongst the set of all integers. The corresponding set of complex zeros of $p(\mathbf{x})$ may be considered geometrically as a projective plane curve C, and so the problem of determining the integral zeros of $p(\mathbf{x})$ is equivalent to finding the rational points of C, a problem of fundamental interest in arithmetic geometry. The general expectation in these problems is that such polynomials should have few primitive integral zeros other than the "obvious" ones. By way of illustration, Wiles [81] has resolved Fermat's notorious conjecture by completing a program of investigation to show that when $n \geq 3$, the equation $x^n + y^n = z^n$ has only the "obvious" integral solutions satisfying $xyz = 0$. In a more general setting, and somewhat earlier, Faltings [29] resolved Mordell's Conjecture by showing that when the curve C defined above has genus exceeding 1, then C has at most finitely many rational points, whence the underlying equation has only finitely many primitive integral solutions. In another rather older direction, when $f(x, y) \in \mathbb{Z}[x, y]$ is a homogeneous polynomial of degree $d \geq 2$ and n is a natural number, the investigation of the integral solutions of the Thue equation $f(x, y) = n$ is fundamental to a whole branch of the theory of diophantine approximations (see, for example, Schmidt [66]). We spend no more space here on this class of diophantine problems, but rather direct the reader to browse the literature wherein papers on this topic proliferate in copious quantities.

(ii) Diophantine problems in many variables. Rather than investigate polynomials which are expected to have few if any integral zeros, one may seek instead to show that a given polynomial has many primitive integral zeros. Consider a homogeneous polynomial $F(\mathbf{x}) \in \mathbb{Z}[x_1, \dots, x_s]$ of degree d. It is a remarkable fact that when d is odd and s is sufficiently large compared to d, there are infinitely many primitive integral zeros of $F(\mathbf{x})$ (see Birch [8]). In consequence, the philosophy underlying investigations concerning the solubility of diophantine equations in many variables takes on an entirely different flavour to the work sketched above. For the purpose of exposition, we characterise two basic problems in this area as follows.

Problem (a). *Existence of solutions.* How large must s be in terms of d so that there exists $\mathbf{x} \in \mathbb{Z}^s \setminus \{\mathbf{0}\}$ such that $F(\mathbf{x}) = 0$?

Problem (b). *Density of solutions.* How small can κ be in terms of s and d so that for each large real number B, one has

$$\text{card}(\{\mathbf{x} \in [-B, B]^s \cap \mathbb{Z}^s : F(\mathbf{x}) = 0\}) \gg B^{s-\kappa}. \tag{1.1}$$

Here, as is usual in analytic number theory, we write $f(t) \gg g(t)$ when for some positive constant c one has $|f(t)/g(t)| > c$ for all t under consideration. Also, we write $f(t) \ll g(t)$ when $g(t) \gg f(t)$.

Fortified with a mild dose of optimism, one might expect that as soon as s is sufficiently large in terms of d, the number of zeros counted by the left hand side of (1.1) should be asymptotic to a suitable product of densities of real solutions and p-adic solutions. In most situations, the truth of such an expectation would imply the validity of the lower bound (1.1) with $\kappa = d$.

Although we restrict attention in this paper primarily to the problems (a) and (b) above, we remark that there are natural generalisations of the latter problems to questions involving inequalities, and the methods described herein can be adapted (with substantial modification) to handle such questions. Consider, for example, the following problem.

Problem (c). *Small values of polynomials.* Consider a homogeneous polynomial $G(\mathbf{x}) \in \mathbb{R}[x_1, \ldots, x_s]$ of degree d. How large must s be in terms of d so that given $\varepsilon > 0$, there are infinitely many $\mathbf{x} \in \mathbb{Z}^s$ such that $|G(x_1, \ldots, x_s)| < \varepsilon$?

Even the simplest cases of the latter problem appear to be formidably difficult. It is a beautiful theorem of Schmidt [58] that when $G(\mathbf{x}) \in \mathbb{R}[x_1, \ldots, x_s]$ is a homogeneous polynomial of odd degree d, there exists an integer $s_0(d)$ such that whenever $s > s_0(d)$ and $\varepsilon > 0$, then there are infinitely many integral solutions of the inequality $|G(\mathbf{x})| < \varepsilon$. At present the only explicit estimate available for $s_0(d)$ is that due to Pitman [53] in the special case where $d = 3$, namely $s_0(3) \leq (1314)^{256} - 2$. Little seems to be known concerning lower bounds on permissible values of $s_0(d)$. For what is known on this and related problems, see Schmidt [63], Baker [6] and Lewis [47].

2. Some simple constraints and observations

Before embarking for more technical territory, we pause to discuss some obvious constraints within the problems (a) and (b) above. In keeping with our initial tack in the introduction, we confine ourselves for the moment to considering single equations, our observations generalising easily to systems of equations.

(i) Real solubility. Plainly, a *definite* polynomial such as $x_1^{2k} + \cdots + x_s^{2k}$ $(k \in \mathbb{N})$ has only the trivial zero $\mathbf{x} = \mathbf{0}$, no matter how large s may be. On the other hand, every homogeneous polynomial $F(\mathbf{x}) \in \mathbb{Z}[x_1, \ldots, x_s]$ of odd degree is indefinite, and necessarily possesses a non-trivial real zero (we leave this as a simple exercise to the reader). We therefore pay particular attention to polynomials of odd degree without further apology.

(ii) p-adic solubility. Since a somewhat detailed discussion at this point is useful in motivating a later argument (see the proof of Theorem 6.2 below), we will be more precise temporarily than would otherwise be warranted by present circumstances. Let d be an integer exceeding 1, let p be a prime number, and write $\mathbb{F}_p$ for the field of p elements. There is a field extension K of $\mathbb{F}_p$ of degree d. Let $\omega_1, \ldots, \omega_d$ be a basis for $K/\mathbb{F}_p$, and consider the norm form $\overline{N}(\mathbf{x}) = N_{K/\mathbb{F}_p}(\omega_1 x_1 + \cdots + \omega_d x_d)$ defined to be the determinant of the linear transformation in K determined by multiplication by $\omega_1 x_1 + \cdots + \omega_d x_d$. When $\boldsymbol{\alpha} \in \mathbb{F}_p^d$ and $\alpha = \alpha_1\omega_1 + \cdots + \alpha_d\omega_d$, we write $\overline{N}(\alpha) = \overline{N}(\boldsymbol{\alpha})$. Plainly $\overline{N}(\mathbf{x})$ is a polynomial of degree d with $\mathbb{F}_p$-rational coefficients. It is easily verified, moreover, that whenever $\alpha, \beta \in K$, one has $\overline{N}(\alpha)\overline{N}(\beta) = \overline{N}(\alpha\beta)$. Consequently, whenever $\boldsymbol{\alpha} \in \mathbb{F}_p^d \setminus \{\mathbf{0}\}$, and $\alpha = \alpha_1\omega_1 + \cdots + \alpha_d\omega_d$, then necessarily one has $\overline{N}(\alpha)\overline{N}(1/\alpha) = 1$, whence $\overline{N}(\boldsymbol{\alpha}) = \overline{N}(\alpha) \neq 0$. So the only zero in $\mathbb{F}_p^d$ of $\overline{N}(\mathbf{x})$ is the trivial one. Identifying now the polynomial $\overline{N}(\mathbf{x})$ with a corresponding polynomial $N(\mathbf{x})$ having integer coefficients, whose reduction modulo p coincides with $\overline{N}(\mathbf{x})$ in the obvious sense, it follows that the polynomial

$$F(\mathbf{x}) = N(x_1, \ldots, x_d) + pN(x_{d+1}, \ldots, x_{2d}) + \ldots + p^{d-1}N(x_{d^2-d+1}, \ldots, x_{d^2}) \tag{2.1}$$

has only the trivial zero $\mathbf{x} = \mathbf{0}$ over $\mathbb{Q}_p$. For whenever this polynomial is divisible by p, it follows from the above argument that $p|x_i$ for $1 \leq i \leq d$. On substituting and dividing by p, an obvious induction shows that when $\mathbf{x}$ is a zero of $F(\mathbf{x})$, then $p^r|x_i$ $(1 \leq i \leq d^2)$ for every $r \in \mathbb{N}$. In particular, there exist forms $F(\mathbf{x}) \in \mathbb{Z}[\mathbf{x}]$ of odd degree d in as many as d^2 variables which fail to possess non-trivial integral zeros.

(iii) Hybrid examples. Motivated by an example described by Swinnerton-Dyer, one may construct hybrid examples less trivial than those above. For example, Cassels and Guy [18] have shown that the equation $5x^3 + 12y^3 - 9z^3 - 10t^3 = 0$ has no non-trivial integral solutions, despite having non-trivial real and p-adic solutions, for every prime p. Consequently the sextic polynomial

$$5(x_1^2 + \cdots + x_s^2)^3 + 12(y_1^2 + \cdots + y_s^2)^3 - 9(z_1^2 + \cdots + z_s^2)^3 - 10(t_1^2 + \cdots + t_s^2)^3$$

has non-trivial real and p-adic zeros, for every prime p, but has no non-trivial integral zeros, no matter how large s may be.

While problem (b) is considerably more subtle than that concerning the mere existence of solutions, the discussion of example (ii) above is nonetheless instructive.

(iv) Density of solutions. Suppose that $F(\mathbf{x}) \in \mathbb{Z}[x_1, \ldots, x_s]$ is a homogeneous polynomial of odd degree d, and suppose that $F(\mathbf{x})$ possesses non-trivial p-adic zeros for every prime p. One may naively expect that when B is large, as we vary $(x_1, \ldots, x_s)$ through the box $[-B, B]^s$, one should find that almost every integer in the convex hull of the set $F([-B, B]^s)$ should receive its fair share of representations. Given this weak probabilistic heuristic, it is to be expected that for a suitable positive real number β, one should have

$$\mathrm{card}(\{\mathbf{x} \in [-B, B]^s \cap \mathbb{Z}^s : F(\mathbf{x}) = 0\}) \gg \frac{\mathrm{card}([-B, B]^s \cap \mathbb{Z}^s)}{\mathrm{card}([-\beta B^d, \beta B^d] \cap \mathbb{Z})} \gg B^{s-d}. \tag{2.2}$$

This lower bound is consistent with the expectation that the number of integral zeros of $F(\mathbf{x})$ in the box $[-B,B]^s$ should be asymptotic to a product of local densities. However, the above heursitic may be far from the truth when the form $F(\mathbf{x})$ is degenerate. Consider, for example, any large natural number s, an odd integer d, and linear polynomials $L_i(\mathbf{x}) \in \mathbb{Z}[x_1,\dots,x_s]$ $(1 \le i \le d^2)$ linearly independent over $\mathbb{Q}$. Recalling the example (2.1) above, we define

$$G(\mathbf{x}) = F(L_1(\mathbf{x}),\dots,L_{d^2}(\mathbf{x})).$$

Then by the argument of example (ii), it follows that whenever $G(\mathbf{x}) = 0$ one necessarily has $L_i(\mathbf{x}) = 0$ $(1 \le i \le d^2)$, and hence the linear independence of the $L_i(\mathbf{x})$ forces us to conclude that

$$\text{card}(\{\mathbf{x} \in [-B,B]^s \cap \mathbb{Z}^s \;:\; G(\mathbf{x}) = 0\}) \ll B^{s-d^2}.$$

We stress that such is the case no matter how large s may be, and when $d > 1$ this estimate sharply contradicts the lower bound (2.2).

We remark that in the current state of knowledge, it remains possible that the following conjecture is true.

Conjecture. *Let s and d be natural numbers with d odd and $s > d^2$. Suppose that $F(\mathbf{x}) \in \mathbb{Z}[x_1,\dots,x_s]$ is a homogeneous polynomial of degree d. Then when B is large, one has*

$$\textit{card}(\{\mathbf{x} \in [-B,B]^s \cap \mathbb{Z}^s \;:\; F(\mathbf{x}) = 0\}) \gg B^{s-d^2}.$$

Similarly, let s and $d_1,\dots,d_r$ be natural numbers with d_i odd $(1 \le i \le r)$ and $s > d_1^2 + \dots + d_r^2$. Suppose that $F_i(\mathbf{x}) \in \mathbb{Z}[x_1,\dots,x_s]$ $(1 \le i \le r)$ is a homogeneous polynomial of degree d_i $(1 \le i \le r)$. Then one might conjecture that

$$\begin{aligned}\text{card}(\{\mathbf{x} \in [-B,B]^s \cap \mathbb{Z}^s \;:\; F_1(\mathbf{x}) = \dots = F_r(\mathbf{x}) = 0\})& \\ \gg B^{s-d_1^2-\dots-d_r^2}&.\end{aligned}$$

3. Approaches to these problems

Except in a few isolated instances, we currently have only two approaches to the problems (a) and (b) above which are guaranteed to achieve some measure of success. While methods from arithmetic geometry and ergodic theory are applicable to special examples (see, for example, Batyrev and Manin [7], Lang [38] and Duke, Rudnick and Sarnak [28]), the applicability of such methods requires detailed knowledge of the geometry and algebraic structure of the examples under consideration. In contrast, the methods we highlight herein are applicable in considerable generality, and make use of only the weakest properties of the underlying polynomials.

(i) Elementary diagonalisation methods. There is a vast body of knowledge available concerning the solubility of additive diophantine equations of the shape

$$a_1 y_1^k + \cdots + a_t y_t^k = 0, \tag{3.1}$$

where the a_i are fixed integers (see, for example, Vaughan [79] and Davenport [22]). Owing to their diagonal structure, such equations are particularly amenable to methods involving exponential sums and the Hardy-Littlewood method. Thus, given a homogeneous polynomial $F(\mathbf{x}) \in \mathbb{Z}[x_1, \ldots, x_s]$ of odd degree d, one may attempt an attack on problem (a) by seeking linear polynomials $L_i(\mathbf{y}) = a_{i1}y_1 + \cdots + a_{it}y_t$ $(1 \le i \le s)$ with $a_{ij} \in \mathbb{Z}$ $(1 \le i \le s, 1 \le j \le t)$, satisfying the property that the equation

$$F(L_1(\mathbf{y}), \ldots, L_s(\mathbf{y})) = 0$$

takes the shape (3.1). Since polynomials of degree exceeding 2 do not, in general, diagonalise under a non-singular substitution, one expects that s need be much larger than t in order that this strategy should stand a chance of success. This approach has been successfully exploited by Brauer [14] and Birch [8] to establish two remarkable theorems about which we will say much more in due course.

Theorem 3.1 (Brauer). *Let d be a natural number. Then there is a number $s_1(d)$ such that whenever p is a prime number and $s > s_1(d)$, and $F(\mathbf{x}) \in \mathbb{Q}_p[x_1, \ldots, x_s]$ is homogeneous of degree d, then the equation $F(\mathbf{x}) = 0$ possesses a solution $\mathbf{x} \in \mathbb{Q}_p^s \setminus \{0\}$.*

Theorem 3.2 (Birch). *Let d be an odd integer. Then there is a number $s_2(d)$ such that whenever $s > s_2(d)$ and $F(\mathbf{x}) \in \mathbb{Q}[x_1, \ldots, x_s]$ is homogeneous of degree d, then it follows that the equation $F(\mathbf{x}) = 0$ possesses a solution $\mathbf{x} \in \mathbb{Q}^s \setminus \{0\}$.*

While these theorems in some sense provide a solution of problem (a) above, neither Brauer nor Birch explicitly computed the dependence of $s_1(d)$ and $s_2(d)$ on d, and with good reason! The arguments used in establishing these theorems involve complicated inductions which lead to bounds so large that they are aptly described by Birch's sarcastic phrase "not even astronomical".

(ii) The Hardy-Littlewood method. Various versions of the Hardy-Littlewood method have been developed in order to discuss the problems (a) and (b) above. All of these versions have rather serious limitations which restrict their use somewhat, and thus we will avoid describing such methods in detail within this paper. Two results typify the kind of conclusions available within this circle of ideas.

Theorem 3.3 (Birch). *Let $F(\mathbf{x}) \in \mathbb{Z}[x_1, \ldots, x_s]$ be homogeneous of degree d, and suppose that the variety defined by the equation $F(\mathbf{x}) = 0$ has a singular locus of dimension at most D. Then whenever $s - D > (d-1)2^d$, one has*

$$\mathit{card}(\{\mathbf{x} \in [-B, B]^s \cap \mathbb{Z}^s : F(\mathbf{x}) = 0\}) \sim CB^{s-d},$$

where C denotes the "product of local densities" within the box $[-B, B]^s$.

Here, in order to save space, we avoid explaining precisely what "product of local densities" means, and instead note merely that this number is positive and uniformly bounded away from 0 whenever the equation $F(\mathbf{x}) = 0$ possesses non-singular real and p-adic solutions for every prime p. In such a situation, it follows from Birch's Theorem that the equation $F(\mathbf{x}) = 0$ possesses infinitely many primitive integral solutions. The difficulty in applying this result of Birch [10] to resolve the problem (a) satisfactorily lies in our failure to adequately understand singular loci. It seems likely that whenever a variety defined as the set of zeros of a polynomial $F(\mathbf{x})$ possesses a singular locus of extremely large dimension, then necessarily that locus contains a subvariety defined by a system of rational equations of small degree. If such were known, then one could apply an inductive procedure in order to infer the existence of non-trivial integral solutions. Presently, however, we do not even understand the singular loci of cubic polynomials in any generality.

There is a second rather general approach due to Schmidt [65]. In order to describe this method, we require some notation. When $F(\mathbf{x}) \in \mathbb{Q}[x_1, \ldots, x_s]$ is a form of degree $d > 1$, write $h(F)$ for the least number h such that F may be written in the form

$$F = A_1B_1 + A_2B_2 + \cdots + A_hB_h,$$

with A_i, B_i forms in $\mathbb{Q}[\mathbf{x}]$ of positive degree $(1 \le i \le h)$. There is an analogous concept for systems of forms which we avoid describing in the interest of saving space.

Theorem 3.4 (Schmidt). *Let d be an integer exceeding* 1, *and write $\chi(d) = d^2 2^{4d} d!$. Let $F(\mathbf{x}) \in \mathbb{Z}[x_1, \ldots, x_s]$ be homogeneous of degree d, and suppose that $h(F) \ge \chi(d)$. Then one has*

$$\mathit{card}(\{\mathbf{x} \in [-B, B]^s \cap \mathbb{Z}^s \ : \ F(\mathbf{x}) = 0\}) \sim CB^{s-d},$$

where C denotes the "product of local densities" within the box $[-B, B]^s$.

While the integer $h(F)$ associated with a form F may be difficult to compute for a specific example, Schmidt's approach has the advantage of leading naturally to an inductive strategy for solving an equation. For suppose that the form $F(\mathbf{x}) \in \mathbb{Z}[x_1, \ldots, x_s]$ has odd degree d, and possesses non-singular real and p-adic solutions for every prime p. Then if $h(F) \ge d^2 2^{4d} d!$, it follows from Theorem 3.4 that the equation $F(\mathbf{x}) = 0$ possesses infinitely many primitive integral solutions. Otherwise we may write

$$F(\mathbf{x}) = A_1(\mathbf{x})B_1(\mathbf{x}) + \cdots + A_h(\mathbf{x})B_h(\mathbf{x})$$

with $h < d^2 2^{4d} d!$, and with $A_1(\mathbf{x}), \ldots, A_h(\mathbf{x}) \in \mathbb{Z}[\mathbf{x}]$ homogeneous of odd degree at most $d-2$. Thus, provided that we can solve the system $A_i(\mathbf{x}) = 0$ $(1 \le i \le h)$, then again we obtain a non-trivial integral solution of the equation $F(\mathbf{x}) = 0$. Moreover, all of the equations occurring in the latter system have degree smaller than that of $F(\mathbf{x})$. We may now, therefore, apply the analogue of Theorem 3.4 for systems of polynomials, and with sufficiently many variables we will either solve the system, or again reduce the degrees of the equations occurring therein. Unfortunately, the number of variables required to establish the existence of solutions using this approach is "not even astronomical" in size, and indeed Birch's elementary approach may be fashioned to do better in this respect.

For more restricted variants of the Hardy-Littlewood method applicable to the solubility of systems of equations in many variables, see also Tartakovsky [76], Davenport [20], [21], [23], Pleasants [54], Schmidt [62], Heath-Brown [33], Hooley [34], [35], [36], Skinner [73], [75] and Vaughan and Wooley [80].

4. Some notation

In order to navigate further our discussion of the problem (a), we require some notation. Despite the unpleasant appearance of this notation, it is best simply to introduce it in the most general form in one clean sweep.

Definition 4.1. *Given an r-tuple of polynomials*

$$\mathbf{F} = (F_1, \dots, F_r)$$

with coefficients in a field k, denote by $\nu(\mathbf{F})$ the number of variables appearing explicitly in $\mathbf{F}$.

We are interested in solution sets, over a field k, of systems of homogeneous polynomial equations with coefficients in k. When such a set contains a linear subspace of the ambient space, we define its dimension to be that when considered as a projective space.

Definition 4.2. *Let k be a field. Denote by $\mathcal{G}_d^{(m)}(r_d, \dots, r_1; k)$ the set of $(r_d + \dots + r_1)$-tuples of homogeneous polynomials, of which r_i have degree i for $1 \le i \le d$, with coefficients in k, which possess no linear space of solutions of dimension m over k. We define $V_d^{(m)}(\mathbf{r}) = V_d^{(m)}(r_d, \dots, r_1; k)$ by*

$$V_d^{(m)}(r_d, \dots, r_1; k) = \sup_{\mathbf{h} \in \mathcal{G}_d^{(m)}(r_d, \dots, r_1; k)} \nu(\mathbf{h}).$$

We abbreviate $V_d^{(m)}(r, 0, \dots, 0; k)$ to $v_{d,r}^{(m)}(k)$, and similarly $v_{d,r}^{(0)}(k)$ to $v_{d,r}(k)$, and $v_{d,1}^{(m)}(k)$ to $v_d^{(m)}(k)$, and $v_d^{(0)}(k)$ to $v_d(k)$.

Notice that this definition simply tells us how many variables we require in order to solve an arbitrary implicit system of equations.

Examples.

(i). We have $v_{1,r}(k) = r$, by familiar linear algebra. In other words, in any field k a system of r linear forms with k-rational coefficients, in $r+1$ or more variables, necessarily possesses a non-trivial k-rational solution. Moreover, there exist systems of r linear forms in r variables which possess only trivial solutions.

(ii). For any prime number p, it follows from the classical theory of quadratic forms that $v_2(\mathbb{Q}_p) = 4$. In other words, any quadratic form with p-adic coefficients in 5 or more variables possesses a non-trivial p-adic solution. Moreover, in view of examples of the type (2.1) above, there are quadratic forms in 4 variables which possess only the trivial p-adic solution.

(iii). We have $v_2(\mathbb{Q}) = +\infty$, because any definite quadratic form has only the trivial zero over $\mathbb{Q}$, no matter how many variables it might have.

The simplest forms of degree d to investigate are diagonal forms of the shape

$$a_1x_1^d + \cdots + a_sx_s^d. \tag{4.1}$$

Definition 4.3. *When k is a field, denote by $\mathcal{D}_{d,r}(k)$ the set of r-tuples of diagonal forms of degree d, with coefficents in a field k, which possess no non-trivial zeros over k. Define*

$$\phi_{d,r}(k) = \sup_{\mathbf{f}\in\mathcal{D}_{d,r}(k)} \nu(\mathbf{f}).$$

We abbreviate $\phi_{d,1}(k)$ to $\phi_d(k)$.

Note that whenever $s > \phi_d(k)$ and $a_i \in k$ $(1 \le i \le s)$, then the polynomial (4.1) possesses a non-trivial k-rational zero.

5. Brauer's method

Equipped with the notation of the previous section, we may restate a quite general version of problem (a) as follows.

Problem (A). *Existence of solutions.* Given a field k, a natural number d, and non-negative integers $r_1, \ldots, r_d$ and m, find an upper bound for $V_d^{(m)}(r_d, \ldots, r_1; k)$.

The methods of Brauer [14] alluded to above in connection with Theorem 3.1 may be applied to provide an elegantly simple solution of problem (A).

Theorem 5.1 (Brauer). *Let k be a field, and suppose that for $i \ge 2$ one has $\phi_i(k) < \infty$. Then for each natural number d and for all non-negative integers $r_1, \ldots, r_d$ and m, one has*

$$V_d^{(m)}(r_d, \ldots, r_1; k) < \infty.$$

It has been known since at least the early part of this century that for every prime number p one has $\phi_d(\mathbb{Q}_p) < \infty$, and thus we see that Theorem 3.1 is an immediate corollary of Theorem 5.1. Unfortunately, since $\phi_t(\mathbb{Q}) = +\infty$ whenever t is even, Theorem 5.1 does not yield insight into bounds for $v_d(\mathbb{Q})$.

As indicated in §3, the method of proof of Theorem 5.1 is a complicated induction, so highly iterative that for decades it was thought that any explicit estimate arising from such methods would surely be too large to be sensibly written down. It was therefore a surpise when Leep and Schmidt [42] were able to obtain a "reasonable" bound by employing a clever variant of Brauer's original method. When p is a prime number and $k = \mathbb{Q}_p$, the conclusions of Leep and Schmidt [42] may be stated reasonably cleanly as follows.

Theorem 5.2 (Leep and Schmidt). *Let p be a prime number, and let d be a natural number. Then for each positive number ε, one has*

$$v_d(\mathbb{Q}_p) \ll_\varepsilon e^{(d!)^2(1+\varepsilon)^d}.$$

Further, when r is a natural number,

$$v_{d,r}(\mathbb{Q}_p) \le \left(\frac{v_2}{2}\right)\left(\frac{v_3}{2}\right)^2 \cdots \left(\frac{v_d}{2}\right)^{2^{d-2}} r^{2^{d-1}} \left(1+O(r^{-1})\right).$$

We will return momentarily to the topic of explicit versions of Brauer's Theorem. In order to better explain the ideas involved in subsequent developments, however, it seems appropriate at this stage to sketch a proof of Theorem 5.1. Let k be a field, and suppose that $\phi_i(k) < \infty$ for $i \ge 2$. When $D \ge 1$, we form the inductive hypothesis that for all non-negative integers $r_D, \ldots, r_1$ and m, one has

$$V_D^{(m)}(r_D, \ldots, r_1; k) < \infty. \tag{5.1}$$

The hypothesis (5.1) is immediate from linear algebra when $D = 1$. We suppose that $d \ge 2$ and that (5.1) holds for all m and $\mathbf{r}$ when $D = d - 1$, and then aim to establish (5.1) for all m and $\mathbf{r}$ when $D = d$.

We start with some simplifications, observing first that for all m and $\mathbf{r}$, one has

$$V_d^{(m)}(r_d, \ldots, r_1; k) \le v_{d,r_d}^{(v)}(k), \tag{5.2}$$

where $v = V_{d-1}^{(m)}(r_{d-1}, \ldots, r_1; k)$. For given any system of r_j forms of degree j $(1 \le j \le d)$ in more than $v_{d,r_d}^{(v)}(k)$ variables, the subsystem of r_d forms of degree d possesses a v-dimensional linear space of k-rational zeros. Writing down a basis for this linear space, and substituting into the remaining forms, we obtain a system of r_i forms of degree i $(1 \le i \le d-1)$ in $v+1$ variables, and by the definition of $V_{d-1}^{(m)}(\mathbf{r}; k)$, this system possesses an m-dimensional linear space of k-rational zeros. The upper bound (5.2) is immediate. Moreover, a similar argument shows that whenever $r \ge 2$, one has

$$v_{d,r}^{(m)}(k) \le v_d^{(w)}(k), \tag{5.3}$$

where $w = v_{d,r-1}^{(m)}(k)$. We therefore deduce that in order to establish (5.1), it suffices to show that $v_d^{(m)}(k) < \infty$ for each m.

Next we indicate how to diagonalise a form. We claim that for each natural number t, there is an integer $n(t)$, depending at most on k and t, such that whenever $s > n(t)$ and $F(\mathbf{x}) \in k[x_1, \ldots, x_s]$ is a form of degree d, then there exist linearly independent k-rational points $\mathbf{y}_1, \ldots, \mathbf{y}_t$ with the property that for every $z_1, \ldots, z_t$ in k one has

$$F(z_1\mathbf{y}_1 + \cdots + z_t\mathbf{y}_t) = F(\mathbf{y}_1)z_1^d + \cdots + F(\mathbf{y}_t)z_t^d. \tag{5.4}$$

In other words, the polynomial $F(\mathbf{x})$ may be reduced non-trivially to a diagonal form in at least t variables. When $t = 1$ this claim is trivial, so we suppose that $t \ge 1$ and that $\mathbf{y}_1, \ldots, \mathbf{y}_t$ satisfy (5.4). Write $\mathbf{u} = z_1\mathbf{y}_1 + \cdots + z_t\mathbf{y}_t$, and consider a point $\mathbf{v} \in k^s$ linearly independent of $\mathbf{y}_1, \ldots, \mathbf{y}_t$. Then for every t and w in k one has

$$F(t\mathbf{u} + w\mathbf{v}) = t^d F(\mathbf{u}) + w^d F(\mathbf{v}) + \sum_{i=1}^{d-1} t^i w^{d-i} G_i(\mathbf{u}, \mathbf{v}), \tag{5.5}$$

where the polynomials $G_i(\mathbf{u},\mathbf{v}) \in k[\mathbf{u},\mathbf{v}]$ are homogeneous of degree i in terms of $\mathbf{u}$, and of degree $d-i$ in terms of $\mathbf{v}$. When $t=1$, so that $\mathbf{u}=z_1\mathbf{y}_1$, the system of equations $G_i(\mathbf{u},\mathbf{v})=0$ $(1\le i\le d-1)$ is equivalent to a system of homogeneous equations in $\mathbf{v}$ of respective degrees $1,2,\dots,d-1$. On recalling that we are choosing $\mathbf{v}$ to be linearly independent of $\mathbf{y}_1$, it follows that whenever $s-1 > V_{d-1}^{(0)}(1,1,\dots,1;k)$, then the latter system possesses a non-trivial k-rational solution. From (5.5) we therefore deduce that the polynomial $F(\mathbf{x})$ may be reduced non-trivially to a diagonal form in at least $t+1$ variables. When $t>1$, we may adopt a similar strategy, examining separately the coefficients of each term of the shape $z_1^{i_1}\dots z_t^{i_t}$, although we emphasise that the number of equations needing to be solved will increase with t. Thus it follows that whenever both $n(t)$ and $V_{d-1}^{(0)}(\mathbf{r};k)$ are finite, for each $\mathbf{r}$, then one has that $n(t+1)$ is finite. Our claim that $n(t)<\infty$ for each t therefore follows by induction.

Consider next a form $F(\mathbf{x}) \in k[x_1,\dots,x_s]$ with $s>n(t)$, where $t=(m+1)\psi$ and $\psi=\phi_d(k)+1$. By the above argument the polynomial $F(\mathbf{x})$ diagonalises via a substitution $\mathbf{x}=z_1\mathbf{y}_1+\dots+z_t\mathbf{y}_t$ to the shape $F(\mathbf{x})=a_1z_1^d+\dots+a_tz_t^d$. But the variables in the latter form may plainly be partitioned into $m+1$ sets each containing ψ variables, and moreover, with the diagonal form underlying each set possessing a non-trivial k-rational zero. Thus it follows that the equation $F(\mathbf{x})=0$ has a k-rational linear space of solutions of dimension m, and hence $v_d^{(m)}(k)\le n(t)<\infty$. In view of (5.2) and (5.3), we may conclude that (5.1) holds with $D=d$, and so our induction is complete.

A cursory examination reveals that the bounds stemming from the above argument will involve highly iterated exponential functions of unpleasant type. In order to establish the respectable bounds embodied in Theorem 5.2, Leep and Schmidt [42] required three new ingredients. First, an efficient new inductive approach is used to generate linear spaces of k-rational solutions to systems of equations, thereby replacing the simplifying bounds (5.2) and (5.3) by ones considerably less wasteful. The idea is to make use of the existence of a linear space of k-rational solutions, via a change of variables, in order to simplify the shape of the equations under consideration, and thence make easier the task of finding a larger linear space of k-rational solutions. By making use of the bound $v_{d,r}(k)\le v_{d,r-1}^{(w)}(k)$, with $w=v_d(k)$, this idea also enables one to bound $v_{d,r}^{(m)}(k)$ simply in terms of $v_j(k)$ $(1\le j\le d)$. Second, Leep and Schmidt diagonalise whole systems of forms simultaneously, rather than just a single form. Then by making use of estimates of Davenport and Lewis [25] for $\phi_{d,r}(\mathbb{Q}_p)$, Leep and Schmidt are able to remove another of the highly iterated inductions from the above argument. Thirdly, and this ingredient should not be underestimated, Leep and Schmidt were brave enough to push the project through to completion! By developing further the theory of simultaneous additive equations, Schmidt [64] was subsequently able to improve the bounds recorded in Theorem 5.2 somewhat, showing that for every prime p and natural number d, one has $v_d(\mathbb{Q}_p)=o(e^{2^d d!})$.

The author recently found that the Leep-Schmidt approach to Brauer's method could be improved further (see Wooley [84]). The key idea is to generate a linear space of k-rational solutions to all but one equation of a system via the Leep-Schmidt process, and at the same time diagonalise the final equation with little additional cost. This amputates another of the iterated inductive steps from the process described above. In addition to providing sharper bounds, this new method offers greater flexibility in its application than that of Leep and Schmidt, for it depends only on the theory of a single additive equation. The latter is by now rather well understood in almost any respectable field. We illustrate the conclusions stemming from these ideas with the following theorem.

Theorem 5.3 (Wooley). *Let m, d and r be non-negative integers with $d \geq 2$ and $r \geq 1$. Write ϕ_i for $\phi_i(k)$ $(2 \leq i \leq d)$. Then whenever k is a field for which $\phi_i < \infty$ $(2 \leq i \leq d)$, one has*

$$v_{d,r}^{(m)}(k) \leq 2(r^2\phi_d + mr)^{2^{d-2}} \prod_{i=2}^{d-1} (\phi_i + 1)^{2^{i-2}}.$$

The conclusion of Theorem 5.3 improves on that of Theorem 5.2 whenever ϕ_j is significantly smaller than v_j $(2 \leq j \leq d)$. Given that we no so much about ϕ_j and little concerning bounds for v_j, the reader will anticipate numerous painless corollaries.

Corollary 1. *Let d be an integer with $d \geq 2$, and let r be a natural number. Then for each prime number p one has $v_{d,r}(\mathbb{Q}_p) \leq (rd^2)^{2^{d-1}}$, and in particular, one has $v_d(\mathbb{Q}_p) \leq d^{2^d}$.*

This follows from Theorem 5.3 via the bound $\phi_d(\mathbb{Q}_p) \leq d^2$ of Davenport and Lewis [24].

Corollary 2. *Let d be an integer with $d \geq 2$, let r be a natural number, and let p be a prime number. Then whenever K is an algebraic extension of $\mathbb{Q}_p$, one has*

$$v_{d,r}(K) \leq r^{2^{d-1}} e^{2^{d+2}(\log d)^2}.$$

Whenever one has a system of equations with coefficients from an algebraic extension of $\mathbb{Q}_p$, these coefficients must all lie in some finite extension K' of $\mathbb{Q}_p$. Thus Corollary 2 follows from Theorem 5.3 via the bound

$$\phi_d(K') \leq d((d+1)^{\max\{2\log d/\log p, 1\}} - 1)$$

of Skinner [74].

It transpires that in purely imaginary field extensions L of $\mathbb{Q}$, the archimedean local solubility condition is automatically satisfied. Consequently, on deriving a bound on $\phi_d(L)$ from work of Siegel [71], [72] and Birch [9], we are able to derive an explicit version of a theorem of Peck [52].

Corollary 3. *Let d be an integer with $d \geq 2$, let r be a natural number, and let L be a purely imaginary field extension of $\mathbb{Q}$. Then $v_{d,r}(L) \leq r^{2^{d-1}} e^{2^d d}$.*

The situation in which Brauer's methods may be expected to be most effective is that in which equations of the shape $ax^d + by^d = 0$ necessarily possess non-trivial solutions. This brings us to the topic of radical solutions of polynomials, an area distinguished since Medieval times and invigorated by the celebrated work of Galois. Such a topic deserves a digression to itself.

6. Solving equations in solvable extensions

Consider a countable field k of characteristic zero, such as $\mathbb{Q}$. We remark that these hypotheses are more a matter of convenience than an essential requirement, the corresponding theory in positive characteristic containing several technical complications. We define the *radical closure* k^{rad} of k as follows. Let $\overline{k}$ denote the algebraic closure of k. Also, let S denote the set of all elements α of $\overline{k}$ for which the field extension $k(\alpha)/k$ is solvable. Since $\overline{k}$ is countable, so too must be S, so that we may write $S = \{\alpha_1, \alpha_2, \dots\}$. We then define

$$k^{\text{rad}} = \bigcup_{n=1}^{\infty} k(\alpha_1, \dots, \alpha_n).$$

The reader will readily verify that k^{rad} satisfies all of the axioms for a field, and moreover, whenever $\alpha \in k^{\text{rad}}$ one has that $k(\alpha)$ is a solvable extension of k, whence α is radical.

The classical theory of equations shows that polynomial equations of the shape

$$f(x) = x^n + a_1 x^{n-1} + \cdots + a_n = 0 \tag{6.1}$$

are solvable over k^{rad} when $n = 2, 3, 4$. Meanwhile, a celebrated consequence of Galois theory shows that when $k = \mathbb{Q}$, for each n with $n \geq 5$ such equations exist with no radical solutions (the only solutions are "irradical"). By homogenising the equation (6.1), therefore, we find that for every countable field k one has $v_d(k^{\text{rad}}) = 1$ when $d = 2, 3, 4$, but that $v_d(\mathbb{Q}^{\text{rad}}) \geq 2$ whenever $d \geq 5$. This classical problem naturally leads one, therefore, to consider upper bounds for $v_d(k^{\text{rad}})$ in general. Since, when $a, b \in k^{\text{rad}}$, the equation $ax^d + by^d = 0$ is always soluble with $x, y \in k^{\text{rad}}$, we have $\phi_d(k^{\text{rad}}) = 1$ for every natural number d. Then as a corollary of Theorem 5.3 we have the following (see Wooley [84]).

Theorem 6.1 (Wooley). *Let k be a countable field of characteristic zero, and let d and r be natural numbers with $d \geq 2$. Then $v_{d,r}(k^{\text{rad}}) \leq (2r^2)^{2^{d-2}}$.*

Fixing attention temporarily on the most familiar case where $k = \mathbb{Q}$, one may ask whether $v_d(\mathbb{Q}^{\text{rad}})$ can be arbitrarily large as d grows. In Wooley [84] we show that for infinitely many integers d one has $v_d(\mathbb{Q}^{\text{rad}}) \geq d^{\frac{\log 2}{\log 5}}$, answering the latter question in the affirmative. But one may, in fact, provide a still larger lower bound for $v_d(\mathbb{Q}^{\text{rad}})$ by use of an example motivated by §2(ii). For background on the necessary Galois theory, see either Garling [30] or Serre [70].

Theorem 6.2. *When $d = 2$, 3, or 4 one has $v_d(\mathbb{Q}^{\text{rad}}) = 1$. But when d is a natural number with $d \geq 5$, one has $v_d(\mathbb{Q}^{\text{rad}}) \geq d$.*

Proof. We have already discussed the first assertion of the theorem. Suppose next that d is an integer with $d \geq 5$, and consider the polynomial $f_d(x) = x^d - x - 1$. By Selmer [69], the polynomial $f_d(x)$ is irreducible over $\mathbb{Q}$. Let K be the splitting field of $f_d(x)$ over $\mathbb{Q}$. Then according to the remarks on p.42 of Serre [70], one can show that the field extension $K/\mathbb{Q}$ has Galois group Γ isomorphic to the full symmetric group S_d. Since the group S_d is not solvable for $d \geq 5$, it follows that the polynomial $f_d(x)$ cannot be solved by radical extensions. We note also that $[K : \mathbb{Q}] = d!$.

Our next step is to show that $f_d(x)$ is irreducible over $\mathbb{Q}^{\mathrm{rad}}$. Suppose that $f_d(x)$ factors over $\mathbb{Q}^{\mathrm{rad}}$ in the form

$$f_d(x) = g_1(x)g_2(x)\dots g_t(x), \tag{6.2}$$

where the $g_i(x)$ are monic polynomials irreducible in $\mathbb{Q}^{\mathrm{rad}}[x]$ of degree d_i $(1 \leq i \leq t)$. Let L_0 denote the field extension of $\mathbb{Q}$ obtained by adjoining the coefficients of the g_i $(1 \leq i \leq t)$. Since each $g_i(x)$ splits over K, it follows that L_0 is contained in K. Consider the Galois group Γ_0 of the field extension $L_0/\mathbb{Q}$. Let σ be an automorphism in Γ_0, and consider its action on the polynomial $g_i(x)$. That is, consider the polynomial $g_i^\sigma(x)$ obtained by replacing the coefficients of $g_i(x)$ by their images under σ. Since $f_d(x)$ is invariant under the action of σ, it follows that $g_i^\sigma(x)$ divides $f_d(x)$. Consequently, one has that $g_i^\sigma(x)$ divides $f_d(x)$ for every $\sigma \in \Gamma_0$, and for each i with $1 \leq i \leq t$. But L_0 is the minimal field extension of $\mathbb{Q}$ containing the coefficients of the g_i, so for each $\sigma \in \Gamma_0$ other than the trivial automorphism, there is at least one g_i which is moved under the action of σ. It follows that Γ_0 is determined by its action on the g_i $(1 \leq i \leq t)$, whence it is isomorphic to some subgroup of the group of permutations on t elements. In particular, one has $|\Gamma_0| \leq t!$, whence also $[L_0 : \mathbb{Q}] \leq t!$.

Next observe that if some subset, say $\{g_{i_1}(x), \dots, g_{i_n}(x)\}$, is left fixed under the action of Γ_0, then the polynomial $\prod_{j=1}^{n} g_{i_j}(x)$ is also invariant under the action of Γ_0, and hence has rational coefficients. Then it follows from the irreducibility of $f_d(x)$ that $n = t$, and moreover there is no loss of generality in supposing that each $g_i(x)$ has the same degree. Thus t divides d, and the degree of each $g_i(x)$ is equal to d/t. Furthermore, one cannot have $t = d$, for then $f_d(x)$ splits over the radical field extension L_0 of $\mathbb{Q}$, and yet $f_d(x)$ has no radical roots. We now construct a tower of field extensions

$$\mathbb{Q} \subseteq L_0 \subseteq L_1 \subseteq \cdots \subseteq L_t = K,$$

as follows. For each i with $0 \leq i \leq t-1$ we take L_{i+1} to be the splitting field of $g_{i+1}(x)$ over L_i. That $K = L_t$ then follows from the factorisation (6.2) together with our earlier observations concerning the field L_0. But we have

$$[L_{i+1} : L_i] \leq (\deg(g_{i+1}))! = (d/t)! \quad (0 \leq i \leq t-1).$$

Consequently,

$$\begin{aligned}[K : \mathbb{Q}] &= [K : L_{t-1}][L_{t-1} : L_{t-2}]\dots[L_1 : L_0][L_0 : \mathbb{Q}] \\ &\leq t!((d/t)!)^t.\end{aligned}$$

By hypothesis, however, we have also $[K : \mathbb{Q}] = d!$, so that necessarily $d! \leq t!((d/t)!)^t$. Moreover, our earlier observation ensures that $t < d$. Since $d \geq 5$, therefore, we are forced to conclude that $t = 1$ and hence that the polynomial $f_d(x)$ is irreducible over $\mathbb{Q}^{\mathrm{rad}}$.

Since $f_d(x)$ is irreducible over $\mathbb{Q}^{\mathrm{rad}}$, if θ is a zero of $f_d(x)$ in the splitting field of $f_d(x)$ over $\mathbb{Q}^{\mathrm{rad}}$, then one has $[\mathbb{Q}^{\mathrm{rad}}(\theta) : \mathbb{Q}^{\mathrm{rad}}] = d$. Write $M = \mathbb{Q}^{\mathrm{rad}}(\theta)$, and let $\omega_1, \ldots, \omega_d$ be a basis for $M/\mathbb{Q}^{\mathrm{rad}}$. Consider the norm form $N(\mathbf{x}) = N_{M/\mathbb{Q}^{\mathrm{rad}}}(\omega_1 x_1 + \cdots + \omega_d x_d)$ defined to be the determinant of the linear transformation in M determined by multiplication by $\omega_1 x_1 + \cdots + \omega_d x_d$. Plainly $N(\mathbf{x})$ is a polynomial of degree d with $\mathbb{Q}^{\mathrm{rad}}$-rational coefficients. By the same multiplicative property discussed in §2(ii), moreoever, it follows that for $\mathbf{y} \in (\mathbb{Q}^{\mathrm{rad}})^d$, one has $N(\mathbf{y}) = 0$ only when $\mathbf{y} = \mathbf{0}$. Thus it follows that $v_d(\mathbb{Q}^{\mathrm{rad}}) \geq d$ whenever $d \geq 5$. This completes the proof of the theorem.

An inspection of the example constructed in the proof of Theorem 6.2 will reveal that the underlying polynomial splits over $\mathbb{C}$, and consequently the solution set of this polynomial is singular. Such is also the case for the example used in establishing the earlier bound $v_d(\mathbb{Q}^{\mathrm{rad}}) \geq d^{\frac{\log 2}{\log 8}}$ of Wooley [84]. It seems natural to ask for absolutely irreducible or even non-singular examples.

Problem. *Is there a curve defined by an absolutely irreducible homogeneous polynomial $p(\mathbf{x}) \in \mathbb{Q}^{\mathrm{rad}}[x_1, x_2, x_3]$ which has no radical points? Further, does such a curve exist which has no singular points?*

To be clear, any homogeneous polynomial $p(\mathbf{x}) \in \mathbb{Z}[x_1, x_2, x_3]$ possessing no radical zeros has the property that whenever $p(\mathbf{x}) = 0$, then either $\mathbf{x} = \mathbf{0}$, or else at least one of x_1, x_2, x_3 does not lie in $\mathbb{Q}^{\mathrm{rad}}$. The work of Rumely [56] may well be relevant to this problem.

One might conclude from the discussion thus far that solving in solvable extensions is a pursuit of wholly artificial nature. Maybe so, but there is at least one application worthy of mention, and indeed this application seems to be what prompted Brauer to investigate the solubility of forms in the first place. We recall the discussion of §3 of Brauer [14]. Let k be a countable field of characteristic zero, and consider the arbitrary algebraic equation of degree n in one variable given by (6.1), with $a_i \in k$ $(1 \leq i \leq n)$. Let the zeros of $f(x)$ be ω_i $(1 \leq i \leq n)$, and when $1 \leq i \leq n$, denote by $\theta_i = \theta_i(\mathbf{u})$ the polynomial

$$\theta_i(\mathbf{u}) = u_0 + u_1\omega_i + \cdots + u_{n-1}\omega_i^{n-1}.$$

It follows that the θ_i are roots of an equation

$$g(x) = x^n + b_1(\mathbf{u})x^{n-1} + \cdots + b_n(\mathbf{u}) = 0, \tag{6.3}$$

where the b_i are homogeneous polynomials in $u_0, \ldots, u_{n-1}$ of degree i for $1 \leq i \leq n$ (this transformation is the classical Tschirnhaus transformation). Since $g(x)$ is invariant under conjugation, moreover, one finds that each $b_i(\mathbf{u})$ has k-rational coefficients for $1 \leq i \leq n$. Suppose that for some fixed d with $1 \leq d \leq n$, one is able to solve the system of equations

$$b_i(\mathbf{u}) = 0 \quad (1 \leq i \leq d) \tag{6.4}$$

non-trivially over k^{rad}. We will sketch below, in fact, an argument which employs the methods described in §5 which establishes the bound

$$V_d^{(0)}(1,1,\dots,1;k^{\mathrm{rad}}) \le 2^{2^{d-1}} - 1. \tag{6.5}$$

Moreover, as is evident from a cursory examination of the underlying arguments, the latter bound guarantees that when $n \ge 2^{2^{d-1}}$, the system (6.4) is soluble non-trivially over a solvable field extension of k whose degree depends at most on d, and such that any prime divisor of this degree is also bounded above by d. Thus it is possible to take $u_0,\dots,u_{n-1}$ in a field obtained from the field of rational functions of $a_1,\dots,a_n$ by adjoining a finite number of radicals. The equation (6.3) then takes the shape

$$x^n + b_{d+1}x^{n-d-1} + \dots + b_n = 0. \tag{6.6}$$

By adjoining a further radical, moreover, there is no loss of generality in supposing that $b_n = 1$. Thus the roots of the equation (6.6) may be considered as algebraic functions of the $n-d-1$ quantities $b_{d+1}, b_{d+2},\dots,b_{n-1}$. Since each ω_i may be expressed in terms of θ_i $(1 \le i \le n)$, it follows that the solution of the general equation of nth degree may be expressed in terms of its coefficients, provided we use radicals and one algebraic function of $n-d-1$ variables.

For each natural number n, let l_n denote the smallest integer l with the property that the roots of the general equation of degree n may be expressed in terms of the coefficients by means of algebraic functions of at most l parameters. Then the discussion above shows that $l_n \le n-d-1$ whenever $n \ge 2^{2^{d-1}}$. Consequently, we have the following theorem.

Theorem 6.3. *Let n and d be natural numbers with $n \ge d \ge 2$. Then $l_n \le n-d$ whenever $n \ge 2^{2^{d-2}}$, and in particular*

$$l_n \le n - 2 - \left[\frac{\log((\log n)/(\log 2))}{\log 2}\right].$$

Proof. We start by establishing the promised bound (6.5). We note that $\phi_d(k^{\mathrm{rad}}) = 1$ for every natural number d, and in order to save effort, we observe that the bound (2.12) of Wooley [84] asserts in particular that

$$v_{d,1}^{(1)}(k^{\mathrm{rad}}) \le 1 + V_d^{(0)}(1,1,\dots,1;k^{\mathrm{rad}}).$$

Since the bound established for $v_{d,1}^{(1)}(k^{\mathrm{rad}})$ in the proof of Theorem 2.4 of Wooley [84] is derived directly from the latter inequality, we may conclude from the proof of Theorem 2.4 of Wooley [84] that

$$\begin{aligned}1 &+ V_d^{(0)}(1,1,\dots,1;k^{\mathrm{rad}})\\ &\le 2(\phi_d(k^{\mathrm{rad}})+1)^{2^{d-2}} \prod_{i=2}^{d-1} (\phi_i(k^{\mathrm{rad}})+1)^{2^{i-2}}\\ &\le 2^{2^{d-1}}.\end{aligned}$$

The upper bound (6.5) is immediate, and the theorem then follows in the manner indicated above.

As far as we are aware, Theorem 6.3 provides the first explicit estimate for l_n as n grows. Brauer [14] had shown that $\lim_{n\to\infty}(n-l_n)=\infty$, and previously Hilbert had shown that $l_n \le n-4$ for $n\ge 5$, and $l_n\le n-5$ for $n\ge 9$, and Segre [68] established that $l_n\le n-6$ for $n\ge 157$. While Theorem 6.3 shows that $l_n\le n-d$ for $n\ge 2^{2^{d-2}}$, it would not be surprising if the lower bound on n in the latter could be replaced by a bound polynomial in d.

7. The truth in local fields

Having discussed upper bounds for $v_{d,r}(\mathbb{Q}_p)$ at some length, it seems appropriate next to discuss lower bounds for the latter quantity, and this permits us to recount one of those epic tales in number theory of Homeric dimensions. We take as our starting point a conjecture of Artin dating from 1936 (see Artin [4, p.x]), usually stated in a form equivalent to the following.

Conjecture (Artin). *For any prime p, whenever d and r_i $(1\le i\le d)$ are integers with $d\ge 2$, one has*

$$V_d^{(0)}(r_d,\dots,r_1;\mathbb{Q}_p)=r_1+4r_2+\dots+d^2r_d. \tag{7.1}$$

In particular, one has $v_d(\mathbb{Q}_p)=d^2$.

In order to establish Artin's Conjecture it suffices to show that $v_d(\mathbb{Q}_p)=d^2$ for each d and each prime p; see Lang [37] and Nagata [51] for details. That

$$V_d^{(0)}(r_d,\dots,r_1;\mathbb{Q}_p)\ge r_1+4r_2+\dots+d^2r_d$$

follows on considering systems of forms of the shape (2.1) discussed in §2(ii), and so the content of Artin's Conjecture lies in the upper bound implicit in (7.1). The evidence in favour of Artin's Conjecture was always weak, but not inconsequential. The classical theory of quadratic forms shows that $v_2(\mathbb{Q}_p)=4$ for every prime p (see Hasse [32]). In the middle of this century, Demyanov [26] (when $p\ne 3$) and Lewis [44] tackled cubic forms, showing that for every prime p one has $v_3(\mathbb{Q}_p)=9$, and subsequently Demyanov [27] considered pairs of quadratic forms, establishing that $v_{2,2}(\mathbb{Q}_p)=8$ for each prime p (see also the treatment of Birch, Lewis and Murphy [13]). Although Artin's Conjecture has never been established in any other instances, strong evidence in its favour has been derived from a number of partial results. Firstly, it was shown by Birch and Lewis [11] and Laxton and Lewis [40], that when $d=5$, 7 or 11, there is a positive number $p_0(d)$ with the property that whenever $p>p_0(d)$, one has $v_d(\mathbb{Q}_p)=d^2$. The arguments used to derive these conclusions make essential use of the Lang-Weil theorem (see Lang and Weil [39]), and thus while no explicit estimate for permissible $p_0(d)$ is provided by these authors, there is in principle no barrier to providing such (see Leep and Yeomans [43], where it is shown that $p_0(5)=43$ is permissible). Moreover, a crucial observation concerning certain factorisations of polynomials dictates that such methods are successful only when d is a prime number not exceeding 11. Also, it follows from Birch and Lewis [12], together with a correction and refinement of Schuur [67], that $v_{2,3}(\mathbb{Q}_p) = 12$ for $p > 7$.

Given the limitations of the above direct approaches to Artin's Conjecture, it is impressive that Ax and Kochen [5], by employing methods from Mathematical Logic, were able to show that Artin's Conjecture is very nearly true in general.

Theorem 7.1 (Ax and Kochen). *For each natural number d, there is a positive number $p_0(d)$, depending at most on d, with the property that whenever $p > p_0(d)$ one has $v_d(\mathbb{Q}_p) = d^2$. More generally, when d and $r_1, \dots, r_d$ are integers with $d \geq 2$, there is a positive number $p_1 = p_1(r_d, \dots, r_1)$ with the property that whenever $p > p_1$, one has*

$$V_d(r_d, \dots, r_1; \mathbb{Q}_p) = r_1 + 4r_2 + \dots + d^2 r_d.$$

Unfortunately, the methods of Ax and Kochen do not enable one to calculate explicit estimates for $p_0(d)$ and $p_1(\mathbf{r})$, and this remains a problem of great interest. The best that is known stems from an alternative treatment due to Cohen [19], which shows that $p_0(d)$ is bounded above by some primitive recursive function of the degree d, with a similar conclusion for $p_1(\mathbf{r})$.

The evidence that we have described thus far shows Artin's Conjecture to be "nearly" true. But in 1966, Terjanian [77] exhibited a homogeneous quartic polynomial with integral coefficients in 18 variables, which surprisingly failed to possess a non-trivial 2-adic solution. Since $18 > 4^2$, it follows that Artin's Conjecture fails when $d = 4$. Subsequently, Terjanian [78] showed that $v_4(\mathbb{Q}_2) \geq 20$ using another explicit example (see also Browkin [15]). While this example, and related ones, showed that Artin's Conjecture was not quite true, later work of Arkhipov and Karatsuba [2], motivated by investigations concerning a problem of Hilbert and Kamke, finally laid a torch to the conjecture. In a sharper form derived more or less simultaneously by Arkhipov and Karatsuba [3], Lewis and Montgomery [48] and Brownawell [16], we may reformulate this crushing of Artin's Conjecture as follows.

Theorem 7.2 (Arkhipov and Karatsuba; Lewis and Montgomery; Brownawell). *When d is a natural number and ε is a positive number, write*

$$\psi(d, \varepsilon) = \exp\left(\frac{d}{(\log d)(\log\log d)^{1+\varepsilon}}\right).$$

Then for each prime number p and positive number ε, there are infinitely many natural numbers d such that $v_d(\mathbb{Q}_p) > \psi(d, \varepsilon)$.

In other words, the number of variables required to guarantee p-adic solubility of a homogeneous equation of degree d may need to be exponentially large in terms of d. Thus, in a certain sense, Artin's Conjecture is spectacularly false! A similar conclusion holds in field extensions of $\mathbb{Q}_p$ (see Alemu [1]). An immediate corollary of Theorem 7.2, which we leave as an exercise to the reader, asserts that for each prime p and positive number ε, and for each natural number r, there are infinitely many d such that $v_{d,r}(\mathbb{Q}_p) > r\psi(d, \varepsilon)$. By only a modest elaboration of the techniques of Arkhipov and Karatsuba [3], Lewis and Montgomery [48] and Brownawell [16], one may sharpen the latter bound as follows (see Wooley [84]).

Theorem 7.3 (Wooley). *Let p be a prime number, and define $q = q(p)$ to be 6 when $p = 2$, and to be $p - 1$ when $p > 2$. Further, let $\alpha_p = (\log p)/(6q)$. There exist positive numbers $d_0(\varepsilon)$ and $r_0(d,\varepsilon)$ with the property that for each $\varepsilon > 0$, whenever d is an integer divisible by q with $d > d_0(\varepsilon)$, and $r > r_0(d,\varepsilon)$, then one has $v_{d,r}(\mathbb{Q}_p) > re^{(\alpha_p-\varepsilon)d}$.*

While the lower bound provided by Theorem 7.3 is larger than that following from the earlier methods, the significance of this conclusion lies in the extent to which Artin's Conjecture may now be said to fail. Thus, while the earlier arguments generated bad failures of Artin's Conjecture for a set of exponents d lying in an exponentially thin set, Theorem 7.3 does so for the prime p for all large exponents d divisible by $p-1$. In particular, bad failures of Artin's Conjecture are essentially ubiquitous, and in particular occur for all large even degrees. A second consequence of the methods used in establishing Theorem 7.3 is a lower bound on the exceptional primes permitted by the Ax-Kochen theorem. For convenience, we abbreviate the notation of Theorem 7.1 by writing $p^*(r_d, d) = p_1(r_d, 0, \dots, 0)$.

Theorem 7.4. *One has $\lim_{D\to\infty} \sup_{1\le d\le D} \sup_{r\in\mathbb{N}} \frac{p^*(r,d)}{d} \ge \frac{1}{30}$.*

Despite the numerous counter-examples to Artin's Conjecture described above, we currently possess none of odd degree. This prompts a conjecture (see Lewis [47]).

Conjecture. *Let p be a prime number. Whenever d is odd and $r_1, r_3, \dots, r_d$ are non-negative integers with $r_d \ge 1$, one has*

$$V_d^{(0)}(r_d, 0, r_{d-2}, 0, \dots, 0, r_3, 0, r_1; \mathbb{Q}_p) = r_1 + 9r_3 + \dots + d^2 r_d.$$

In particular, one has $v_d(\mathbb{Q}_p) = d^2$.

Some workers believe this conjecture to be more likely for prime exponents d.

8. Concluding remarks on solubility over the p-adic numbers

Despite discussing at length both upper and lower bounds for $v_{d,r}(\mathbb{Q}_p)$, we have neglected a number of topics worthy of investigation. Greater effort has been expended on bounds for $v_{d,r}(\mathbb{Q}_p)$ when d is small. Martin [50], improving on earlier work of Leep [41], has shown that $v_{2,r}(\mathbb{Q}_p) \le 2r^2$ when r is even, and $v_{2,r}(\mathbb{Q}_p) \le 2r^2 + 2$ when r is odd. Also, Schmidt [59], [60], [61] has applied analytic methods to establish that for each natural number r one has $v_{3,r}(\mathbb{Q}_p) \le 5300r(3r+1)^2$. Notice that the bounds described thus far have all been non-linear in r, and in particular those from §5 take the shape $v_{d,r}(\mathbb{Q}_p) \ll_d r^{2^{d-1}}$ when r is large. This raises the problem of determining the true rate of growth of $v_{d,r}(\mathbb{Q}_p)$ in terms of r. For all we know, the following could be true.

Conjecture. *For each prime p, and each natural number d, one has $v_{d,r}(\mathbb{Q}_p) \ll_d r$.*

Finally, we observe that it would be desirable to know that a given system possesses non-singular p-adic solutions in order to successfully apply the Hardy-Littlewood method in some generality. For example, suppose that a system of equations has a singular locus of very high dimension, but nonetheless possesses non-singular real and p-adic solutions for every prime p. Then one might hope that a suitable version of the circle method would establish an asymptotic formula for the number of rational points, up to a given height, in a neighbourhood away from the singular points. Unfortunately, the issue of existence of non-singular points is complicated by degeneracy. In any field k, for example, when s and d are natural numbers with $d > 1$, all solutions of the equation $(a_1x_1 + \cdots + a_sx_s)^d = 0$ are singular. Although we have no space to describe such matters herein, the author has made some progress on this problem by showing that given sufficiently many variables in terms of the degree, a form possesses non-singular solutions provided that it is not badly degenerate. We refer the reader to forthcoming work (Wooley [86]) for details.

9. Birch's method

Before discussions of Birch's method begin in earnest, it is useful to record some further notation relevant to systems of forms of odd degree. Let k be a field. When d is an odd number, we abbreviate

$$V_d^{(m)}(r_d, 0, r_{d-2}, 0, \ldots, 0, r_3, 0, r_1; k)$$

to

$$w_d^{(m)}(r_d, r_{d-2}, \ldots, r_3, r_1; k).$$

Next, when $m \geq 2$, we define $\mathcal{H}_d^{(m)}(r; k)$ to be the set of r-tuples, $(F_1, \ldots, F_r)$, of homogeneous polynomials of degree d, with coefficients in k, for which no linearly independent k-rational vectors $\mathbf{e}_1, \ldots, \mathbf{e}_m$ exist such that $F_i(t_1\mathbf{e}_1 + \cdots + t_m\mathbf{e}_m)$ is a diagonal form in $t_1, \ldots, t_m$ for $1 \leq i \leq r$. We then define $\tilde{w}_d^{(m)}(r) = \tilde{w}_d^{(m)}(r; k)$ by

$$\tilde{w}_d^{(m)}(r; k) = \sup_{\mathbf{h} \in \mathcal{H}_d^{(m)}(r;k)} \nu(\mathbf{h}).$$

Further, we adopt the convention that $\tilde{w}_d^{(1)}(r; k) = 0$. Note that $\tilde{w}_d^{(m)}(r; k)$ is an increasing function of the arguments m and r. Moreover, when $s > \tilde{w}_d^{(m)}(r; k)$ and $F_1, \ldots, F_r$ are homogeneous polynomials of degree d with coefficients in k possessing s variables, then there exist linearly independent k-rational vectors $\mathbf{e}_1, \ldots, \mathbf{e}_m$ with the property that $F_i(t_1\mathbf{e}_1 + \cdots + t_m\mathbf{e}_m)$ is a diagonal form in $t_1, \ldots, t_m$ for $1 \leq i \leq r$.

The historical sequence of events leading to the first bounds on $v_{d,r}(\mathbb{Q})$, for odd exponents d, illustrates the diverse nature of the ideas in this area. The interested reader will find a commentary on such matters provided by the editor of volume 4 of Mathematika. It seems that independently, Lewis, Birch and Davenport more or less simultaneously showed that a cubic form with rational coefficients in sufficiently many variables possesses a non-trivial rational solution (it seems that priority runs in the order indicated).

Lewis [46] observed that if $F(\mathbf{x}) \in \mathbb{Q}[x_1, \ldots, x_s]$ is a cubic form, then it possesses non-trivial $\mathbb{Q}$-rational zeros provided only that it possesses non-trivial $\mathbb{Q}(\sqrt{-1})$-rational zeros. For suppose that the latter is the case, and that α is a non-trivial

$\mathbb{Q}(\sqrt{-1})$-rational zero. If, under conjugation, this solution remains fixed (considered as a projective point on the hypersurface defined by $F = 0$), then this $\mathbb{Q}(\sqrt{-1})$-rational solution is equivalent to a non-trivial $\mathbb{Q}$-rational one. Otherwise, the two points $\boldsymbol{\alpha}$ and $\overline{\boldsymbol{\alpha}}$ are distinct, and by Bézout's theorem, the line passing through $\boldsymbol{\alpha}$ and $\overline{\boldsymbol{\alpha}}$ intersects the cubic hypersurface at some new point $\boldsymbol{\beta}$. But $\boldsymbol{\beta}$ is fixed under conjugation, so again we have a non-trivial $\mathbb{Q}$-rational zero. However, if s is sufficiently large, then it follows from Peck's theorem (see Peck [52], or Corollary 3 to Theorem 5.3 above) that F possesses a non-trivial $\mathbb{Q}(\sqrt{-1})$-rational zero, and we are done. What is astonishing about this idea of Lewis is that it generalises to show that for any number field K, and given any non-negative integers m and r with $r \geq 1$, one has $v_{3,r}^{(m)}(K) < \infty$. Unfortunately, however, the bounds stemming from such a method are weak, and indeed Lewis did not explicitly calculate a bound even for $v_3(\mathbb{Q})$.

Birch [8] cleverly adapted Brauer's elementary diagonalisation method, which we described in §5, in order to handle quite general systems.

Theorem 9.1 (Birch). *Let k be a field, and suppose that for each odd natural number i with $i \geq 3$ one has $\phi_i(k) < \infty$. Then for each odd natural number d, and for all non-negative integers $r_1, r_3, \dots, r_d$ and m, one has*

$$w_d^{(m)}(r_d, r_{d-2}, \dots, r_3, r_1; k) < \infty.$$

Since when j is odd the bound $\phi_j(k) < \infty$ holds in any algebraic number field k, as a consequence of Siegel's version (Siegel [71], [72]) of the Hardy-Littlewood method, it follows that $w_d^{(m)}(\mathbf{r}; k) < \infty$ for every number field k. Owing to the highly iterated nature of the induction leading to Birch's theorem, however, in general no explicit bounds were available for $w_d^{(m)}(\mathbf{r}; k)$ until very recently. We will discuss such matters further, and sketch a proof of Birch's theorem, later in this section.

Davenport's approach to bounding $v_3(\mathbb{Q})$ was through the use of the Hardy-Littlewood method (see, in particular, Davenport [20]). In a remarkable tour de force which has stimulated much subsequent work, Davenport initially succeeded in establishing that $v_3(\mathbb{Q}) \leq 31$ (see Davenport [20]), subsequently improving this bound first to $v_3(\mathbb{Q}) \leq 28$ (see Davenport [21]), and then $v_3(\mathbb{Q}) \leq 15$ (see Davenport [23]). The latter bound has never been improved, although in view of the local condition implicit in the conclusion $v_3(\mathbb{Q}_p) = 9$ for each prime p, one strongly suspects that $v_3(\mathbb{Q}) = 9$ (see Heath-Brown [33] and Hooley [34], [35], [36] for progress on non-singular cubic forms). In general the circle method is extremely difficult to apply to higher degree forms, although, as the strength of Davenport's bound suggests, a successful treatment should yield impressive consequences.

The solubility of a single cubic form over $\mathbb{Q}$ is the simplest problem along these lines to consider, and regrettably our knowledge extends hardly any further than this. Pleasants [54] has shown that in any number field K one has $v_3(K) \leq 15$ (see also Ramanujam [55] and Ryavec [57] for earlier results, and Skinner [73] for non-singular cubic forms). Schmidt [62] has studied systems of cubic forms via an impressive technical development of Davenport's methods. In particular, Schmidt obtains the bound $v_{3,2}(\mathbb{Q}) \le 5139$, and in general he shows that $v_{3,r}(\mathbb{Q}) \le (10r)^5$. In an interesting twist of fate, the emergence of Birch's quite general methods (Birch [8]) seems to have obscured Lewis' work [46] of the same time. In any case, by employing the latter's ideas, the author (Wooley [82]) has shown that in any algebraic extension K of $\mathbb{Q}$ (possibly $\mathbb{Q}$ itself), one has $v_{3,2}(K) \le 855$. Finally, Schmidt [65] has shown that as a consequence of the precise version of his Theorem 3.4 above, one may obtain a bound of the shape

$$v_{5,r}(\mathbb{Q}) < A\exp(\exp(Br)), \tag{9.1}$$

for suitable positive constants A and B.

Recently the author (Wooley [83], [85]) found that a rather efficient diagonalisation procedure could be engineered by exploiting ideas of Lewis and Schulze-Pillot [49]. So far as estimates for $v_{3,r}(\mathbb{Q})$ are concerned, this refinement of Birch's method yields weaker conclusions than those obtained by Schmidt [62]. However, the methods can be employed in any field k for which bounds are available for $\phi_{3,r}(k)$. Since a simple argument shows that for each natural number r one has $\phi_{3,r}(k)+1 \le (\phi_3(k)+1)^r$ (see, for example, the argument of the proof of Lemma 10.4 below), it suffices simply to have an upper bound for $\phi_3(k)$.

Theorem 9.2 (Wooley). *Let k be a field, let m and r be non-negative integers with $r \ge 1$, and suppose that $\phi_{3,r}(k)$ is finite. Then*

$$v_{3,r}^{(m)}(k) \le r^3(m+1)^5(\phi_{3,r}(k)+1)^5.$$

In particular, one has $v_{3,r}^{(m)}(k) \le r^3(m+1)^5(\phi_3(k)+1)^{5r}$.

The methods described in Wooley [83] are already more effective than those available hitherto for systems of quintic forms. Thus the bound (9.1) due to Schmidt [65] may be improved as follows.

Theorem 9.3 (Wooley). *Let m and r be non-negative integers with $r \ge 1$. Then*

$$v_{5,r}^{(m)}(\mathbb{Q}) < \exp\left(10^{32}((m+1)r\log(3r))^{\kappa}\log(3r(m+1))\right),$$

where $\kappa = \frac{\log 3430}{\log 4} = 5.87199\ldots$. In particular, $v_{5,r}(\mathbb{Q}) = o(e^{r^6})$.

For systems of forms of degree higher than 5, the bounds provided by the methods of Wooley [83], [85] are embarrassingly weak, but currently such are the only explicit bounds available in Birch's theorem. In order to describe these bounds we will require some notation. Suppose that A is a subset of $\mathbb{R}$ and Ψ is a function mapping A into A. When α is a real number, write $[\alpha]$ for the largest integer not exceeding α. Then we adopt the notation that whenever x and y are real numbers with $x \ge 1$, then $\Psi_x(y)$ denotes the real number $a_{[x]}$, were $(a_n)_{n=1}^{\infty}$ is the sequence defined by taking $a_1 = \Psi(y)$, and $a_{i+1} = \Psi(a_i)$ $(i \ge 1)$. Finally, when n is a non-negative integer we define the functions $\psi^{(n)}(x)$ by taking $\psi^{(0)}(x) = \exp(x)$, and when $n > 0$ by putting

$$\psi^{(n)}(x) = \psi^{(n-1)}_{42\log x}(x). \tag{9.2}$$

Theorem 9.4 (Wooley). *Let d be an odd integer exceeding 5, and let r and m be non-negative integers with $r \geq 1$. Then*

$$v_{d,r}^{(m)}(\mathbb{Q}) < \psi^{((d-5)/2)}(dr(m+1)).$$

More generally, when $r_1, r_3, \dots, r_d$ are non-negative integers with $r_d \geq 1$, one has

$$w_d^{(m)}(r_d, \dots, r_1; \mathbb{Q}) \leq \psi^{((d-5)/2)}(d(r_1 + r_3 + \dots + r_d)(m+1)).$$

We note that the number 42 occurring in the definition (9.2) could certainly be reduced with greater effort, especially for large values of the parameters d, r and $m+1$. However, the level of iteration involved in the function $\psi^{(n)}$ seems difficult to improve, and in particular present methods seem unable to reduce the function $\log x$ in (9.2) with any function of significantly smaller rate of growth.

We now sketch the basic ideas underlying the proof of Birch's theorem (Theorem 9.1). Suppose that $w_D^{(M)}(\mathbf{r}; k) < \infty$ for each odd D with $D < d$, and for every $\mathbf{r}$ and m. In broad outline we follow the strategy for establishing Brauer's theorem as sketched in §5. In particular, it suffices to bound $v_d^{(m)}(k)$ for odd exponents d, and such is possible provided that $\tilde{w}_d^{(m)}(1; k) < \infty$ for each natural number m. We establish the latter bound by induction on m, noting trivially that $\tilde{w}_d^{(1)}(1; k) = 0$. Suppose then that $m \geq 1$ and that $\tilde{w}_d^{(m)}(1; k)$ is finite, and write $n = \tilde{w}_d^{(m)}(1; k)$. We take q to be a natural number sufficiently large in terms of d, n and k, and we take p to be a natural number sufficiently large in terms of d, n, q and k. Let $s = p + q$, and consider a linear subspace U of k^s of affine dimension p. We take V to be the complementary linear space of affine dimension q, so that $k^s = U \oplus V$. Let $\mathbf{v}_1, \dots, \mathbf{v}_q$ be a k-rational basis for V, and consider an arbitrary element $\mathbf{v} = c_1\mathbf{v}_1 + \dots + c_q\mathbf{v}_q$ of V. Also, let $\mathbf{u}$ be an arbitrary element of U, and consider a form $F(\mathbf{x}) \in k[x_1, \dots, x_s]$ of odd degree d. Then for every $t, w \in k$,

$$F(t\mathbf{u} + w\mathbf{v}) = t^d F(\mathbf{u}) + w^d F(\mathbf{v}) + \sum_{\substack{i=1 \\ i \text{ odd}}}^{d-1} t^i w^{d-i} G_i(\mathbf{u}, \mathbf{v}) + \sum_{\substack{j=1 \\ j \text{ even}}}^{d-1} t^j w^{d-j} H_j(\mathbf{u}, \mathbf{v}), \tag{9.3}$$

where $G_i(\mathbf{u}, \mathbf{v}) \in k[\mathbf{u}, \mathbf{v}]$ is a homogeneous polynomial of odd degree i in $\mathbf{u}$ and even degree $d - i$ in $\mathbf{v}$ ($i = 1, 3, \dots, d-2$), and $H_j(\mathbf{u}, \mathbf{v}) \in k[\mathbf{u}, \mathbf{v}]$ is a homogeneous polynomial of odd degree $d - j$ in $\mathbf{v}$ and even degree j in $\mathbf{u}$ ($j = 2, 4, \dots, d-1$). On examining the coefficient of each term $c_1^{i_1} \dots c_q^{i_q}$ in (9.3), we find that the system of equations

$$G_i(\mathbf{u}, \mathbf{v}) = 0 \quad (i = 1, 3, \dots, d-2) \tag{9.4}$$

is satisfied for every $\mathbf{v} \in V$ provided only that a certain system of equations in $\mathbf{u}$ is satisfied. The latter system consists of homogeneous equations of odd degree at most $d-2$, say r_j of degree j ($j = 1, 3, \dots, d-2$), with r_j bounded in terms of d and q for each j. Since we may suppose that $\dim(U) = p > w_{d-2}^{(0)}(\mathbf{r}; k)$, it follows that the system (9.4) is satisfied for every $\mathbf{v} \in V$ for some fixed non-trivial $\mathbf{u} \in U$. Consider next the system of equations

$$H_j(\mathbf{u}, \mathbf{v}) = 0 \quad (j = 2, 4, \dots, d-1). \tag{9.5}$$

Since $\mathbf{u}$ is now fixed, this is a system of equations of odd degree at most $d-2$ in $\mathbf{v}$, say with r_j' equations of degree j ($j = 1, 3, \dots, d-2$). Since we may suppose that $\dim(V) = q > w_{d-2}^{(n)}(\mathbf{r}'; k)$, it follows that the system (9.5) possesses a k-rational linear space of solutions of affine dimension $n+1 > \tilde{w}_d^{(m)}(1; k)$. But for each solution $\mathbf{v}$ lying in the latter linear space one has, for each $t, w \in k$, that $F(t\mathbf{u} + w\mathbf{v}) = t^d F(\mathbf{u}) + w^d F(\mathbf{v})$, and so it follows that the polynomial $F(\mathbf{x})$ may be reduced non-trivially to a diagonal form in at least $m+1$ variables, whence $\tilde{w}_d^{(m+1)}(1; k) < s < \infty$. Thus it follows that whenever both $\tilde{w}_d^{(m)}(1; k)$ and $w_{d-2}^{(M)}(\mathbf{r}; k)$ are finite, for each M and $\mathbf{r}$, then one has that $\tilde{w}_d^{(m+1)}(1; k)$ is finite. Our claim that $\tilde{w}_d^{(m)}(1; k) < \infty$ for each m therefore follows by induction.

We conclude by remarking that the above strategy may be improved, firstly by diagonalising a whole system of equations simultaneously, removing one of the unpleasant iterated steps suppressed above. Also, and this is an idea whose origins lie in work of Lewis and Schulze-Pillot [49], one may diagonalise in such a way that rather than iterating from $\tilde{w}_d^{(m)}(1; k)$ to $\tilde{w}_d^{(m+1)}(1; k)$, as sketched above, one may instead iterate from $\tilde{w}_d^{(m)}(1; k)$ to $\tilde{w}_d^{(2m)}(1; k)$. Such a procedure plainly has the potential to dramatically accelerate the diagonalisation process, and it is this idea which underlies the work of Wooley [83], [85] described above.

10. Birch's Theorem in algebraic number fields

The only major obstruction to establishing a strong analogue of Theorem 9.4 in algebraic number fields lies in our lack of knowledge concerning the solubility of simultaneous additive equations. Thus, while the bound

$$\phi_{d,r}(\mathbb{Q}) + 1 \le 48rd^3 \log(3rd^2) \tag{10.1}$$

due to Brüdern and Cook [17], or earlier bounds due to Davenport and Lewis [25], provide a suitable foundation for Birch's method over $\mathbb{Q}$, we have no strong analogue of the bound (10.1) when the field $\mathbb{Q}$ is replaced by an algebraic number field. Thus, while there is no difficulty in principle to establishing such bounds, the significant quantity of work required to establish suitable estimates has thus far deterred detailed investigations. Such difficulties, moreover, are compounded when one is interested in bounds independent of the degree of the field extension. Since there appears to be some interest in this topic, we take the opportunity herein to conclude our discussion of diophantine problems in many variables by establishing an explicit version of Birch's theorem in algebraic number fields, albeit a weak one.

We begin by recalling some technical lemmata from Wooley [83], starting with a lemma which provides information concerning the number of variables required to diagonalise a system of forms. In what follows, we denote by k an arbitrary field.

Lemma 10.1. *Let d be an odd integer with $d \geq 3$, and let r, n and m be natural numbers. Then*

$$\tilde{w}_d^{(n+m)}(r;k) \leq s + w_{d-2}^{(M)}(\mathbf{R};k),$$

where

$$M = \tilde{w}_d^{(n)}(r;k), \quad s = 1 + w_{d-2}^{(N)}(\mathbf{S};k), \quad N = \tilde{w}_d^{(m)}(r;k),$$

and for $0 \leq u \leq (d-1)/2$,

$$R_{2u+1} = r\binom{s+d-2u-2}{d-2u-1} \quad \text{and} \quad S_{2u+1} = r\binom{n+d-2u-2}{d-2u-1}.$$

Proof. This is Lemma 2.1 of Wooley [83].

We next bound $v_{d,r}^{(m)}(k)$ in terms of $\tilde{w}_d^{(M)}(r;k)$.

Lemma 10.2. *Let d be an odd positive number, let r be a natural number, and let m be a non-negative integer. Then*

$$v_{d,r}^{(m)}(k) \leq \tilde{w}_d^{(M)}(r;k),$$

where $M = (m+1)(\phi_{d,r}(k)+1)$.

Proof. This is Lemma 2.2 of Wooley [83].

Finally, we provide a bound for $w_d^{(m)}(\mathbf{r};k)$ in terms of the quantities $w_{d-2}^{(M)}(\mathbf{r};k)$ and $v_{d,r}^{(m)}(k)$.

Lemma 10.3. *Let $d \geq 3$ be an odd positive number, and let $r_1, r_3, \ldots, r_d$ be non-negative integers with $r_d > 0$. Then for each non-negative integer m one has*

$$w_d^{(m)}(r_d, r_{d-2}, \ldots, r_1; k) \leq w_{d-2}^{(M)}(r_{d-2}, \ldots, r_1; k),$$

where $M = v_{d,r_d}^{(m)}(k)$.

Proof. This is Lemma 2.3 of Wooley [83].

Henceforth we restrict attention to a fixed algebraic extension K of $\mathbb{Q}$, not necessarily of finite degree, and for the sake of concision, when it is convenient so to do, we drop explicit mention of this field from our various notations. In order to make use of Lemma 10.2 in our argument we require an estimate for $\phi_{d,r}(K)$, and here we are interested in simple estimates independent of K. The problem of bounding $\phi_{d,r}(K)$ is substantially more difficult, in general, than that of bounding $\phi_{d,r}(\mathbb{Q})$, in large part because the local solubility problem for systems of forms over K may be significantly more complicated than the corresponding problem over $\mathbb{Q}$. We circumvent such issues by employing only estimates of simple type depending on our knowledge of $\phi_d(K)$.

Lemma 10.4. *Let d be an odd positive number with $d \geq 3$, and let r be a natural number. Then one has $\phi_{d,r}(K) + 1 \leq e^{2dr}$.*

Proof. We consider the algebraic extension K of $\mathbb{Q}$. Let $s \geq 2^d + 1$, and suppose that $b_1, \ldots, b_s \in K$. Note first that there is a finite extension L of $\mathbb{Q}$ with the property that $b_1, \ldots, b_s \in L$, and moreover that whenever the equation

$$b_1 x_1^d + \cdots + b_s x_s^d = 0 \tag{10.2}$$

possesses a non-trivial solution in L, then it does so also in K. Next we recall that by using Siegel's version of the circle method (see Siegel [71], [72]), Birch [9, Theorem 3] was able to show that whenever $s > 2^d$, the equation (10.2) has a non-trivial solution in L provided that it has a non-trivial solution in every real and $\mathfrak{p}$-adic completion of L. The condition that the equation (10.2) be soluble in the real completion of L is, of course, trivially satisfied when that completion is $\mathbb{C}$. When that completion is $\mathbb{R}$, meanwhile, the equation (10.2) has a non-trivial solution in the real completion of L whenever d is odd. On the other hand, when M is a finite extension of $\mathbb{Q}_p$ and d is an integer with $d \geq 2$, it follows from Skinner [74] that

$$\phi_d(M) \leq d\left((d+1)^{\max\{2\log d/\log p, 1\}} - 1\right).$$

Consequently, when d is odd the equation (10.2) is soluble non-trivially over L, and hence also over K, provided only that

$$s > \max\{2^d, d((d+1)^{\max\{2\log d/\log p, 1\}} - 1)\}. \tag{10.3}$$

The simple bound $\phi_d(K) + 1 \leq e^{2d}$ follows from (10.3) with a modicum of computation.

Next suppose that $r > 1$, and write $m = \phi_{d,r-1}(K)$. Suppose that $s \geq (m+1)(\phi_d(K)+1)$, and consider elements $b_{ij} \in K$ $(1 \leq i \leq r,\ 1 \leq j \leq s)$, and the system of equations

$$\sum_{j=1}^{s} b_{ij} x_j^d = 0 \quad (1 \leq i \leq r). \tag{10.4}$$

Since $m + 1 > \phi_{d,r-1}(K)$, it follows that each of the systems

$$\sum_{j=t(m+1)+1}^{(t+1)(m+1)} b_{ij} x_j^d = 0 \quad (2 \leq i \leq r)$$

has a solution $\mathbf{x} = (a_{t(m+1)+1}, \ldots, a_{(t+1)(m+1)}) \in K^{m+1} \setminus \{\mathbf{0}\}$ for $0 \leq t \leq \phi_d(K)$. On substituting

$$x_{t(m+1)+j} = a_{t(m+1)+j} y_t \quad (1 \leq j \leq m+1,\ 0 \leq t \leq \phi_d(K)),$$

we find that the system (10.4) is soluble provided that there is a non-trivial K-rational solution to the equation

$$c_0 y_0^d + \cdots + c_T y_T^d = 0, \tag{10.5}$$

where $T = \phi_d(K)$, and

$$c_t = \sum_{j=t(m+1)+1}^{(t+1)(m+1)} b_{1j} a_{t(m+1)+j}^d \quad (0 \le t \le \phi_d(K)).$$

But $T + 1 > \phi_d(K)$, and so the equation (10.5) does indeed possess a non-trivial K-rational solution. We therefore conclude that whenever $r > 1$, one has

$$\phi_{d,r}(K) + 1 \le (\phi_{d,r-1}(K) + 1)(\phi_d(K) + 1),$$

and hence by induction one obtains $\phi_{d,r}(K) + 1 \le (\phi_d(K) + 1)^r$. On recalling the conclusion of the previous paragraph, we therefore conclude that $\phi_{d,r}(K)+1 \le e^{2dr}$. This completes the proof of the lemma.

Having negotiated the preliminaries, we begin the main body of our investigation, and make no apology for following closely the argument of Wooley [85]. We start with bounds for $v_{3,r}(K)$ and $v_{5,r}(K)$, the former quantity being bounded easily by means of Theorem 9.2 above.

Theorem 10.5. *Let m and r be non-negative integers with $r \ge 1$. Then one has*

$$v_{3,r}^{(m)}(K) \le (m+1)^5 r^3 3^{10r}.$$

Proof. Since it is readily available from the literature, we make use of a sharper bound for $\phi_{3,r}(K)$ than is immediately available from Lemma 10.4. By Lewis [45], it follows that for any algebraic extension K_p of $\mathbb{Q}_p$, one has $\phi_3(K_p) \le 6$. Consequently, the argument of the proof of Lemma 10.4 shows that

$$\phi_3(K) + 1 \le \max\{7, 2^3 + 1\} = 9,$$

whence $\phi_{3,r}(K) + 1 \le 9^r$. On substituting the latter bound into the conclusion of Theorem 9.2, the desired conclusion is immediate.

We require a simplification of Theorem 10.5 of use in our main inductive process, and for this purpose the following lemma suffices.

Lemma 10.6. *Suppose that r_3, r_1 and m are non-negative integers with $r_1 < 3r_3^2$. Then*

$$w_3^{(m)}(r_3, r_1; K) < \exp(21 r_3(m+1)).$$

Proof. Whenever $r_1 < 3r_3^2$, one finds by elimination of the implicit linear equations that

$$w_3^{(m)}(r_3, r_1) = r_1 + v_{3,r_3}^{(m)} < 3r_3^2 + (m+1)^5 r_3^3 3^{10r_3},$$

and a modest calculation therefore leads to the desired conclusion.

We must now dispose of systems of quintic forms, beginning with the diagonalisation process.

Lemma 10.7. *Suppose that m and r are natural numbers. Then*

$$\tilde{w}_5^{(m)}(r;K) < \exp_{5\log(5m)}(5rm).$$

Proof. When m and r are natural numbers, write

$$\overline{w}_5^{(m)}(r) = \exp_{5\log(5m)}(5rm). \tag{10.6}$$

We aim to show that for each R and M one has

$$\tilde{w}_5^{(M)}(R) < \overline{w}_5^{(M)}(R), \tag{10.7}$$

and from this the conclusion of the lemma is immediate. Note that by definition, for each natural number R one has $\tilde{w}_5^{(1)}(R) = 0$, so that (10.7) certainly holds when $M = 1$. Next suppose that $m > 1$, and that for each R the inequality (10.7) holds whenever $M < m$. We will establish that (10.7) holds for each R when $M = m$, whence (10.7) follows for all R and M by induction.

Let m and r be natual numbers with $m \geq 2$. Write $n = [(m+1)/2]$, and note that $n < m$. By Lemma 10.1 one has

$$\tilde{w}_5^{(m)}(r) \leq \tilde{w}_5^{(2n)}(r) \leq s + w_3^{(N)}(\mathbf{R}), \tag{10.8}$$

where

$$N = \tilde{w}_5^{(n)}(r), \quad s = 1 + w_3^{(N)}(\mathbf{S}), \tag{10.9}$$

and for $0 \leq u \leq 2$,

$$R_{2u+1} = r\binom{s+3-2u}{4-2u} \quad \text{and} \quad S_{2u+1} = r\binom{n+3-2u}{4-2u}.$$

We first bound s. Write $\overline{N} = [\overline{w}_5^{(n)}(r)]$, and note that since $n < m$, the inductive hypothesis shows that $N \leq \overline{N}$. Note also that for $0 \leq u \leq 2$ one has $S_{2u+1} \leq rn^{4-2u}$. Then the hypotheses required for the application of Lemma 10.6 to bound $w_3^{(N)}(\mathbf{S})$ are satisfied, and we may conclude from (10.9) that

$$s \leq 1 + w_3^{(\overline{N})}(rn^2, rn^4) \leq w_3^{(\overline{N})}(rn^2, 2rn^4),$$

whence

$$s < \exp(21rn^2(\overline{N}+1)). \tag{10.10}$$

But by (10.6) one has

$$\overline{N} \leq \overline{w}_5^{(n)}(r) = \exp_{5\log(5n)}(5rn) \tag{10.11}$$

and

$$\overline{N} = [\exp_{5\log(5n)}(5rn)] \geq 5rn. \tag{10.12}$$

Also, plainly, for each $m \geq 2$ it follows from (10.6) that $\overline{N} \geq \exp_5(5)$. Then by combining (10.10) and (10.12), we obtain

$$s \leq \exp((\overline{N}+1)^3) < \exp(\overline{N}^4). \tag{10.13}$$

Finally we bound $\tilde{w}_5^{(m)}(r)$ by substituting (10.13) into (10.8). Note that for $0 \leq u \leq 2$ one has $R_{2u+1} \leq rs^{4-2u}$. Then the hypotheses required for the application of Lemma 10.6 to bound $w_3^{(\overline{N})}(\mathbf{R})$ are satisfied, and we may conclude that

$$\tilde{w}_5^{(m)}(r) \leq s + w_3^{(\overline{N})}(rs^2, rs^4) \leq w_3^{(\overline{N})}(rs^2, 2rs^4),$$

whence

$$\tilde{w}_5^{(m)}(r) < \exp(21rs^2(\overline{N}+1)).$$

Write $\overline{s} = \exp(\overline{N}^4)$. Then by (10.12) and (10.13), one has

$$\tilde{w}_5^{(m)}(r) < \exp((s(\overline{N}+1))^3) \leq \exp(\overline{s}^6) = \exp_2(6\overline{N}^4) < \exp_2(\overline{N}^5).$$

We therefore deduce from (10.11) that

$$\tilde{w}_5^{(m)}(r) < \exp_3(5\log\overline{N}) < \exp_3(\exp_{5\log(5n)-1}(5rm)).$$

But on noting that whenever $m \geq 2$ one has

$$\left[5\log\left(5\left[\frac{m+1}{2}\right]\right)\right] \leq [5\log(5m)-2],$$

we may conclude from (10.6) that

$$\tilde{w}_5^{(m)}(r) < \exp_{5\log(5m)}(5rm) = \overline{w}_5^{(m)}(r),$$

thereby establishing the inequality (10.7) with $M = m$ and $R = r$. Thus, on recalling the comments concluding the first paragraph of the proof, the proof of the lemma is complete.

A bound for $v_{5,r}^{(m)}(K)$ follows on substituting the conclusion of Lemma 10.7 into Lemma 10.2.

Theorem 10.8. *Suppose that m and r are non-negative integers with $r \geq 1$. Then*

$$v_{5,r}^{(m)}(K) < \exp_{67r(m+1)}(5r(m+1)).$$

Proof. By combining Lemmata 10.2 and 10.4 with Lemma 10.7, one obtains

$$v_{5,r}^{(m)}(K) < \exp_{5\log(5M)}(5rM), \tag{10.14}$$

where $M = (m+1)e^{10r}$. But a little calculation reveals that

$$\log(5M) < \log(5\exp(10r+(m+1))) < 13r(m+1),$$

and

$$\log(5rM) < 10r + \log(5r(m+1)) < \exp(5r(m+1)),$$

and hence (10.14) provides the estimate

$$v_{5,r}^{(m)} < \exp_{65r(m+1)+2}(5r(m+1)).$$

The conclusion of the lemma follows immediately.

We will require a slightly more general conclusion within our main induction.

Lemma 10.9. *Suppose that r_1, r_3, r_5 and m are non-negative integers with $r_1 \le 3r_3^2$ and $r_3 < 3r_5^2$. Then*

$$w_5^{(m)}(r_5, r_3, r_1) < \exp_{69r_5(m+1)}(5r_5(m+1)).$$

Proof. By Lemma 10.3 one has $w_5^{(m)}(\mathbf{r}) \le w_3^{(v)}(r_3, r_1)$, where $v = v_{5,r_5}^{(m)}$. But in view of the hypotheses concerning r_1 and r_3, we may apply Lemma 10.6 together with Theorem 10.8 to conclude that

$$\log w_5^{(m)}(\mathbf{r}) < 63r_5^2\left(1 + \exp_{67r_5(m+1)}(5r_5(m+1))\right),$$

whence

$$\log_2 w_5^{(m)}(\mathbf{r}) < \exp_{67r_5(m+1)}(5r_5(m+1)).$$

The desired conclusion is essentially immediate from the latter inequality.

We have now established the basis for our induction. In order to establish the main inductive step, we require some further notation. We say that the function $\Psi : [1,\infty) \to [1,\infty)$ satisfies the exponential growth condition if it has derivatives of all orders on $[1,\infty)$, and moreover for each non-negative integer n, one has for each $x \in [1,\infty)$ that

$$\frac{d^n\Psi(x)}{dx^n} \ge e^x.$$

When D is an integer exceeding 3, we make use of the following hypothesis.

Hypothesis $\mathcal{H}_D(\Psi)$. *For all natural numbers M, and all $\frac{1}{2}(D+1)$-tuples $\mathbf{R} = (R_D, R_{D-2}, \dots, R_1)$ of non-negative integers satisfying $R_{D-2} < 3R_D^2$ and $R_i \le 3R_{i+2}^2$ $(i = 1, 3, \dots, D-4)$, one has*

$$w_D^{(M)}(\mathbf{R}) < \Psi(DR_D(M+1)). \tag{10.15}$$

Fortunately, the diagonalisation process is largely independent of the ambient field under consideration.

Lemma 10.10. *Let d be an odd integer exceeding 5. Suppose that Ψ is a function satisfying the exponential growth condition, and suppose further that the hypothesis $\mathcal{H}_{d-2}(\Psi)$ holds. Then whenever m and r are natural numbers, one has*

$$\tilde{w}_d^{(m)}(r) < \Psi_{5\log(dm)}(drm).$$

Proof. This is essentially Lemma 4.1 of Wooley [85].

On combining Lemma 10.10 with Lemma 10.2, we are able to bound $v_{d,r}^{(m)}(K)$ on the hypothesis $\mathcal{H}_{d-2}(\Psi)$.

Lemma 10.11. *Let d be an odd integer exceeding 5. Suppose that Ψ is a function satisfying the exponential growth condition, and suppose further that the hypothesis $\mathcal{H}_{d-2}(\Psi)$ holds. Then whenever m and r are non-negative integers with $r \geq 1$, one has*

$$v_{d,r}^{(m)}(K) < \Psi_{16dr(m+1)}(dr(m+1)).$$

Proof. On combining Lemmata 10.2 and 10.4 with Lemma 10.10, one obtains

$$v_{d,r}^{(m)}(K) < \Psi_{5\log(dM)}(drM), \tag{10.16}$$

where $M = (m+1)e^{2dr}$. But a modicum of computation reveals that

$$\log(dM) = 2dr + \log(d(m+1)) < 3dr(m+1),$$

and

$$\log(drM) = 2dr + \log(dr(m+1)) < 3dr(m+1) < \exp(dr(m+1)),$$

and hence (10.16) leads to the upper bound

$$v_{d,r}^{(m)}(K) < \Psi_{15dr(m+1)+2}(dr(m+1)).$$

The conclusion of the lemma follows immediately.

In order to complete the main inductive step we must combine the conclusion of Lemma 10.11 with the hypothesis $\mathcal{H}_{d-2}(\Psi)$ in order to bound $w_d^{(m)}(\mathbf{r};K)$.

Lemma 10.12. *Let d be an odd integer exceeding 5. Suppose that Ψ is a function satisfying the exponential growth condition, and suppose further that the hypothesis $\mathcal{H}_{d-2}(\Psi)$ holds. Then whenever r_{2u+1} $(0 \leq u \leq \frac{1}{2}(d-1))$ and m are non-negative integers with $r_i \leq 3r_{i+2}^2$ $(i = 1, 3, \ldots, d-4)$ and $r_{d-2} < 3r_d^2$, one has*

$$w_d^{(m)}(\mathbf{r};K) < \Psi_{17dr_d(m+1)}(dr_d(m+1)).$$

Proof. By Lemma 10.3 one has

$$w_d^{(m)}(\mathbf{r}) \leq w_{d-2}^{(v)}(r_{d-2}, \ldots, r_1),$$

where $v = v_{d,r_d}^{(m)}$. The hypotheses concerning r_i for $i = 1, 3, \ldots, d-2$ permit us the use of the hypothesis $\mathcal{H}_{d-2}(\Psi)$ in order to bound the quantity $w_{d-2}^{(v)}(r_{d-2}, \ldots, r_1)$, and thus on employing Lemma 10.11 to bound v, we deduce that

$$\begin{aligned} w_{d-2}^{(v)}(r_{d-2}, \ldots, r_1) &< \Psi(3(d-2)r_d^2(v+1)) \\ &< \Psi_2(2\Psi_{16dr_d(m+1)-1}(dr_d(m+1))) \\ &\leq \Psi_{16dr_d(m+1)+2}(dr_d(m+1)). \end{aligned}$$

The conclusion of the lemma is now immediate.

We have now reached the crescendo of our argument, and this demands further notation. We define the functions $\phi^{(n)}(x)$ by taking $\phi^{(0)}(x) = \exp(x)$, and when $n > 0$ by putting

$$\phi^{(n)}(x) = \phi_{17x}^{(n-1)}(x). \tag{10.17}$$

Theorem 10.13. *Let K be an algebraic extension of $\mathbb{Q}$, let d be an odd integer exceeding 3, and let r and m be non-negative integers with $r \geq 1$. Then*

$$v_{d,r}^{(m)}(K) < \phi^{((d-3)/2)}(dr(m+1)).$$

Proof. The conclusion of the theorem is immediate when $d = 5$, in view of Theorem 10.8. Note next that by Lemma 10.9, the hypothesis $\mathcal{H}_5(\phi^{(1)})$ holds, where $\phi^{(1)}$ is defined by (10.17). Moreover $\phi^{(1)}$ plainly satisfies the exponential growth condition. Suppose now that d is an odd integer exceeding 5, and that the hypothesis $\mathcal{H}_{d-2}(\phi^{((d-5)/2)})$ holds. Since plainly $\phi^{((d-5)/2)}$ also satisfies the exponential growth condition, it follows from Lemma 10.12 that the hypothesis $\mathcal{H}_d(\phi^{((d-3)/2)})$ holds. We therefore deduce, by induction, that the hypothesis $\mathcal{H}_d(\phi^{((d-3)/2)})$ holds for every odd integer d exceeding 3. Consequently, on applying Lemma 10.11, we conclude that the inequality

$$v_{d,r}^{(m)}(K) < \phi_{16dr(m+1)}^{((d-5)/2)}(dr(m+1)) < \phi^{((d-3)/2)}(dr(m+1))$$

holds for every odd integer exceeding 3. This completes the proof of the theorem.

REFERENCES

1. Y. Alemu, *On zeros of forms over local fields*, Acta Arith. **65** (1985), 163–171.
2. G. I. Arkhipov and A. A. Karatsuba, *Local representation of zero by a form*, Izv. Akad. Nauk SSSR Ser. Mat. **45** (1981), 948–961. (Russian)
3. G. I. Arkhipov and A. A. Karatsuba, *A problem of comparison theory*, Uspekhi Mat. Nauk **37** (1982), 161–162. (Russian)
4. E. Artin, *The collected papers of Emil Artin*, Addison-Wesley, 1965.
5. J. Ax and S. Kochen, *Diophantine problems over local fields, I*, Amer. J. Math. **87** (1965), 605–630.
6. R. C. Baker, *Diophantine inequalities*, Clarendon Press, Oxford, 1986.
7. V. Batyrev and Yu. Manin, *Sur le nombre des points rationnels de hauteur bornée des variétés algébriques*, Math. Ann. **286** (1990), 27–43.
8. B. J. Birch, *Homogeneous forms of odd degree in a large number of variables*, Mathematika **4** (1957), 102–105.
9. B. J. Birch, *Waring's problem in algebraic number fields*, Proc. Cambridge Philos. Soc. **57** (1961), 449–459.
10. B. J. Birch, *Forms in many variables*, Proc. Roy. Soc. Ser. A **265** (1961), 245–263.
11. B. J. Birch and D. J. Lewis, *$\mathfrak{p}$-adic forms*, J. Indian Math. Soc. **23** (1959), 11–31.
12. B. J. Birch and D. J. Lewis, *Systems of three quadratic forms*, Acta Arith. **10** (1965), 423–442.
13. B. J. Birch, D. J. Lewis and T. G. Murphy, *Simultaneous quadratic forms*, Amer. J. Math. **84** (1962), 110–115.
14. R. Brauer, *A note on systems of homogeneous algebraic equations*, Bull. Amer. Math. Soc. **51** (1945), 749–755.
15. J. Browkin, *On forms over p-adic fields*, Bull. Acad. Polon. Sci. Sér. Sci. Math. Astronom. Phys. **14** (1966), 489–492.
16. W. D. Brownawell, *On p-adic zeros of forms*, J. Number Theory **18** (1984), 342–349.
17. J. Brüdern and R. J. Cook, *On simultaneous diagonal equations and inequalities*, Acta Arith. **62** (1992), 125–149.
18. J. W. S. Cassels and M. J. T. Guy, *On the Hasse principle for cubic surfaces*, Mathematika **13** (1966), 111–120.

19. P. J. Cohen, *Decision procedures for real and p-adic fields*, Comm. Pure Appl. Math. **22** (1969), 131–151.
20. H. Davenport, *Cubic forms in thirty-two variables*, Philos. Trans. Roy. Soc. London Ser. A **251** (1959), 193–232.
21. H. Davenport, *Cubic forms in 29 variables*, Proc. Roy. Soc. Ser. A **266** (1962), 287–298.
22. H. Davenport, *Analytic methods for Diophantine equations and Diophantine inequalities*, Ann Arbor Publishers, Ann Arbor, 1962.
23. H. Davenport, *Cubic forms in 16 variables*, Proc. Roy. Soc. Ser. A **272** (1963), 285–303.
24. H. Davenport and D. J. Lewis, *Homogeneous additive equations*, Proc. Roy. Soc. Ser. A **274** (1963), 443–460.
25. H. Davenport and D. J. Lewis, *Simultaneous equations of additive type*, Philos. Trans. Roy. Soc. London Ser. A **264** (1969), 557–595.
26. V. B. Dem'yanov, *On cubic forms in discretely normed fields*, Dokl. Akad. Nauk SSSR **74** (1950), 889–891. (Russian)
27. V. B. Dem'yanov, *Pairs of quadratic forms over a complete field with discrete norm with a finite field of residue classes*, Izv. Akad. Nauk SSSR Ser. Mat. **20** (1956), 307–324. (Russian)
28. W. Duke, Z. Rudnick and P. Sarnak, *Density of integer points on affine homogeneous varieties*, Duke. Math. J. **71** (1993), 143–179.
29. G. Faltings, *Endlichkeitssätze für abelsche Varietäten über Zahlkörpen*, Invent. Math. **73** (1983), 349–366.
30. D. J. H. Garling, *A course in Galois theory*, Cambridge University Press, Cambridge-New York, 1986.
31. M. J. Greenberg, *Lectures on forms in many variables*, W. A. Benjamin and Co., New York, 1969.
32. H. Hasse, *Über die Darstellbarkeit von Zahlen durch quadratische Formen in Körper der rationalen Zahlen*, J. Reine Angew. Math. **152** (1923), 129–148.
33. D. R. Heath-Brown, *Cubic forms in ten variables*, Proc. London Math. Soc. (3) **47** (1983), 225–257.
34. C. Hooley, *On nonary cubic forms*, J. Reine Angew. Math. **386** (1988), 32–98.
35. C. Hooley, *On nonary cubic forms. II*, J. Reine Angew. Math. **415** (1991), 95–165.
36. C. Hooley, *On nonary cubic forms. III*, J. Reine Angew. Math. **456** (1994), 53–63.
37. S. Lang, *On quasi algebraic closure*, Ann. of Math. **55** (1952), 373–390.
38. S. Lang, ed., *Number Theory III: Diophantine Geometry, Encyclopaedia Math. Sci.*, vol. 60, Springer-Verlag, Berlin, 1991.
39. S. Lang and A. Weil, *Number of points of varieties in finite fields*, Amer. J. Math. **76** (1954), 819–827.
40. R. R. Laxton and D. J. Lewis, *Forms of degrees* 7 *and* 11 *over* $\mathfrak{p}$*-adic fields*, Proceedings of Symposia in Pure Mathematics, vol. 8, American Mathematical Society, 1965, pp. 16–21.
41. D. B. Leep, *Systems of quadratic forms*, J. Reine Angew. Math. **350** (1984), 109–116.
42. D. B. Leep and W. M. Schmidt, *Systems of homogeneous equations*, Invent. Math. **71** (1983), 539–549.
43. D. B. Leep and C. C. Yeomans, *Quintic forms over p-adic fields*, J. Number Theory **57** (1996), 231–241.
44. D. J. Lewis, *Cubic homogeneous polynomials over p-adic number fields*, Ann. of Math. (2) **56** (1952), 473–478.
45. D. J. Lewis, *Cubic congruences*, Michigan Math. J. **4** (1957), 85–95.
46. D. J. Lewis, *Cubic forms over algebraic number fields*, Mathematika **4** (1957), 97–101.
47. D. J. Lewis, *Diophantine problems: solved and unsolved*, Number Theory and Applications (Banff, AB, 1988), NATO Adv. Sci. Inst. Ser. C Math. Phys. Sci. (R. A. Mollin, ed.), vol. 265, Kluwer Academic Publishers, Dordrecht, 1989, pp. 103–121.
48. D. J. Lewis and H. L. Montgomery, *On zeros of p-adic forms*, Michigan Math. J. **30** (1983), 83–87.
49. D. J. Lewis and R. Schulze-Pillot, *Linear spaces on the intersection of cubic hypersurfaces*, Monatsh. Math. **97** (1984), 277–285.

50. G. Martin, *Solubility of systems of quadratic forms*, Bull. London Math. Soc. **29** (1997), 385–388.
51. M. Nagata, *Note on a paper of Lang concerning quasi algebraic closure*, Mem. Coll. Sci. Univ. Kyoto Ser. A. Math. **30** (1957), 237–241.
52. L. G. Peck, *Diophantine equations in algebraic number fields*, Amer. J. Math. **71** (1949), 387–402.
53. J. Pitman, *Cubic inequalities*, J. London Math. Soc. **43** (1968), 119–126.
54. P. A. B. Pleasants, *Cubic polynomials over algebraic number fields*, J. Number Theory **7** (1975), 310–344.
55. C. P. Ramanujam, *Cubic forms over algebraic number fields*, Proc. Cambridge Philos. Soc. **59** (1963), 683–705.
56. R. S. Rumely, *Arithmetic over the ring of all algebraic integers*, J. Reine Angew. Math. **368** (1986), 127–133.
57. C. Ryavec, *Cubic forms over algebraic number fields*, Proc. Cambridge Philos. Soc. **66** (1969), 323–333.
58. W. M. Schmidt, *Diophantine inequalities for forms of odd degree*, Adv. in Math. **38** (1980), 128–151.
59. W. M. Schmidt, *On cubic polynomials I. Hua's estimate of exponential sums*, Monatsh. Math. **93** (1982), 63–74.
60. W. M. Schmidt, *On cubic polynomials II. Multiple exponential sums*, Monatsh. Math. **93** (1982), 141–168.
61. W. M. Schmidt, *On cubic polynomials III. Systems of p-adic equations*, Monatsh. Math. **93** (1982), 211–223.
62. W. M. Schmidt, *On cubic polynomials IV. Systems of rational equations*, Monatsh. Math. **93** (1982), 329–348.
63. W. M. Schmidt, *Analytic methods for congruences, Diophantine equations and approximations*, Proceedings of the International Congress of Mathematicians, Vol. 1 (Warszaw, 1983), PWN, Warszaw, 1984, pp. 515–524.
64. W. M. Schmidt, *The solubility of certain p-adic equations*, J. Number Theory **19** (1984), 63–80.
65. W. M. Schmidt, *The density of integer points on homogeneous varieties*, Acta Math. **154** (1985), 243–296.
66. W. M. Schmidt, *Diophantine approximations and Diophantine equations, Lecture Notes in Mathematics*, vol. 1467, Springer-Verlag, Berlin, 1991.
67. S. E. Schuur, *On systems of three quadratic forms*, Acta Arith. **36** (1980), 315–322.
68. B. Segre, *The algebraic equations of degrees* $5, 9, 157, \ldots$, *and the arithmetic upon an algebraic variety*, Ann. of Math. (2) **46** (1945), 287–301.
69. E. S. Selmer, *On the irreducibility of certain trinomials*, Math. Scand. **4** (1956), 287–302.
70. J.-P. Serre, *Topics in Galois theory. Lecture notes prepared by Henri Darmon. With a foreword by Darmon and the author. Research Notes in Mathematics, 1*, Jones and Bartlett Publishers, Boston, MA, 1992.
71. C. L. Siegel, *Generalization of Waring's problem to algebraic number fields*, Amer. J. Math. **66** (1944), 122–136.
72. C. L. Siegel, *Sums of m-th powers of algebraic integers*, Ann. of Math. **46** (1945), 313–339.
73. C. M. Skinner, *Rational points on non-singular cubic hypersurfaces*, Duke Math. J. **75** (1994), 409–466.
74. C. M. Skinner, *Solvability of* $\mathfrak{p}$*-adic diagonal equations*, Acta Arith. **75** (1996), 251–258.
75. C. M. Skinner, *Forms over number fields and weak approximation*, Compositio Math. **106** (1997), 11–29.
76. W. Tartakovsky, *Über asymptotische Gesetze der allgemeinen Diophantischen Analyse mit vielen Unbekannten*, Bull. Acad. Sci. USSR (1935), 483–524.
77. G. Terjanian, *Un contre-example à une conjecture d'Artin*, C. R. Acad. Sci. Paris Ser. AB **262** (1966), A612.
78. G. Terjanian, *Formes p-adiques anisotropes*, J. Reine Angew. Math. **313** (1980), 217–220.

79. R. C. Vaughan, *The Hardy–Littlewood Method, second edition*, Cambridge University Press, 1997.
80. R. C. Vaughan and T. D. Wooley, *On a certain nonary cubic form and related equations*, Duke Math. J. **80** (1995), 669–735.
81. A. Wiles, *Modular elliptic curves and Fermat's last theorem*, Ann. of Math. (2) **141** (1995), 443–551.
82. T. D. Wooley, *Linear spaces on cubic hypersurfaces, and pairs of homogeneous cubic equations*, Bull. London Math. Soc. **29** (1997), 556–562.
83. T. D. Wooley, *Forms in many variables*, Analytic Number Theory: Proceedings of the 39th Taniguchi International Symposium, Kyoto, May 1996 (Y. Motohashi, ed.), London Mathematical Society Lecture Notes 247, Cambridge University Press, Cambridge, 1997, pp. 361–376.
84. T. D. Wooley, *On the local solubility of diophantine systems*, Compositio Math. **111** (1998), 149–165.
85. T. D. Wooley, *An explicit version of Birch's Theorem*, Acta Arith. **85** (1998), 79–96.
86. T. D. Wooley, *On the existence of non-singular local solutions to diophantine systems* (in preparation).

Department of Mathematics, University of Michigan, East Hall, 525 East University Avenue, Ann Arbor, MI 48109-1109, U.S.A.

E-mail address: `wooley@math.lsa.umich.edu`

EQUATIONS $F(x, y, \alpha^x) = 0$

SCOTT AHLGREN

ABSTRACT. We consider equations $F(x, y, \alpha^x) = 0$ in integer unknowns x and y, where $F(X, Y, Z) \in \mathbb{C}[X, Y, Z]$ is an irreducible polynomial with degree 2 in Z and $\alpha \in \mathbb{C}^\times$ is not a root of unity. We prove (under an additional hypothesis) that if such an equation has solutions with arbitrarily large values of $|x|$, then F and α must be of a particularly simple form.

1. INTRODUCTION AND STATEMENT OF RESULTS

In [5], Schmidt studies equations

$$\alpha^x = R(x, y) \tag{1.1}$$

in unknowns x, $y \in \mathbb{Z}$, where $\alpha \in \mathbb{C}^\times$ is not a root of unity and $R(X, Y) \in \mathbb{C}(X, Y)$. It is assumed that (as a rational function of Y), R has an expansion at infinity of the type

$$R(X, Y) = r_0 Y^\ell + r_1(X) Y^{\ell-1} + \dots \tag{1.2}$$

where $r_0 \neq 0$ is constant and r_1, $r_2, \dots$ are rational functions of X. The following result is obtained. (Let $\overline{\mathbb{Q}}$ be the field of algebraic numbers.)

Theorem 1.1 ([5, Thm. 1]). *Suppose that* (1.1) *has solutions with arbitrarily large values of* $|x|$. *Then* R *is of the type*

$$R(X, Y) = r_0 (Y - h(X))^\ell$$

with $\ell \neq 0$ *and* $h(X) \in \mathbb{Q}[X]$, *and* $\alpha^u \in \mathbb{Z}$ *for some* $u \in \mathbb{Z} \setminus \{0\}$.

Schmidt [5, Thm. 2] goes on to give a detailed description of the set of solutions.

Equations (1.1) are examples of more general equations

$$F(x, y, \alpha^x) = 0 \qquad (x, y) \in \mathbb{Z}^2, \tag{1.3}$$

where $F(X, Y, Z) \in \mathbb{C}[X, Y, Z]$ is an irreducible polynomial with positive degree in Z, and $\alpha \in \mathbb{C}^\times$ is not a root of unity. A general theorem of Laurent [3, Théorème 1] shows that many polynomial-exponential equations have only finitely many solutions; it is of interest to study (1.3) as it is one of the simplest equations which can indeed possess infinitely many solutions. In this case we can still hope that (1.3) has solutions with arbitrarily large $|x|$ only when F and α are of a particularly simple form; this idea is substantiated by the first theorem.

A key step towards Theorem 1 is to show that if (1.1) has solutions with infinitely many values of x, then α is algebraic. In [2] the present author extended this result on the algebraicity of α to a wider range of polynomial–exponential equations. In particular we know the following

S.D. Ahlgren et al. (eds.), Topics in Number Theory, 85–100.

Theorem 1.2 ([2, Thm. 1]). *If* (1.3) *has solutions with infinitely many x then α is algebraic.*

In (1.3) there may be a fixed value of x such that $F(x, Y, \alpha^x)$ vanishes identically as a polynomial in Y. For this reason we speak of "solutions with infinitely many values of x" rather than "infinitely many solutions".

A chapter of the author's Ph. D. thesis [1] is devoted to the study of the equation (1.3). Since this undertaking is rather involved (it requires, among other things, the development of a Newton Polygon attached to such an equation), we shall be content here to consider the case when F has degree 2 in Z.

In the present situation we may write $F(X,Y,Z) = f_0(X,Y) + f_1(X,Y)Z + f_2(X,Y)Z^2$ with polynomials f_0, f_1, and f_2 (where $f_2 \neq 0$); then the equation (1.3) becomes

$$f_0(x,y) + f_1(x,y)\alpha^x + f_2(x,y)\alpha^{2x} = 0. \tag{1.4}$$

For $i = 0, 1, 2$, let d_i be the degree of f_i in Y. We will assume that

$$|d_0 + d_2 - 2d_1| \geq 3. \tag{a}$$

(This condition may, with some loss of simplicity, be slightly weakened.) We also suppose that if $i, j \in \{0,1,2\}$ and $f_j \neq 0$, then

$$\frac{f_i(X,Y)}{f_j(X,Y)} \text{ has an expansion at infinity of the form (1.2).} \tag{b}$$

Our aim in this paper will be to prove the following result.

Theorem 1.3. *Suppose that F is irreducible, and operate under assumptions* (a) *and* (b). *Suppose that* (1.4) *has solutions $(x,y) \in \mathbb{Z}^2$ with infinitely many values of x. Then one of the following is true:*

(1) *There exist $k \in \mathbb{Z} \setminus \{0\}$, $r \in \overline{\mathbb{Q}}$, and $h(X) \in \mathbb{Q}[X]$ such that*

$$\frac{F(X,Y,Z)}{f_2(X,Y)} = Z^2 - r(Y - h(X))^k,$$

and there exists $p \in \mathbb{Z} \setminus \{0\}$ such that $\alpha^p \in \mathbb{Z}$, or

(2) *There exist $P(X) \in \overline{\mathbb{Q}}[X]$ and $\gamma \in \overline{\mathbb{Q}}$ such that, up to a constant multiple,*

$$F(X,Y,Z) = Z^2 + P(Y - h(X))Z + \gamma,$$

and there exists a positive integer r such that α^r is a real quadratic unit.

Example. The theorem shows that for any α, the equation $\alpha^{2x} + y^2\alpha^x + y = 0$ has solutions for only finitely many x. On the other hand, there are equations of the form described in each case of the theorem which do have solutions for arbitrarily large $|x|$. For example, let β be a real quadratic unit. Then $\beta^x + \beta^{-x}$ is the trace of β^x, and so is an integer for any integer value of x. Therefore the equation

$y^2 = (\beta^x + \beta^{-x})^2 = \beta^{2x} + \beta^{-2x} + 2$ has infinitely many solutions. Setting $\alpha = \beta^2$, and multiplying through by β^{2x}, we see that the equation $\alpha^{2x} + (2-y^2)\alpha^x + 1 = 0$ has solutions with infinitely many values of x.

Before proceeding with the proof, we make the following remarks:

(1) In case (1) we see that, up to a constant multiple, either $F = Z^2 - r(Y - h(X))^k$ or $F = r(Y-h(X))^k Z^2 - 1$ with $k > 0$.
(2) Theorem 2 of [5] characterizes the set of solutions in case (1).
(3) The polynomial P appearing in case (2) is described completely in Section 6 below.

Our method of proof follows [5]; new difficulties arise from the fact that our equation includes more than one power of α^x. This forces the consideration of certain partitions of the set of archimedian places of a number field; this is undertaken in Section 4. Our main tool will be Schlickewei's $\mathfrak{p}$-adic generalization of Schmidt's Subspace Theorem.

2. Reduction to the Algebraic Case

Set $\log^+ z = \max(\log z, 1)$. A general theorem of Laurent [3, Théorème 1] shows that if x, $y \in \mathbb{Z}$ satisfy (1.4) then

$$|x| \ll \log^+ |y|, \tag{2.1}$$

where the implied constant depends only on the equation. When $a(X)$ is a non-zero rational function and $\epsilon > 0$, pairs (x,y) with (2.1) have $|y|^{-\epsilon} \ll_\epsilon |a(x)| \ll_\epsilon |y|^\epsilon$ for large $|x|$; this fact will be used throughout. Suppose that $f(X,Y) = a_0(X)Y^d + a_1(X)Y^{d-1} + \dots$ is a non-zero polynomial with degree d in Y. If (x,y) has (2.1), then for large $|x|$ with $a_0(x) \neq 0$, we have

$$|y|^d \ll |f(x,y)| \ll |y|^{d+1/8}. \tag{2.2}$$

(Note that 1/8 could be replaced by any positive constant.) In particular, by (2.2) we see that f has only finitely many zeros (x,y) satisfying (2.1).

Lemma 2.1. *Suppose that P, $Q \in \mathbb{C}(X,Y)[Z]$, and that for infinitely many $x \in \mathbb{Z}$ there exist $y \in \mathbb{Z}$ and $z \in \mathbb{C}$ with (2.1) and*

$$P(x,y,z) = Q(x,y,z) = 0. \tag{2.3}$$

Then P and Q have a common factor of positive degree in Z.

Proof. Let $R(X,Y) \in \mathbb{C}(X,Y)$ be the resultant of P and Q. Whenever there is a triple (x,y,z) with (2.3), we have $R(x,y) = 0$. Therefore $R(x,y) = 0$ for infinitely many values of x with (2.1), whence $R(X,Y) = 0$. We conclude that P and Q have a factor of positive degree in Z. □

Let V be the vector space generated over $\mathbb{Q}(\alpha)$ by the coefficients of F, and let $\beta_1, \dots, \beta_t$ be a basis for V over $\mathbb{Q}(\alpha)$. For $i = 0, 1, 2$ write $f_i = \sum_{j=1}^t \beta_j f_i^{(j)}$, where the $f_i^{(j)}$ are uniquely determined polynomials with coefficients in $\mathbb{Q}(\alpha)$. The equation (1.4) with $x, y \in \mathbf{Z}$ is equivalent to the system of equations

$$f_0^{(j)}(x,y) + f_1^{(j)}(x,y)\alpha^x + f_2^{(j)}(x,y)\alpha^{2x} = 0 \qquad (1 \le j \le t).$$

For each j, define $F^{(j)}(X,Y,Z) = \sum_{i=0}^{2} f_i^{(j)}(X,Y)Z^i$. Then $F^{(j)}(x,y,\alpha^x) = F(x,y,\alpha^x) = 0$ for infinitely many values of x. By Lemma 2.1, F and $F^{(j)}$ have a factor of positive degree in Z. Since F is assumed to be irreducible, this factor must have degree 2; therefore there exists $U^{(j)} \in \mathbb{C}[X,Y]$ (possibly zero), with $F^{(j)} = U^{(j)}F$. As the total degree of $F^{(j)}$ does not exceed that of F, each $U^{(j)}$ is a constant. Some $F^{(j)} \neq 0$, whence $U^{(j)} \neq 0$, and $F = \frac{1}{U^{(j)}}F^{(j)}$ where $F^{(j)}$ has coefficients in $\mathbb{Q}(\alpha)$. We may therefore assume that the coefficients of the polynomials f_i lie in $\mathbb{Q}(\alpha)$; by Theorem 1.2 we know that $\mathbb{Q}(\alpha)$ is an algebraic number field.

3. The first estimates

In the next lemma we obtain some estimates which are essential for our application of the Subspace Theorem. (The implied constants depend on the equation (1.4), and to ease notation we will sometimes write f_i to mean $f_i(x,y)$.)

Lemma 3.1. *Let (x,y) be a solution of (1.4) with $|x|$ sufficiently large, and recall that d_i denotes the degree of f_i in Y.*

(a) *If $d_0 + d_2 - 2d_1 \ge 3$ then*

$$(3.1) \qquad |\alpha^{2x} + \frac{f_0}{f_2}| \ll |y|^{d_0-d_2-\frac{5}{4}} \quad \text{and} \quad |\alpha^{-2x} + \frac{f_2}{f_0}| \ll |y|^{d_2-d_0-\frac{5}{4}}.$$

(b) *If $d_0 + d_2 - 2d_1 \le -3$ then either*

$$(3.2) \qquad |\alpha^{x} + \frac{f_1}{f_2}| \ll |y|^{d_1-d_2-\frac{5}{4}} \quad \text{and} \quad |\alpha^{-x} + \frac{f_2}{f_1}| \ll |y|^{d_2-d_1-\frac{5}{4}},$$

or

$$(3.3) \qquad |\alpha^{-x} + \frac{f_1}{f_0}| \ll |y|^{d_1-d_0-\frac{5}{4}} \quad \text{and} \quad |\alpha^{x} + \frac{f_0}{f_1}| \ll |y|^{d_0-d_1-\frac{5}{4}}.$$

Proof. To expose some symmetry, write (1.4) as

$$(3.4) \qquad f_0(x,y)\alpha^{-x} + f_1(x,y) + f_2(x,y)\alpha^x = 0.$$

Suppose first that

$$(3.5) \qquad d_0 + d_2 - 2d_1 \ge 3.$$

If

$$(3.6) \qquad |f_1| \ge \frac{1}{2}|f_2\alpha^x|,$$

then (2.2) shows that $|\alpha^{x}| \ll |y|^{d_1-d_2+\frac{1}{8}}$, so that, using (3.5) and (2.2),

$$|f_0\alpha^{-x}| \gg |y|^{d_0+d_2-d_1-\frac{1}{8}} \geq |y|^{d_1+2} \gg |f_1||y|.$$

With (3.6) this shows that the first summand in (3.4) has larger order of magnitude than the others, which is impossible.

Therefore (3.6) fails to hold, and it must be the case that $|f_0\alpha^{-x}| \geq \frac{1}{2}|f_2\alpha^{x}|$. By (2.2), we see that $|\alpha^{x}| \ll |y|^{\frac{1}{2}(d_0-d_2+\frac{1}{8})}$. Then (2.2) and (3.5) give

$$\begin{aligned}|\alpha^{2x} + \frac{f_0}{f_2}| = |\frac{f_1}{f_2}\alpha^{x}| &\ll |y|^{d_1-d_2+\frac{1}{8}+\frac{1}{2}(d_0-d_2+\frac{1}{8})} \\ &= |y|^{(d_0-d_2)+(d_1-\frac{1}{2}d_0-\frac{1}{2}d_2)+\frac{3}{16}} \ll |y|^{d_0-d_2-\frac{3}{2}+\frac{3}{16}},\end{aligned}$$

which yields the first assertion in (3.1). The second holds by symmetry.

Suppose, then, that

$$d_0 + d_2 - 2d_1 \leq -3. \tag{3.7}$$

By (3.4) one of $|f_2\alpha^{x}| \geq \frac{1}{2}|f_1|$ or $|f_0\alpha^{-x}| \geq \frac{1}{2}|f_1|$ must hold. If $|f_2\alpha^{x}| \geq \frac{1}{2}|f_1|$ then (2.2) shows that $|\alpha^{x}| \gg |\frac{f_1}{f_2}| \gg |y|^{d_1-d_2-\frac{1}{8}}$. Therefore, using (3.7),

$$\begin{aligned}|\alpha^{x} + \frac{f_1}{f_2}| = |\frac{f_0}{f_2}\alpha^{-x}| &\ll |y|^{(d_0-d_2+\frac{1}{8})+(d_2-d_1+\frac{1}{8})} \\ &\ll |y|^{d_0-d_1+\frac{1}{4}} \leq |y|^{d_1-d_2-\frac{11}{4}},\end{aligned}$$

which gives the first estimate in (3.2). For the second, notice that

$$\begin{aligned}|\alpha^{-x} + \frac{f_2}{f_1}| = |\frac{f_0}{f_1}\alpha^{-2x}| &\ll |y|^{d_0-d_1+\frac{1}{8}+2(d_2-d_1+\frac{1}{8})} \\ &\ll |y|^{(d_2-d_1)+(d_0+d_2-2d_1)+\frac{3}{8}} \ll |y|^{d_2-d_1-\frac{21}{8}}.\end{aligned}$$

If $|f_0\alpha^{-x}| \geq \frac{1}{2}|f_1|$, then, appealing to symmetry, we conclude that (3.3) must be true. This establishes the lemma. □

Let m be a non-zero integer, and K an algebraic number field. Suppose that f, $g \in K[X,Y]$ are polynomials such that $\frac{f}{g}$ has an expansion of the type (1.2). Set $k = \deg_Y f - \deg_Y g$, and suppose that $k \geq 0$. In view of Lemma 3.1, we have cause to consider estimates of the form

$$|\alpha^{mx} + \frac{f}{g}| \ll |y|^{k-\frac{5}{4}}. \tag{3.8}$$

We introduce the notation $\boldsymbol{x} = (x, y)$.

Lemma 3.2. *Let K be embedded in $\mathbb{C}$. Define*

$$\ell = \begin{cases} m & \text{if } k = 0 \\ m/k, & \text{if } k > 0. \end{cases}$$

There exists a function $\mathfrak{y}(X,Y)$ where $\mathfrak{y}(X,Y) = 1$ or $\mathfrak{y}(X,Y) = Y - h(X)$ with $h(X) \in K(X)$, such that the following holds: if x satisfies (2.1) and (3.8) then there exists $\rho(x) \in \overline{\mathbb{Q}}^{\times}$ such that

$$|\alpha^{\ell x} - \rho(x)\mathfrak{y}(x)| \ll |y|^{-\frac{1}{4}}. \tag{3.9}$$

Further, the values $\rho(x)$ all lie in a finite subset of $\overline{\mathbb{Q}}$.

Proof. By (1.2) and (2.1) we may write

$$\frac{f(x,y)}{g(x,y)} = -r_0 y^k - r_1(x) y^{k-1} + O(|y|^{k-3/2}), \tag{3.10}$$

where $r_0 \in K^{\times}$ and $r_1(X) \in K(X)$. If $k = 0$ then (3.8) and (3.10) yield $|\alpha^{mx} - r_0| \ll |y|^{-\frac{1}{4}}$, which has the form (3.9), with $\ell = m$, $\mathfrak{y}(X,Y) = 1$, and $\rho(x) = r_0$.

If $k > 0$ then $-r_0 y^k - r_1(x) y^{k-1} = -r_0(y - h(x))^k + O(|y|^{k-3/2})$, where $h(x) = -r_1(x)/r_0 k$. This, together with (3.8) and (3.10), gives

$$|\alpha^{mx} - r_0(y - h(x))^k| \ll |y|^{k-\frac{5}{4}}.$$

Let $\alpha^{1/k}$ be a kth root of α, and let $\rho_1, \ldots, \rho_k$ be the kth roots of r_0. The last inequality becomes

$$\prod_{i=1}^{k} |\alpha^{(m/k)x} - \rho_i(y - h(x))| \ll |y|^{k-\frac{5}{4}}.$$

All but at most one factor on the left is $\gg |y|$ (to see this consider the difference of any two factors). Therefore there must be one small factor; in fact there is an index $i = i(x)$ for which $|\alpha^{(m/k)x} - \rho_{i(x)}(y - h(x))| \ll |y|^{-\frac{1}{4}}$. This again has the form (3.9), with $\ell = m/k$, $\mathfrak{y}(X,Y) = Y - h(X)$, and $\rho(x) = \rho_{i(x)}$. The lemma follows. □

4. The heart of the proof

When K is a number field, let $V = V(K)$ denote the set of places of K. Write $V = V_\infty \cup V_0$ where V_∞ consists of archimedean, and V_0 non-archimedean places. For $v \in V$, let $|\cdot|_v$ be the associated absolute value, normalized to extend the usual, or a p-adic absolute value of $\mathbb{Q}$, and set $\|\cdot\|_v = |\cdot|_v^{d_v/d}$, where d is the degree of K and d_v is the local degree. The product formula asserts that $\prod_{v \in V} \|\xi\|_v = 1$ for any $\xi \in K^{\times}$. Finally, when $n \geq 2$ and $\alpha = (\alpha_1, \ldots, \alpha_n) \in K^n$ we define the absolute multiplicative height $H(\alpha) = \prod_{v \in V} \|\alpha\|_v$, where $\|\alpha\|_v = \max(\|\alpha_1\|_v, \ldots, \|\alpha_n\|_v)$.

We will use Lemmas 3.1 and 3.2 in order to prove the following.

Lemma 4.1. *Operate under assumptions (a) and (b), and suppose that* (1.4) *has solutions with infinitely many values of* x. *Then there exists a number field* K *containing* α *such that one of the following options holds:*

(1) *There exist* $\ell = m/k \in \mathbb{Q}^\times$ *with* $|m| \le 2$, *a function* $\mathfrak{y}(X,Y)$ *where* $\mathfrak{y}(X,Y) = 1$ *or* $\mathfrak{y}(X,Y) = Y - h(X)$ *with* $h(X) \in K(X)$, *and a collection* $\{\rho(v)\}_{v \in V_\infty}$ *of non-zero elements of* K *such that there exist solutions with arbitrarily large* $|x|$ *which satisfy*

$$|\alpha^{\ell x} - \rho(v)\mathfrak{y}(x)|_v \ll |y|^{-1/4} \qquad (v \in V_\infty). \tag{4.1}$$

(2) *There is a fixed nontrivial partition* $V_1 \cup V_2$ *of* V_∞ *and there are fixed non-zero rational numbers* $\ell_1 \neq \ell_2$ *with numerators* ± 1, *such that there exist solutions with arbitrarily large* $|x|$ *which satisfy*

$$|\alpha^{\ell_i x} - \rho(v)\mathfrak{y}_i(x)|_v \ll |y|^{-1/4} \qquad (v \in V_i,\ i = 1,2), \tag{4.2}$$

where $\mathfrak{y}_i(X,Y) = 1$ *or* $\mathfrak{y}_i(X,Y) = Y - h_i(X)$ *with* $h_i(X) \in K(X)$, *and* $\rho(v) \in K^\times$ *for each* $v \in V_\infty$.

Proof. Let K be a number field containing $\mathbb{Q}(\alpha)$. Suppose that $d_0 + d_2 - 2d_1 \le -3$ and that x is a solution of (1.4) with $|x|$ large. By Lemma 3.1 we have either (3.2) or (3.3); suppose to begin that (3.2) holds. Choose the inequality in (3.2) which has the form (3.8) with $k \ge 0$ and $m = \pm 1$; then Lemma 3.2 shows that

$$|\alpha^{\ell_1 x} - \rho(x)\mathfrak{y}_1(x)| \ll |y|^{-1/4}, \tag{4.3}$$

where $\ell_1 \in \mathbb{Q}^\times$ has numerator ± 1 and $\mathfrak{y}_1 = 1$ or $\mathfrak{y}_1 = Y - h_1(X)$ with $h_1 \in K(X)$. Similarly, if (3.3) holds we obtain

$$|\alpha^{\ell_2 x} - \rho(x)\mathfrak{y}_2(x)| \ll |y|^{-1/4}. \tag{4.4}$$

Enlarging K if necessary, we may suppose that it contains all of the finitely many values $\rho(x)$ appearing in (4.3), (4.4).

This argument may be applied whenever K is embedded in $\mathbb{C}$. So for every $v \in V_\infty$ we have either (4.3) or (4.4); i.e. we have

$$|\alpha^{\ell_{i(v,x)} x} - \rho(v,x)\mathfrak{y}_{i(v,x)}(x)|_v \ll |y|^{-1/4} \qquad (v \in V_\infty),$$

where $i(v,x) = 1$ or 2, and the quantities $\rho(v,x)$ all lie in a finite subset of $K^\times$. For each $v \in V_\infty$ we may fix quantities $i(v)$, $\rho(v)$ such that infinitely many solutions satisfy

$$|\alpha^{\ell_{i(v)} x} - \rho(v)\mathfrak{y}_{i(v)}(x)|_v \ll |y|^{-1/4}.$$

For $i = 1,2$ let V_i consist of $v \in V_\infty$ such that $i(v) = i$. If one of V_1 or V_2 is empty the first assertion of our lemma holds, while if neither is empty the second is true.

We are left to consider the situation when $d_0 + d_2 - 2d_1 \ge 3$, in which case (3.1) holds. Since one of the estimates therein has the form (3.8) with $k \ge 0$, a slight modification of the argument above yields the first assertion of our lemma. □

The Subspace Theorem will show that the inequalities of the last lemma can only be the result of a linear relation between the quantities $\alpha^{\ell_i x}$ and $\mathfrak{y}_i(x)$; this is the content of the next lemma.

Lemma 4.2. *Operate under assumptions (a) and (b), and suppose that* (1.4) *has solutions with infinitely many values of x. Then there exists a number field K containing α such that one of the following options holds:*

(1) *There exist $\ell = m/k \in \mathbb{Q}^\times$ with $|m| \le 2$, $\rho \in K^\times$, and $h(X) \in K(X)$ such that there exist solutions with arbitrarily large $|x|$ which satisfy*

$$\alpha^{\ell x} = \rho(y - h(x)).$$

(2) *There exist fixed non-zero rational numbers $\ell_1 \ne \ell_2$ with numerators ± 1, ρ_1, $\rho_2 \in K^\times$, and $h(X) \in K(X)$ such that there exist solutions with arbitrarily large $|x|$ which satisfy*

$$\rho_1^{-1}\alpha^{\ell_1 x} + \rho_2^{-1}\alpha^{\ell_2 x} = y - h(x).$$

Proof. We show that these follow from the corresponding assertions of Lemma 4.1; we begin with the most difficult case. Suppose that (2) of Lemma 4.1 holds; we may then assume with no loss of generality that $\ell_1 > \ell_2$. With $\mathfrak{x}_i = \alpha^{\ell_i x}$ $(i = 1, 2)$, (4.2) becomes

$$|\mathfrak{x}_i - \rho(v)\mathfrak{y}_i|_v \ll |y|^{-1/4} \qquad (v \in V_i,\ i = 1, 2). \tag{4.5}$$

We first assume that $\mathfrak{y}_i(X, Y) \ne 1$ $(i = 1, 2)$. By (2.1) we see that

$$|y| \ll |\mathfrak{y}_i|_v \ll |y| \qquad (v \in V_\infty,\ i = 1, 2). \tag{4.6}$$

Therefore, (4.5) gives

$$|y| \ll |\mathfrak{x}_i|_v \ll |y| \qquad (v \in V_i,\ i = 1, 2). \tag{4.7}$$

Since $\mathfrak{x}_i^{\ell_j} = \mathfrak{x}_j^{\ell_i}$ $(1 \le i, j \le 2)$, we have

$$|y|^{\ell_j/\ell_i} \ll |\mathfrak{x}_j|_v \ll |y|^{\ell_j/\ell_i} \qquad (v \in V_i,\ \ 1 \le i, j \le 2). \tag{4.8}$$

We now make the further assumption that $\mathfrak{y}_1 \ne \mathfrak{y}_2$ (i.e., $h_1(X) \ne h_2(X)$). Set $\mathfrak{x} = (\mathfrak{x}_1, \mathfrak{x}_2, \mathfrak{y}_1, \mathfrak{y}_2)$, let $S_0 \subseteq V_0$ consist of $v \in V_0$ for which $|\alpha|_v \ne 1$, and set $S = S_0 \cup V_\infty$. Define linear forms for $v \in S$ in the variables $\mathfrak{x}$ as follows:

$$L_i^v(\mathfrak{x}) = \begin{cases} \mathfrak{x}_i - \rho(v)\mathfrak{y}_i, & \text{if } v \in V_i \\ \mathfrak{x}_i, & \text{if } v \in S \setminus V_i \end{cases} \qquad (i = 1, 2),$$
$$L_3^v(\mathfrak{x}) = \mathfrak{y}_1,$$
$$L_4^v(\mathfrak{x}) = \mathfrak{y}_1 - \mathfrak{y}_2.$$

Then L_1^v, L_2^v, L_3^v, L_4^v are linearly independent forms for any $v \in S$. We must make some estimates in order to apply the Subspace Theorem.

Define $\delta_i = \sum_{v\in V_i} \frac{d_v}{d}$; then $\delta_1 + \delta_2 = \sum_{v\in V_\infty} \frac{d_v}{d} = 1$. For $i = 1, 2$ we have, by (4.5),

$$\prod_{v\in V_i} ||L_i(\mathfrak{x})||_v = \prod_{v\in V_i} ||\mathfrak{x}_i - \rho(v)\mathfrak{y}_i||_v \ll |y|^{-\frac{1}{4}\delta_i}. \tag{4.9}$$

Using (4.7), the product formula, and the fact that α is an S-unit, we obtain, for $i = 1, 2$,

$$\begin{aligned} \prod_{v\in S\setminus V_i} ||L_i(\mathfrak{x})||_v &= \prod_{v\in S_0} ||\mathfrak{x}_i||_v \prod_{v\in V_\infty\setminus V_i} ||\mathfrak{x}_i||_v = \prod_{v\in V_0} ||\mathfrak{x}_i||_v \prod_{v\in V_\infty\setminus V_i} ||\mathfrak{x}_i||_v \\ &= \prod_{v\in V_\infty} ||\mathfrak{x}_i||_v^{-1} \prod_{v\in V_\infty\setminus V_i} ||\mathfrak{x}_i||_v = \prod_{v\in V_i} ||\mathfrak{x}_i||_v^{-1} \ll |y|^{-\delta_i}. \end{aligned} \tag{4.10}$$

Combining (4.9) and (4.10) gives

$$\prod_{v\in S} ||L_i^v(x)||_v \ll |y|^{-\frac{5}{4}\delta_i} \qquad (i = 1, 2). \tag{4.11}$$

Write $h_1(x) = s_1(x)/t_1(x)$ where s_1, t_1 are polynomials with algebraic integer coefficients. Then $y - h_1(x) = \dfrac{yt_1(x) - s_1(x)}{t_1(x)}$, so that for large $|x|$ with (2.1), the product formula gives

$$\begin{aligned} \prod_{v\in S_0} ||L_3^v(\mathfrak{x})||_v &= \prod_{v\in S_0} ||y - h_1(x)||_v \le \prod_{v\in S_0} ||t_1(x)||_v^{-1} \\ &\le \prod_{v\in V_0} ||t_1(x)||_v^{-1} = \prod_{v\in V_\infty} ||t_1(x)||_v \ll |y|^{\frac{1}{16}}. \end{aligned} \tag{4.12}$$

By (4.6) we have $\prod_{v\in V_\infty} ||L_3^v(\mathfrak{x})||_v = \prod_{v\in V_\infty} ||\mathfrak{y}_1||_v \ll |y|$. Combining this with (4.12), we obtain $\prod_{v\in S} ||L_3^v(\mathfrak{x})||_v \ll |y|^{1+\frac{1}{16}}$, which, together with (4.11) and the fact that $\delta_1 + \delta_2 = 1$, shows that

$$\prod_{v\in S} \prod_{i=1}^{3} ||L_i^v(\mathfrak{x})||_v \ll |y|^{-3/16}. \tag{4.13}$$

Recall that $L_4^v(\mathfrak{x}) = h_1(x) - h_2(x)$. Using (2.1) and arguing as in (4.12) gives $\prod_{v\in S} ||L_4(\mathfrak{x})||_v \ll |y|^{\frac{1}{16}}$, and we conclude that

$$\prod_{v\in S} \prod_{i=1}^{4} ||L_i^v(\mathfrak{x})||_v \ll |y|^{-1/8}. \tag{4.14}$$

It is now necessary to consider the height of the vector $\mathfrak{x}$. Since α is an S-unit,

$$\begin{aligned} \prod_{v\notin S} ||\mathfrak{x}||_v &= \prod_{v\notin S} \max(||\mathfrak{x}_1||_v, ||\mathfrak{x}_2||_v, ||\mathfrak{y}_1||_v, ||\mathfrak{y}_2||_v) \\ &= \prod_{v\notin S} \max(1, ||\mathfrak{y}_1||_v, ||\mathfrak{y}_2||_v) \le \prod_{v\notin S} \max(1, ||\mathfrak{y}_1||_v) \max(1, ||\mathfrak{y}_2||_v). \end{aligned}$$

Modifying the argument in (4.12) shows that for any $\epsilon > 0$,

$$\prod_{v \notin S} \max(1, \|\mathfrak{y}_1\|_v) \le \prod_{v \notin S} \max(1, \|t_1(x)\|_v^{-1})$$
$$= \prod_{v \notin S} \|t_1(x)\|_v^{-1} \le \prod_{v \in V_0} \|t_1(x)\|_v^{-1} \ll |y|^{\epsilon}.$$

Since this holds also with $\mathfrak{y}_1$ replaced by $\mathfrak{y}_2$, we may conclude that for any $\epsilon > 0$,

$$\prod_{v \notin S} \|\mathfrak{x}\|_v \ll |y|^{\epsilon}. \tag{4.15}$$

Since $|x| \ll \log^+ |y|$, for each $v \in V$ there exists a constant D_v so that

$$\|\mathfrak{x}\|_v = \max(\|\alpha^{\ell_1 x}\|_v, \|\alpha^{\ell_2 x}\|_v, \|y - h_1(x)\|_v, \|y - h_2(x)\|_v) \ll |y|^{D_v}.$$

Therefore (since S is a finite set), $\prod_{v \in S} \|\mathfrak{x}\|_v$ is bounded by some power of $|y|$, which together with (4.15) shows that for some D we have

$$H(\mathfrak{x}) \ll |y|^D. \tag{4.16}$$

Notice that $\prod_{v \in S} \|\mathfrak{x}\|_v^{-1} = H(\mathfrak{x})^{-1} \prod_{v \notin S} \|\mathfrak{x}\|_v$. Choosing $\epsilon = 1/64$ in (4.15), and using (4.14) and (4.16), we conclude that

$$\begin{aligned} \prod_{v \in S} \prod_{i=1}^{4} \|L_i^v(\mathfrak{x})\|_v \|\mathfrak{x}\|_v^{-1} &\ll |y|^{-\frac{1}{8}} H(\mathfrak{x})^{-4} \prod_{v \notin S} \|\mathfrak{x}\|_v^4 \\ &\ll \frac{|y|^{-\frac{1}{16}}}{H(\mathfrak{x})^4} \ll \frac{1}{H(\mathfrak{x})^{4+\frac{1}{16D}}}. \end{aligned} \tag{4.17}$$

We are now in a position to apply Schlickewei's $\mathfrak{p}$-adic Subspace Theorem [4] (in fact, the version best suited for our present use is recorded as Theorem 1D′ of [6, Chapter V]). The Subspace Theorem asserts that the solutions $\mathfrak{x} \in K^4$ of (4.17) lie in finitely many proper subspaces of K^4. Therefore some subspace contains infinitely many solutions; that is, infinitely many solutions satisfy a non-trivial linear relation

$$a_1 \mathfrak{x}_1 + a_2 \mathfrak{x}_2 + b_1 \mathfrak{y}_1 + b_2 \mathfrak{y}_2 = 0. \tag{4.18}$$

If b_1 and b_2 are both zero, then, since the relation is nontrivial, neither a_1 nor a_2 may be zero, so that $|\mathfrak{x}_1|_v \gg\ll |\mathfrak{x}_2|_v$ for all v, against (4.8). Therefore some $b_i \neq 0$, and, since the rational functions $Y - h_1(X)$ and $Y - h_2(X)$ are linearly independent over K (recall that $h_1(X) \neq h_2(X)$), (4.18) has the form

$$a_1 \mathfrak{x}_1 + a_2 \mathfrak{x}_2 + by + j(x) = 0, \tag{4.19}$$

with a non-zero rational function $bY + j(X)$.

Up to now we have assumed that $h_1(X) \neq h_2(X)$. If, however, $h_1 = h_2$, then $\mathfrak{y}_1 = \mathfrak{y}_2$, so that for each $v \in S$ the forms L_1^v, L_2^v, L_3^v defined above are three linearly independent forms in the variables $\mathfrak{x} = (\mathfrak{x}_1, \mathfrak{x}_2, \mathfrak{y}_1)$. The estimates (4.15) and (4.16) continue to hold, and by (4.13), we have, in analogy with (4.17), $\prod_{v \in S} \prod_{i=1}^3 \|L_i^v(\mathfrak{x})\|_v \|\mathfrak{x}\|_v^{-1} \ll \frac{1}{H(\mathfrak{x})^{3+\lambda}}$ for some positive λ. The Subspace Theorem shows that infinitely many solutions satisfy an equation of the form (4.18), in this case with $b_2 = 0$. Then $b_1 \neq 0$ (if not we reach the contradiction $|\mathfrak{x}_1|_v \gg\ll |\mathfrak{x}_2|_v$ as before). So here also we obtain an equation of the form (4.19). In either case, therefore, we may suppose that (4.19) holds for infinitely many solutions.

Suppose that $bY + j(X)$ is a non-zero rational function. If $b \neq 0$ then

$$|y| \ll |by + j(x)|_v \ll |y| \qquad (v \in V_\infty), \tag{4.20}$$

while if $b = 0$ then $j(X)$ is non-zero, so that for any $\epsilon > 0$, (2.1) yields

$$|y|^{-\epsilon} \ll |by + j(x)|_v \ll |y|^\epsilon \qquad (v \in V_\infty). \tag{4.21}$$

Either (4.20) holds for all large $|x|$, or (4.21) holds for all large $|x|$. We may now conclude that in (4.19) neither a_1 nor a_2 is zero: if, say a_1 were zero then (4.8) and (4.19) would give

$$|y|^{\ell_2/\ell_1} \gg\ll |\mathfrak{x}_2|_v \gg\ll |by + j(x)|_v \qquad (v \in V_1), \tag{4.22}$$

which is impossible as it violates each of (4.20), (4.21) (recall that $\ell_1 \neq \ell_2$).

Since $a_1 \neq 0$ we have $\mathfrak{x}_1 = \frac{-1}{a_1}(a_2\mathfrak{x}_2 + by + j(x))$, and (4.5) yields

$$\left| \frac{-1}{a_1}(a_2\mathfrak{x}_2 + by + j(x)) - \rho(v)(y - h_1(x)) \right|_v \ll |y|^{-1/4} \qquad (v \in V_1). \tag{4.23}$$

The quantity on the left has the form $\frac{-a_2}{a_1}\mathfrak{x}_2 + b'y + j'(x)$. Suppose that $b'Y + j'(X)$ is not identically zero. Then for $v \in V_1$, we have either $|b'Y + j'(X)|_v \gg\ll |y|$ or $|y|^{-\epsilon} \ll |b'Y + j'(X)|_v \ll |y|^\epsilon$ for any $\epsilon > 0$. But each of these inequalities violates (4.23), in view of the fact that $|\mathfrak{x}_2|_v \gg\ll |y|^{\ell_2/\ell_1}$.

We may therefore conclude that $b'Y + j'(X) = 0$. In other words, infinitely many solutions satisfy

$$\frac{-1}{a_1}(by + j(x)) = \rho(v)(y - h_1(x)). \tag{4.24}$$

Since $|x| \ll \log^+ |y|$, we conclude that $\rho(v) = -b/a_1$ for all $v \in V_1$. By symmetry, we see that for $i = 1, 2$ and $v \in V_i$ we have $\rho(v) = -b/a_i = \rho_i$, say. Then (4.19) becomes $a_1\mathfrak{x}_1 + a_2\mathfrak{x}_2 = a_1\rho_1 y - j(x)$, and dividing by $a_1\rho_1$ gives assertion (2) of Lemma 4.2, with $h(x) = \frac{j(x)}{a_1\rho_1}$.

We have established assertion (2) in the case when neither $\mathfrak{y}_i(X, Y)$ is identically 1, and the remaining case is more easily treated. Suppose for example that

$\mathfrak{y}_1(X,Y)=1$ (if $\mathfrak{y}_2=1$ the situation is identical). Let $\mathfrak{x}_1=\alpha^{\ell_1 x}$ and $\mathfrak{x}=(\mathfrak{x}_1,\mathfrak{y}_1)$, let S be as above, and define linear forms

$$L_1^v(\mathfrak{x})=\begin{cases}\mathfrak{x}_1-\rho(v)\mathfrak{y}_1, & \text{if } v\in V_1\\ \mathfrak{x}_1, & \text{if } v\in S\setminus V_1,\end{cases}\qquad L_2^v(\mathfrak{x})=\mathfrak{y}_1.$$

In analogy with (4.11) we have $\prod_{v\in S}\|L_1^v(\mathfrak{x})\|_v\ll|y|^{-\frac{4}{3}\delta_1}\le|y|^{-\delta}$ for some $\delta>0$. We still have (4.15) and (4.16), and, since L_2^v is identically 1, we conclude that there is a positive λ such that

$$\prod_{v\in S}\|L_1^v(\mathfrak{x})\|_v\|L_2^v(\mathfrak{x})\|_v\|\mathfrak{x}\|_v^{-2}\ll\frac{1}{H(\mathfrak{x})^{2+\lambda}}.$$

The Subspace Theorem asserts that all solutions lie on finitely many lines $\mathfrak{x}_1=\gamma\mathfrak{y}_1$; that is, $\alpha^{\ell_1 x}=\gamma$. In this case there can be only finitely many solutions, for if any line contains two solutions then α is a root of unity, against the hypothesis.

Suppose finally that (1) of Lemma 4.1 holds (here we find ourselves in a situation similar to [5]). If $\mathfrak{y}(X,Y)=1$ then, as above, we reach a contradiction. If, however, $\mathfrak{y}(X,Y)=Y-h(X)$, then let S be as before and define linear forms in the variables $\mathfrak{x}=(\mathfrak{x},\mathfrak{y})$ in the following way:

$$L_1^v(\mathfrak{x})=\begin{cases}\mathfrak{x}-\rho(v)\mathfrak{y}, & \text{if } v\in V_\infty\\ \mathfrak{x}, & \text{if } v\in S_0,\end{cases}\qquad L_2^v(\mathfrak{x})=\mathfrak{y}.$$

Using the Subspace Theorem, we proceed to the conclusion that the variables $\mathfrak{x}$, $\mathfrak{y}$ lie in finitely many subspaces $\mathfrak{x}=\rho\mathfrak{y}$; i.e. $\alpha^{\ell x}=\rho(y-h(x))$. Since one such subspace must contain infinitely many solutions, the first assertion of Lemma 4.2 follows. □

5. The Equation $\rho_1^{-1}\alpha^{\ell_1 x}+\rho_2^{-1}\alpha^{\ell_2 x}=y-h(x)$

In this section we examine further the consequences of assertion (2) of Lemma 4.1. For the duration, we shall write $a=\rho_1^{-1}$ and $b=\rho_2^{-1}$ for simplicity.

Lemma 5.1. *Suppose that (2) of Lemma 4.1 (whence (2) of Lemma 4.2) holds, and therefore that infinitely many solutions of* (1.4) *satisfy*

$$a\alpha^{\ell_1 x}+b\alpha^{\ell_2 x}=y-h(x). \tag{5.1}$$

Then $\ell_1=-\ell_2$, $h(X)\in\mathbb{Q}[X]$, $ab\in\mathbb{Q}$, and there exists a positive integer r such that $\alpha^{\ell_1 r}$ is a real quadratic unit.

Proof. When τ is an automorphism of $\mathbb{C}$ and $\omega\in\mathbb{C}$, write ω_τ for the image of ω under τ. Apply such a τ to (5.1) and write the resulting equation as

$$a_\tau\alpha_\tau^{\ell_1 x}+b_\tau\alpha_\tau^{\ell_2 x}=y-h_\tau(x). \tag{5.2}$$

Eliminating y from (5.1) and (5.2) gives

$$a\alpha^{\ell_1 x} + b\alpha^{\ell_2 x} - a_\tau \alpha_\tau^{\ell_1 x} - b_\tau \alpha_\tau^{\ell_2 x} + h(x) - h_\tau(x) = 0. \tag{5.3}$$

Writing $h(x) = s(x)/t(x)$ where s, t are polynomials, this becomes

$$\begin{aligned} t(x)t_\tau(x)(a\alpha^{\ell_1 x} + b\alpha^{\ell_2 x} - a_\tau \alpha_\tau^{\ell_1 x} - b_\tau \alpha_\tau^{\ell_2 x}) \\ + s(x)t_\tau(x) - s_\tau(x)t(x) = 0. \end{aligned} \tag{5.4}$$

This is an equation of the form

$$\sum_{i=1}^{m} p_i(x)\alpha_i^x = 0, \tag{5.5}$$

where for each i, p_i is a polynomial with complex coefficients and $\alpha_i \in \mathbb{C}^\times$. As another special consequence of the previously mentioned theorem of Laurent [3, Théorème 1], we know the following:

If (5.5) ***has infinitely many solutions for which no proper subsum vanishes, then there exists a positive integer*** z ***such that*** $\alpha_i^z = \alpha_j^z$ $(1 \le i, j \le m)$.

In (5.4) the exponential functions are $\alpha^{\ell_1 x}$, $\alpha^{\ell_2 x}$, $\alpha_\tau^{\ell_1 x}$, $\alpha_\tau^{\ell_2 x}$, and 1^x. Since α is not a root of unity, we see that the subsum $s(x)t_\tau(x) - s_\tau(x)t(x)$ vanishes for all but finitely many solutions. Therefore $h(x) = h_\tau(x)$ for infinitely many x, and so $h(X) = h_\tau(X)$. Since this is true for any automorphism of $\mathbb{C}$, we must have $h(X) \in \mathbb{Q}(X)$.

Since V_1 is non-empty, (4.7) shows that we may embed K into $\mathbb{C}$ in such a way that

$$|\alpha^{\ell_1 x}| \gg\ll |y|. \tag{5.6}$$

Then, since also V_2 is non-empty, there exists an automorphism τ of $\mathbb{C}$ for which

$$|\alpha_\tau^{\ell_2 x}| \gg\ll |y|. \tag{5.7}$$

For this τ, (5.3) becomes

$$a\alpha^{\ell_1 x} + b\alpha^{\ell_2 x} - a_\tau \alpha_\tau^{\ell_1 x} - b_\tau \alpha_\tau^{\ell_2 x} = 0. \tag{5.8}$$

Since α is not a root of one, no positive integer z satisfies $\alpha^{\ell_1 z} = \alpha^{\ell_2 z}$. If $\alpha^{\ell_1 z} = \alpha_\tau^{\ell_1 z}$, then $|\alpha| = |\alpha_\tau|$, from which the contradiction $\ell_1 = \ell_2$ by (5.6), (5.7). In view of these facts, Laurent's theorem implies that a subsum of (5.8) vanishes; in fact, we see that for all but finitely many solutions we have

$$a\alpha^{\ell_1 x} = b_\tau \alpha_\tau^{\ell_2 x}, \qquad b\alpha^{\ell_2 x} = a_\tau \alpha_\tau^{\ell_1 x}. \tag{5.9}$$

Applying Laurent's theorem to these equations, we see that there exist positive integers z, z' such that

$$\alpha^{\ell_1 z} = \alpha_\tau^{\ell_2 z}, \qquad \alpha^{\ell_2 z'} = \alpha_\tau^{\ell_1 z'},$$

from which $\alpha_1^{\ell_1^2 zz'} = \alpha_\tau^{\ell_1\ell_2 zz'} = \alpha_2^{\ell_2^2 zz'}$. Therefore $\ell_1^2 = \ell_2^2$, so that $\ell_1 = -\ell_2 = \ell$, say (recall that $\ell_1 \neq \ell_2$).

For simplicity, set $\beta = \alpha^\ell$; then (5.9) becomes

$$a\beta^x = b_\tau \beta_\tau^{-x}, \qquad b\beta^{-x} = a_\tau \beta_\tau^x. \tag{5.10}$$

If, on the other hand, σ is an automorphism for which $|\alpha_\sigma^{\ell x}| = |\beta_\sigma^x| \gg\ll |y|$, a similar argument shows that for all but finitely many solutions we have

$$a\beta^x = a_\sigma \beta_\sigma^x, \qquad b\beta^{-x} = b_\sigma \beta_\sigma^{-x}. \tag{5.11}$$

If (5.10) and (5.11) hold for x_1 and x_2, then

$$(\beta\beta_\tau)^{x_1-x_2} = (\beta/\beta_\sigma)^{x_1-x_2} = 1.$$

Let r be the least positive integer such that for all automorphisms σ, τ as above, we have $(\beta\beta_\tau)^r = (\beta/\beta_\sigma)^r = 1$. Since by (4.8), every automorphism of $\mathbb{C}$ is one of these, we conclude that β^r is quadratic, having only itself and β^{-r} as conjugates. Together, (5.10) and (5.11) give $ab = a_\sigma b_\sigma = a_\tau b_\tau$ for all σ, τ, showing that $ab \in \mathbb{Q}$.

To show that β is a unit, write $h = s/t$, and suppose that $|\beta|_v = \omega > 1$ for some $v \in V_0$. If there are solutions with infinitely many positive x, then (5.1) gives

$$\begin{aligned}\omega^x \ll |a\beta^x|_v &= |y - b\beta^{-x} - h(x)|_v \le \max(|y|_v, |b\beta^{-x}|_v, |h(x)|_v)\\ &\ll |t(x)|_v^{-1} \le \prod_{v\in V_0} \|t(x)\|_v^{-d} = \prod_{v\in V_\infty} \|t(x)\|_v^d \ll |x|^C\end{aligned}$$

for some constant C, which is impossible. Therefore $|\beta|_v \le 1$ for all $v \in V_0$. By symmetry, $|\beta^{-1}|_v \le 1$ for all $v \in V_0$, proving that β is a unit. (If there are solutions for infinitely many negative x, a similar argument holds.)

It remains to show that $h(x)$ is a polynomial. Write $h(x) = h_0(x) + c_1/x + c_2/x^2 + \ldots$, where $h_0(X) \in \mathbb{Q}[X]$ and $c_1, c_2, \cdots \in \mathbb{Q}$ (this series converges when $|x|$ is sufficiently large). Let s be a common denominator of a, b, and the coefficients of h_0. By (5.1) we have

$$\begin{aligned}sa\beta^x + sb\beta^{-x} &= sy - sh_0(x) - s(c_1/x + c_2/x^2 + \ldots)\\ &= sy - sh_0(x) - r(x), \quad \text{say.}\end{aligned}$$

We have $sa\beta^x + sb\beta^{-x} - sy + sh_0(x) = -r(x)$, so the left side is rational, and is an algebraic integer since β is a unit. Therefore $r(x) \in \mathbb{Z}$ for infinitely many x. Suppose now that some $c_i \neq 0$, and let i be the smallest index for which this is true. Then $r(x) = \frac{sc_i}{x^i} + O(|x|^{-i-1})$. When $|x|$ is large, we see that $\frac{1}{2}\left|\frac{sc_i}{x^i}\right| < |r(x)| < \frac{3}{2}\left|\frac{sc_i}{x^i}\right|$, from which $0 < |r(x)| < 1$, so that $r(x) \notin \mathbb{Z}$. Therefore each $c_i = 0$, which shows that $h(X) \in \mathbb{Q}[X]$, and establishes Lemma 5.1. □

6. Proof of Theorem 1.3

When (1.4) has infinitely many solutions, one of the two conclusions of Lemma 4.2 must hold. If the first holds, then infinitely many solutions have $\alpha^{\ell x} = \rho(y - h(x))$, where $\ell = m/k$ with $k > 0$ and $|m| \le 2$. Raising both sides to the kth power gives $\alpha^{mx} = r(y - h(x))^k$, with $r = \rho^k$. Set

$$\tilde{F}(X,Y,Z) = \begin{cases} Z^m - r(Y - h(X))^k, & \text{if } m > 0 \\ r(Y - h(X))^k Z^{-m} - 1, & \text{if } m < 0. \end{cases}$$

Then $\tilde{F}(x, y, \alpha^x) = F(x, y, \alpha^x) = 0$ for infinitely many values of x. Since F is irreducible and $|m| \le 2$, Lemma 2.1 shows that, up to a constant multiple, $\tilde{F}(X,Y,Z) = F(X,Y,Z)$. Schmidt's theorem shows that $h(X) \in \mathbb{Q}[X]$ and $\alpha^p \in \mathbb{Z}$ for some $p \in \mathbb{Z} \setminus \{0\}$, and in this case we obtain the first assertion of our theorem.

If, on the other hand, the last assertion of Lemma 4.2 holds, then, as in (5.1), infinitely many solutions satisfy

$$a\alpha^{\ell x} + b\alpha^{-\ell x} = y - h(x), \tag{6.1}$$

where $\ell = 1/k$ with $k > 0$, and, by Lemma 5.1, $ab \in \mathbb{Q}$ and $h(X) \in \mathbb{Q}[X]$. There exists a positive integer r such that $\alpha^{\ell r}$ is a real quadratic unit; since $\ell = 1/k$ we see that α^r is also a real quadratic unit.

When V is an indeterminate and k a natural number, we have an identity

$$V^k + (1/V)^k = P_k(V + 1/V), \tag{6.2}$$

where P_k is a polynomial with integer coefficients depending only on k (to see this, consider the expansion of $(V + 1/V)^k$). From this identity we obtain, for any A,

$$V^k + (A/V)^k = A^{k/2} P_k\left(\frac{1}{\sqrt{A}}(V + A/V)\right).$$

Applying this to (6.1) with $V = a\alpha^{\ell x}$ and $A = ab$, we obtain

$$a^k\alpha^x + b^k\alpha^{-x} = (ab)^{k/2} P_k\left(\frac{1}{\sqrt{ab}}(y - h(x))\right). \tag{6.3}$$

Define

$$\tilde{F}(X,Y,Z) = Z^2 - \left(\frac{b}{a}\right)^{k/2} P_k\left(\frac{1}{\sqrt{ab}}(Y - h(X))\right) Z + \left(\frac{b}{a}\right)^k.$$

Then every solution to (6.1) has $\tilde{F}(x, y, \alpha^x) = 0$. There are infinitely many values of x for which $\tilde{F}(x, y, \alpha^x) = F(x, y, \alpha^x) = 0$; since F is irreducible, Lemma 2.1 shows that F divides $\tilde{F}$ in $\mathbb{C}(X,Y)[Z]$. Comparing degrees we see that, up to a constant multiple, $\tilde{F}(X,Y,Z) = F(X,Y,Z)$. Setting $\gamma = \left(\frac{b}{a}\right)^k$ and $\rho = \frac{1}{ab} \in \mathbb{Q}$, we have, again up to a constant multiple,

$$F(X,Y,Z) = Z^2 - \sqrt{\gamma} P_k\left(\sqrt{\rho}(Y - h(X))\right) Z + \gamma. \tag{6.4}$$

The proof is finished; for completeness, we record the following.

Supplement to case (2) of Theorem 1.3. *$F(X, Y, Z)$ is as in (6.4), where P_k is defined in (6.2).*

Remark. When $\deg_Z F = n > 2$, one can prove a result similar to Theorem 1.3, under an assumption generalizing (a) on the Newton polygon attached to F. In the general case, we must consider partitions of V_∞ (as in Lemma 4.1) into n sets, and additional technical difficulties are encountered. For a full account, see [1].

References

1. S. Ahlgren, *Equations of polynomial–exponential type*, Ph. D. thesis, University of Colorado (1996).
2. ———, *Polynomial–exponential equations in two variables*, J. Number Theory **62** (1997), 428-438.
3. M. Laurent, *Équations exponentielles–polynômes et suites récurrentes linéaires, II*, J. Number Theory **31** (1989), 24–53.
4. H. P. Schlickewei, *p-adic Thue-Siegel-Roth-Schmidt theorem*, Arch. Math. (Basel) **29** (1977), 267-270.
5. W. Schmidt, *Equations* $\alpha^x = R(x, y)$, J. Number Theory **47** (1994), 348–358.
6. ———, *Diophantine approximations and diophantine equations*, Springer Lecture Notes in Math. **1467**, 1992.

Department of Mathematics, The Pennsylvania State University, University Park, PA 16802

E-mail address: ahlgren@math.psu.edu

A FUNDAMENTAL INVARIANT IN THE THEORY OF PARTITIONS

KRISHNASWAMI ALLADI

To my teacher and friend, Basil Gordon, on his sixty fifth birthday.

ABSTRACT. One of the important invariants under conjugation of Ferrers graphs is the number of different parts. But this invariant has not been exploited in the literature. In this paper, utilizing this invariance, we obtain a new proof and a variant of Cauchy's q−binomial theorem, a six parameter extension of Heine's fundamental transformation, and an extension of a fifth order mock-theta function identity due to Ramanujan.

1. INTRODUCTION

Given a partition $\pi : b_1 + b_2 + \cdots + b_\nu$ into parts $b_1 \geq b_2 \geq \ldots b_\nu$, its Ferrers graph is an array of nodes equally spaced such that there are b_i nodes in the i−th row. The nodes are arranged such that the left most node of each row will lie on a common vertical line. We use the symbol π to denote a partition as well as its Ferrers graph.

Next, given a partition π, its conjugate π^* is obtained by interchanging the rows and columns of the Ferrers graph. For example, the Ferrers graph of the partition $\pi : 7+7+5+4+2+2+1$ is given in figure 1 and the Ferrers graph of its conjugate $\pi^* : 7+6+4+4+3+2+2$ is given in figure 2.

Conjugation has been used extensively in the study of partitions not only to enhance combinatorial understanding but to simplify proofs as well.

Conjugation is an involution. That is

$$(\pi^*)^* = \pi \tag{1.1}$$

for any partition π. Under conjugation the largest part $\lambda(\pi)$ and the number of parts $\nu(\pi)$ are interchanged. That is

$$\lambda(\pi^*) = \nu(\pi), \quad \nu(\pi^*) = \lambda(\pi). \tag{1.2}$$

Therefore

$$\lambda(\pi) + \nu(\pi) = \lambda(\pi^*) + \nu(\pi^*)$$

1991 *Mathematics Subject Classification.* Primary 05A15, 05A17, 11P83. Secondary 05A19.
Key words and phrases. partitions, Ferrers graph, conjugation, invariance.
Research supported in part by the National Science Foundation grant DMS-9400191.

S.D. Ahlgren et al. (eds.), Topics in Number Theory, 101–113.

is invariant under conjugation. Another invariant is $D(\pi)$, the Durfee square of a partition π. This is the largest square of nodes starting from the upper left hand corner of the Ferrers graph of π. That is

$$D(\pi) = D(\pi^*).$$

In figures 1 and 2 both Durfee squares are of size 4×4.

π :

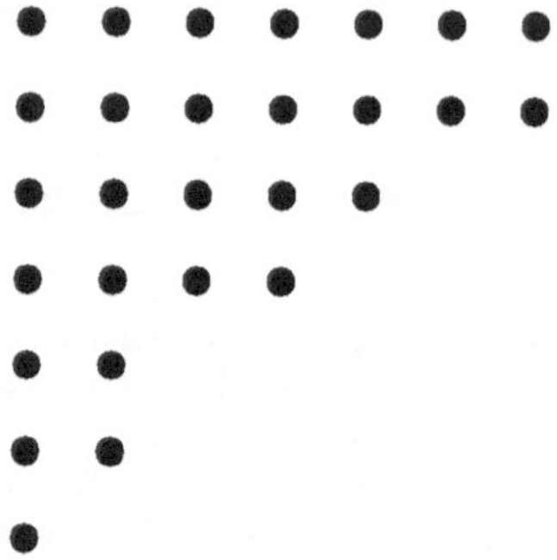

FIG. 1 Ferrers graph of π

π^* :

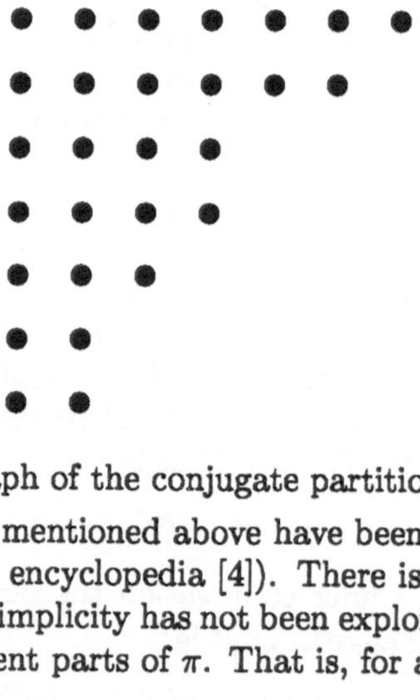

FIG. 2 Ferrers graph of the conjugate partition π^*

Conjugation and the invariants mentioned above have been used in a variety of ways over the years (see Andrews' encyclopedia [4]). There is however one fundamental invariant which despite its simplicity has not been exploited in the literature. This is $\nu_d(\pi)$, the number of different parts of π. That is, for all partitions π

$$\nu_d(\pi) = \nu_d(\pi^*). \tag{1.3}$$

In figures 1 and 2 we have $\nu_d(\pi) = \nu_d(\pi^*) = 5$.

Although the partition statistic $\nu_d(\pi)$ has been studied in several situations, the usefulness of the invariance (1.3) seems to have escaped attention. In this paper we indicate briefly how (1.3) could be used to obtain new combinatorial proofs of Cauchy's q−binomial theorem and Heine's transformation (see §4 and §5). The main consequence of this approach is that it leads to a new six parameter extension of Heine's transformation (see §6). Finally, in §7, (1.3) is used to obtain an extension of a fifth–order mock theta function identity that Ramanujan communicated in his last letter to Hardy.

2. Cuts

Given the Ferrers graph of a partition π, a (horizontal) cut is a horizontal line drawn between the nodes such that the number of nodes in the row above this line is greater than the number of nodes in the row below this line. If the number of nodes on the row below this line is k, this line is called a horizontal k–cut.

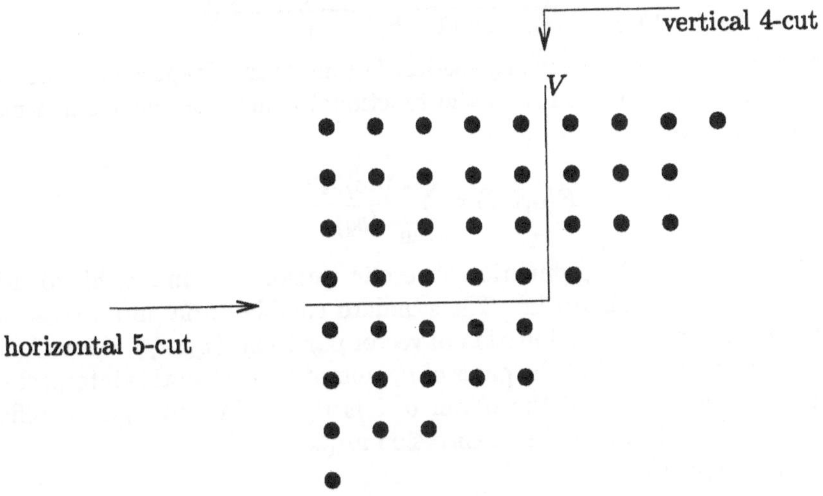

FIG. 3 Horizontal and vertical cuts

Given a horizontal cut, consider a line orthogonal to it but just to the right of the last node on the row below this horizontal cut. This line V is a vertical cut in the sense that the number of nodes in the column to the left of V is greater than the number of nodes in the column to the right of V. Since every horizontal cut generates a vertical cut and vice–versa, the number of horizontal and vertical cuts are equal. The invariance (1.3) follows from the equality of the number of horizontal and vertical cuts.

3. Notation and combinatorial interpretation

In the theory of q–series, the symbols $(a)_n = (a;q)_n$ are defined as follows. For a positive integer n,

$$(a)_n = \prod_{j=0}^{n-1}(1-aq^j) \tag{3.1}$$

and

$$(a)_\infty = \lim_{n\to\infty}(a)_n = \prod_{j=0}^{\infty}(1-aq^j), \text{ for } |q|<1. \tag{3.2}$$

If $|q| < 1$, then (3.2) can be used to define the symbols $(a)_n$ for all real n by means of the relation

$$(a)_n = \frac{(a)_\infty}{(aq^n)_\infty}. \tag{3.3}$$

From this definition it follows that

$$\frac{1}{(q)_n} = 0, \text{ for } n = -1, -2, -3, \ldots \tag{3.4}$$

The rational function

$$\frac{(aq)_n}{(bq)_n} = \frac{(1-aq)(1-aq^2)\ldots(1-aq^n)}{(1-bq)(1-bq^2)\ldots(1-bq^n)} \tag{3.5}$$

arises quite naturally in the study of q–series. For instance, Chapter 1 of Fine's book [5] is devoted to a detailed study of the functional equations and transformation properties of the series

$$F(a,b;t) = \sum_{n=0}^{\infty} \frac{(aq)_n t^n}{(bq)_n}. \tag{3.6}$$

Although Fine states striking partition theoretic versions of some of his identities, his methods are not combinatorial. The standard combinatorial interpretation for (3.5) is that it is the generating function of vector partitions $(\pi_1; \pi_2)$, where $\lambda(\pi_1) \leq n$ and $\lambda(\pi_2) \leq n$, and such that the parts of π_1 cannot repeat. In this interpretation, the power of $-a$ is $\nu(\pi_1)$, and the power of b is $\nu(\pi_2)$. We now give a different combinatorial interpretation for the expression in (3.5).

Consider the expansion

$$\begin{aligned}\frac{1-abq^j}{1-bq^j} &= \frac{1-bq^j+bq^j-abq^j}{1-bq^j} \\ &= 1+(1-a)\{bq^j+b^2q^{2j}+b^3q^{3j}+\ldots\}.\end{aligned} \tag{3.7}$$

We interpret this as an expansion involving only one part, namely j, where the power of b indicates how often the part occurs, while the power of $(1-a)$ indicates whether the part occurs or not. Thus, we interpret

$$\frac{(abq)_n}{(bq)_n} = \sum_{\lambda(\pi)\leq n} (1-a)^{\nu_d(\pi)} b^{\nu(\pi)} q^{\sigma(\pi)} \tag{3.8}$$

as the generating function of partitions π into parts $\leq n$, such that the power of b is $\nu(\pi)$ and the power of $(1-a)$ is $\nu_d(\pi)$. In (3.8) and in what follows, $\sigma(\pi)$ is the sum of the parts of π; that is, $\sigma(\pi)$ is the integer being partitioned. The interpretation given in (3.8) is the one that is crucial in this paper.

4. Cauchy's identity and a variant

The q–binomial theorem or Cauchy's identity is

$$\sum_{n=0}^{\infty} \frac{(a)_n t^n}{(q)_n} = \frac{(at)_\infty}{(t)_\infty}. \tag{4.1}$$

Various proofs of (4.1) are known (see Andrews [4]). Here we give a proof using the invariance (1.3) and in that process obtain a variant of (4.1).

Consider the three parameter generating function for all unrestricted partitions, namely, the function $f(a,b,c;q)$ defined by

$$f(a,b,c;q)=\sum_{\pi}(1-a)^{\nu_d(\pi)}b^{\nu(\pi)}c^{\lambda(\pi)}q^{\sigma(\pi)}. \tag{4.2}$$

From the combinatorial interpretation underlying (3.8) it follows that

$$f(a,b,c;q)=1+\sum_{n=1}^{\infty}\frac{(1-a)(abq)_{n-1}bc^nq^n}{(bq)_n}. \tag{4.3}$$

To justify this, consider all partitions π for which $\lambda(\pi)=n$. This accounts for the term c^nq^n in (4.3). Since π contains n as a part, we have the factor b in the numerator of (4.3) for the part n enumerated by $\nu(\pi)$ and a factor $(1-a)$ for n enumerated by $\nu_d(\pi)$. The part n may repeat. This is given by the factor $(1-bq^n)$ in the denominator. The repetition of n will not contribute to $\nu_d(\pi)$ and so there is no further power of $(1-a)$ contributed by the part n. The parts $1,2,\ldots,n-1$ could repeat and their contribution to the generating function is given by the term

$$\frac{(abq)_{n-1}}{(bq)_{n-1}}.$$

Thus

$$\frac{(abq)_{n-1}(1-a)bc^nq^n}{(bq)_{n-1}(1-bq^n)}=\frac{(1-a)(abq)_{n-1}bc^nq^n}{(bq)_n} \tag{4.4}$$

is the contribution to the sum in (4.2) arising from all partitions π with $\lambda(\pi)=n>1$. Summing the terms in (4.4) from $n=1$ to ∞ and interpreting 1 as representing the null partition, yields (4.3).

The three parameter generating function (4.3) does not have a product representation which is perhaps the reason it has not attracted much attention. Under conjugation we have the relations (1.2) and (1.3). Thus we have

$$f(a,b,c;q)=f(a,c,b;q), \tag{4.5}$$

although this is not obvious from the series representation (4.3).

Next take $c=1$ in (4.3). This means we are enumerating only $\nu_d(\pi)$ and $\nu(\pi)$, but not $\lambda(\pi)$. It is clear from the interpretation underlying (3.8) that

$$f(a,b,1;q)=\sum_{\pi}(1-a)^{\nu_d(\pi)}b^{\nu(\pi)}q^{\sigma(\pi)}=\frac{(abq)_\infty}{(bq)_\infty}. \tag{4.6}$$

Thus from (4.3) and (4.6) we have proved the following <u>variant</u> of Cauchy's identity:

$$f(a,b,1;q)=1+\sum_{n=1}^{\infty}\frac{(1-a)(abq)_{n-1}bq^n}{(bq)_n}=\frac{(abq)_\infty}{(bq)_\infty}. \tag{4.7}$$

Finally to get (4.1) from (4.7), take $b=1$ in (4.3). Notice that in this case

$$(1-a)(aq)_{n-1}=(a)_n. \tag{4.8}$$

Using (4.8), (4.6) and exploiting the symmetry given by (4.5) we get

$$\sum_{n=0}^{\infty} \frac{(a)_n c^n q^n}{(q)_n} = f(a,1,c;q) = f(a,c,1;q) = \frac{(acq)_\infty}{(cq)_\infty} \tag{4.9}$$

which is equivalent to (4.1) if we replace t by cq. This completes the combinatorial proof of Cauchy's identity.

Remark: The exact relationship between our function $f(a,b,c;q)$ and Fine's function $F(a,b;t)$ is given by

$$\frac{1-ab}{(1-a)b}(f(a,b,c;q)-1) = F(abq^{-1},b;cq)-1.$$

We preferred to study $f(a,b,c;q)$ here because of its combinatorial interpretation as the generating function of unrestricted partitions.

Finite and bilateral versions: The Cauchy variant (4.7) has the advantage over the standard version (4.1) in the sense that (4.7) is true even in a finite form. More precisely consider the generating functions

$$f_j(a,b,1;q) = \sum_{\lambda(\pi)=j} (1-a)^{\nu_d(\pi)} b^{\nu(\pi)} q^{\sigma(\pi)}, \tag{4.10}$$

and

$$f^{(n)}(a,b,1;q) = \sum_{0\le j\le n} f_j(a,b,1;q) = \sum_{\lambda(\pi)\le n} (1-a)^{\nu_d(\pi)} b^{\nu(\pi)} q^{\sigma(\pi)}. \tag{4.11}$$

We know from (4.4) that

$$f_j(a,b,1;q) = \frac{(1-a)(abq)_{j-1} bq^j}{(bq)_j}, \quad \text{for} \quad j \ge 1, \tag{4.12}$$

and from (3.7) that

$$f^{(n)}(a,b,1;q) = \frac{(abq)_n}{(bq)_n}. \tag{4.13}$$

Thus (4.11) - (4.13) yield

$$1 + \sum_{j=1}^{n} \frac{(1-a)(abq)_{j-1} bq^j}{(bq)_j} = \frac{(abq)_n}{(bq)_n} \tag{4.14}$$

which is a finite version of the Cauchy variant (4.7).

Actually (4.14) is true in a more general finite form which then leads to a bilateral version. To see this, formally consider the term on the left in (4.14) corresponding to $j=0$. Using (3.3) we get

$$\frac{(1-a)(abq)_{-1} bq^0}{(bq)_0} = \frac{(1-a)b}{1-ab}. \tag{4.15}$$

Thus from (4.14) and (4.15) we get

$$\sum_{j=0}^{n} \frac{(1-a)(abq)_{j-1}bq^j}{(bq)_j} = \frac{(abq)_n}{(bq)_n} - 1 + \frac{(1-a)b}{1-ab}$$
$$= \frac{(abq)_n}{(bq)_n} - \frac{1-b}{1-ab} = \frac{(abq)_n}{(bq)_n} - \frac{(abq)_{-1}}{(bq)_{-1}}.$$

Pursuing this idea further we get

$$\sum_{j=M}^{N} \frac{(1-a)(abq)_{j-1}bq^j}{(bq)_j} = \frac{(abq)_N}{(bq)_N} - \frac{(abq)_{M-1}}{(bq)_{M-1}} \tag{4.16}$$

for all integers M and N positive or negative. We call this the ***universal Cauchy identity*** since there are no restrictions on M or N.

Finally we note that if in addition to $|q| < 1$ we also have $|a| < 1$, then

$$\lim_{M\to-\infty} \frac{(abq)_{M-1}}{(bq)_{M-1}} = 0.$$

This leads to the following bilateral form of the Cauchy variant:

$$\sum_{j=-\infty}^{\infty} \frac{(1-a)(abq)_{j-1}bq^j}{(bq)_j} = \frac{(abq)_\infty}{(bq)_\infty}. \tag{4.17}$$

In concluding this section we note that although there is no finite form of the classical version of Cauchy's identity like (4.14), there is (trivially) a bilateral version because of (3.4). That is,

$$\sum_{n=0}^{\infty} \frac{(a)_n t^n}{(q)_n} = \sum_{n=-\infty}^{\infty} \frac{(a)_n t^n}{(q)_n} = \frac{(at)_\infty}{(t)_\infty}. \tag{4.18}$$

Both bilateral forms (4.17) and (4.18) are special cases of Ramanujan's ${}_1\psi_1$ summation formula [6].

5. Heine's transformation

Heine's fundamental transformation for basic hypergeometric series is

$$\sum_{n=0}^{\infty} \frac{(a)_n(\gamma)_n c^n}{(\alpha)_n(q)_n} = \frac{(\gamma)_\infty(ac)_\infty}{(\alpha)_\infty(c)_\infty} \sum_{n=0}^{\infty} \frac{(\alpha/\gamma)_n(c)_n\gamma^n}{(ac)_n(q)_n}. \tag{5.1}$$

Andrews [2] gave an enumerative proof of (5.1). We follow Andrews in recasting (5.1) into a suitable symmetric form but we interpret this differently. Andrews proved the symmetric version by interpreting it in terms of certain vector partitions. We will establish the symmetric form by appealing to the invariance of $\nu_d(\pi)$ under conjugation. The main advantage in our approach is that it leads naturally to a six parameter extension of Heine's transformation (see §6).

Following Andrews, replace α by $\alpha\gamma$ in (5.1) to get

$$\sum_{n=0}^{\infty} \frac{(a)_n(\gamma)_n c^n}{(\alpha\gamma)_n(q)_n} = \frac{(\gamma)_\infty(ac)_\infty}{(\alpha\gamma)_\infty(c)_\infty} \sum_{n=0}^{\infty} \frac{(\alpha)_n(c)_n\gamma^n}{(ac)_n(q)_n}. \tag{5.2}$$

Rewrite (5.2) as

$$\sum_{n=0}^{\infty}\frac{(a)_n(\alpha\gamma q^n)_\infty c^n}{(\gamma q^n)_\infty(q)_n}=\sum_{n=0}^{\infty}\frac{(\alpha)_n(acq^n)_\infty\gamma^n}{(cq^n)_\infty(q)_n}.$$

At this stage we depart from Andrews [2]. First we replace c by cq and γ by γq to get

$$\sum_{n=0}^{\infty}\frac{(a)_n(\alpha\gamma q^{n+1})_\infty c^n q^n}{(\gamma q^{n+1})_\infty(q)_n}=\sum_{m=0}^{\infty}\frac{(\alpha)_m(acq^{m+1})_\infty\gamma^m q^m}{(cq^{m+1})_\infty(q)_m}. \tag{5.3}$$

This is equivalent to (5.1) and is in a symmetric form, namely,

$$h(a,c,\gamma,\alpha)=h(\alpha,\gamma,c,a). \tag{5.4}$$

We now prove (5.3) combinatorially.

Consider Ferrers graphs of a partition π with an n-cut.

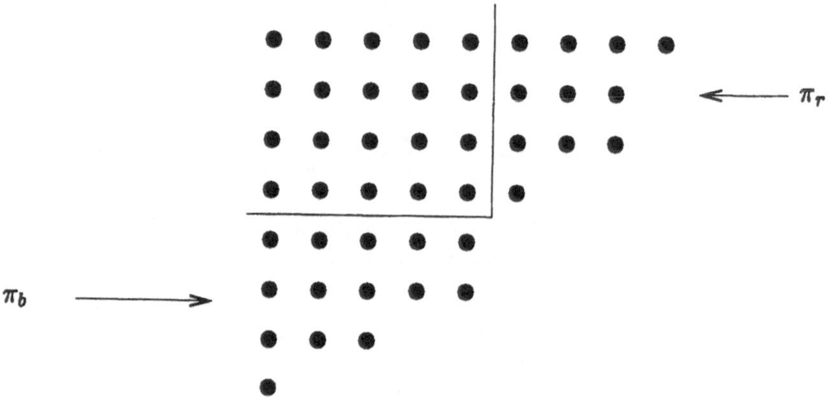

FIG. 4

Let π_b denote the portion of the Ferrers graph below the n-cut, and π_r denote the portion of the Ferrers graph to the right of the vertical cut generated by this horizontal n-cut. Let the vertical cut be an m-cut. Also let π_a denote the portion of the graph above the $n - cut$ and π_l the portion of the graph to the left of the $m - cut$.

Next, consider the generating function given by

$$h_n(a,c,\gamma,\alpha)=\sum_{\lambda(\pi_b)=n}(1-a)^{\nu_d(\pi_b)}c^{\lambda(\pi_b)}q^{\sigma(\pi_b)}(1-\alpha)^{\nu_d(\pi_r)}\gamma^{\nu(\pi_r)}q^{\sigma(\pi_a)} \tag{5.5}$$

for a fixed n, where the summation is over all Ferrers graphs that can be generated by an n-cut. Note that since $\nu(\pi_a)=\nu(\pi_r)$ and $\nu_d(\pi_a)=\nu_d(\pi_r)$, it follows that

$$h_n(a,c,\gamma,\alpha)=\frac{(a)_n c^n q^n}{(q)_n}\cdot\frac{(\alpha\gamma q^{n+1})_\infty}{(\gamma q^{n+1})_\infty} \tag{5.6}$$

is the n-th term on the left in (5.3). Summing h_n over all n yields the left hand side of (5.3) and the function $h(a,c,\gamma,\alpha)$ in (5.4). Now consider the conjugates of these Ferrers graphs generated by a vertical cut of a specific size $\lambda(\pi_r^*)=m=\nu(\pi_r)$.

Under conjugation $\lambda(\pi_b) = \nu(\pi_b^*)$ and $\nu(\pi_r) = \lambda(\pi_r^*)$, but ν_d functions remains invariant. Thus

$$(5.7)\qquad \begin{aligned} &h_n(\alpha,\gamma,c,a)\\ &= \sum_{\lambda(\pi_r^*)=m} (1-\alpha)^{\nu_d(\pi_r^*)}\gamma^{\lambda(\pi_r^*)}q^{\sigma(\pi_r^*)}(1-a)^{\nu_d(\pi_b^*)}c^{\nu(\pi_b^*)}q^{\sigma(\pi_1^*)}\\ &= \frac{(\alpha)_m\gamma^m q^m}{(q)_m}\cdot\frac{(acq^{m+1})_\infty}{(cq^{m+1})_\infty}\end{aligned}$$

which is the m–th term of the sum on the right in (5.3). Summing this over m establishes (5.3) and completes the combinatorial proof of Heine's transformation.

Bilateral form: By utilizing the results of the previous section, the symmetric form of Heine's transformation (5.3) can be cast in a bilateral form as we show presently. First we use (4.17) to rewrite (5.3) in the form

$$(5.8)\qquad \begin{aligned} &\sum_{n=0}^{\infty}\frac{(a)_n c^n q^n}{(q)_n}\left(\sum_{t=-\infty}^{\infty}\frac{(1-\alpha)(\alpha\gamma q^{n+1})_{t-1}\gamma q^{n+t}}{(\gamma q^{n+1})_t}\right)\\ &= \sum_{m=0}^{\infty}\frac{(\alpha)_m\gamma^m q^m}{(q)_m}\left(\sum_{t=-\infty}^{\infty}\frac{(1-a)(acq^{m+1})_{t-1}cq^{m+1}}{(cq^{m+1})_t}\right).\end{aligned}$$

Since (3.4) holds, (5.8) may be rewritten in the form

$$(5.9)\qquad \begin{aligned} &\sum_{n=-\infty}^{\infty}\frac{(a)_n c^n q^n}{(q)_n}\left(\sum_{t=-\infty}^{\infty}\frac{(1-\alpha)(\alpha\gamma q^{n+1})_{t-1}\gamma q^{n+t}}{(\gamma q^{n+1})_t}\right)\\ &= \sum_{m=-\infty}^{\infty}\frac{(\alpha)_m\gamma^m q^m}{(q)_m}\left(\sum_{t=-\infty}^{\infty}\frac{(1-a)(acq^{m+1})_{t-1}cq^{m+1}}{(cq^{m+1})_t}\right).\end{aligned}$$

This is the symmetric bilateral form of Heine's transformation in which the bilateral forms of the classical version of Cauchy identity and Cauchy's variant are both present.

6. A SIX PARAMETER EXTENSION

The combinatorial proof of Heine's transformation given in §5 suggests the following question: Why not also keep track of the statistics $\nu(\pi_b)$ and $\lambda(\pi_r)$ and consider the six parameter generating function

$$(6.1)\qquad \begin{aligned} &H_n(a,b,c,\gamma,\beta,\alpha)\\ &= \sum_{\lambda(\pi_b)=n} (1-a)^{\nu_d(\pi_b)}b^{\nu(\pi_b)}c^{\lambda(\pi_b)}q^{\sigma(\pi_b)}(1-\alpha)^{\nu_d(\pi_r)}\beta^{\lambda(\pi_r)}\gamma^{\nu(\pi_r)}q^{\sigma(\pi_a)}?\end{aligned}$$

Using the combinatorial interpretations given in §5 and the ideas underlying the Cauchy variant (4.7) we see that

$$H_n(a,b,c,\gamma,\beta,\alpha) = \frac{(1-a)(abq)_{n-1}bc^nq^n}{(bq)_n}\left(1+\sum_{t=1}^{\infty}\frac{(1-\alpha)(\alpha\gamma q^{n+1})_{t-1}\gamma\beta^tq^{n+t}}{(\gamma q^{n+1})_t}\right) \tag{6.2}$$

for $n \geq 1$, while

$$H_0(a,b,c,\gamma,\beta,\alpha) = 1+\sum_{t=1}^{\infty}\frac{(1-\alpha)(\alpha\gamma q)_{t-1}\gamma\beta^tq^t}{(\gamma q)_t} \tag{6.3}$$

does not involve the parameters a, b, c, because $n = 0$. Summing H_n over all $n \geq 0$, and considering conjugates of all such Ferrers graphs we get the following six parameter extension of Heine's transformation by exploiting invariants under conjugation:

$$\begin{aligned}
&1+\sum_{t=1}^{\infty}\frac{(1-\alpha)(\alpha\gamma q)_{t-1}\gamma\beta^tq^t}{(\gamma q)_t} \\
&\quad+\sum_{n=1}^{\infty}\frac{(1-a)(abq)_{n-1}bc^nq^n}{(bq)_n}\left(1+\sum_{t=1}^{\infty}\frac{(1-\alpha)(\alpha\gamma q^{n+1})_{t-1}\gamma\beta^tq^{n+t}}{(\gamma q^{n+1})_t}\right) \\
&=1+\sum_{t=1}^{\infty}\frac{(1-a)(acq)_{t-1}cb^tq^t}{(cq)_t} \\
&\quad+\sum_{m=1}^{\infty}\frac{(1-\alpha)(\alpha\beta q)_{m-1}\beta\gamma^mq^m}{(\beta q)_m}\left(1+\sum_{t=1}^{\infty}\frac{(1-a)(acq^{m+1})_{t-1}cb^tq^{m+1}}{(cq^{m+1})_t}\right)
\end{aligned} \tag{6.4}$$

This is a symmetric identity in the form

$$H(a,b,c,\gamma,\beta,\alpha) = H(\alpha,\beta,\gamma,c,b,a). \tag{6.5}$$

What I have given here is a combinatorial proof of (6.4) by exploiting the invariance of ν_d under conjugation. I do not have a purely q–series proof of (6.4) and it would be worthwhile to obtain such a proof.

Setting $b = \beta = 1$ in (6.4) yields (5.3), the symmetric form of Heine's transformation. With these choices of b and β in (6.4), the variant of Cauchy's identity yields

$$\begin{aligned}
&1+\sum_{t=1}^{\infty}\frac{(1-\alpha)(\alpha\gamma q)_{t-1}\gamma q^t}{(\gamma q)_t}+\sum_{n=1}^{\infty}\frac{(a)_nc^nq^n(\alpha\gamma q^{n+1})_\infty}{(q)_n(\gamma q^{n+1})_\infty} \\
&=1+\sum_{t=1}^{\infty}\frac{(1-a)(acq)_{t-1}cq^t}{(cq)_t}+\sum_{m=1}^{\infty}\frac{(\alpha)_m\gamma^mq^m(acq^{m+1})_\infty}{(q)_m(cq^{m+1})_\infty}.
\end{aligned} \tag{6.6}$$

By another use of the variant of Cauchy's identity in (6.6) we get

$$\frac{(\alpha\gamma q)_\infty}{(\gamma q)_\infty}+\frac{(\alpha\gamma q)_\infty}{(\gamma q)_\infty}\sum_{n=1}^{\infty}\frac{(a)_n(\gamma q)_n c^n q^n}{(q)_n(\alpha\gamma q)_n}$$
$$=\frac{(acq)_\infty}{(cq)_\infty}+\frac{(acq)_\infty}{(cq)_\infty}\sum_{m=1}^{\infty}\frac{(\alpha)_m(cq)_m\gamma^m q^m}{(q)_m(acq)_m}$$

This yields

$$\frac{(\alpha\gamma q)_\infty}{(\gamma q)_\infty}\sum_{n=0}^{\infty}\frac{(a)_n(\gamma q)_n c^n q^n}{(q)_n(\alpha\gamma q)_n}=\frac{(acq)_\infty}{(cq)_\infty}\sum_{m=0}^{\infty}\frac{(\alpha)_m(cq)_m\gamma^m q^m}{(q)_m(acq)_m}\tag{6.7}$$

which is clearly equivalent to (5.3).

7. Ramanujan's mock-theta identity

In his last letter to Hardy dated 12 January 1920, Ramanujan stated the following fifth order mock–theta function identity:

$$\sum_{n=0}^{\infty}\frac{q^n}{(q^{n+1})_n}=1+\sum_{m=0}^{\infty}\frac{q^{2m+1}}{(q^{m+1})_{m+1}}.\tag{7.1}$$

Identity (7.1) has the following partition interpretation as noted by Andrews [2]: *The number of partitions of N where the smallest part does not repeat and the largest part is at most twice the smallest part, equals the number of partitions of N where the largest part is odd and the smallest part is larger than half the largest part.*

Andrews [2] gave a combinatorial proof this partition identity. It goes as follows. Consider a Ferrers graph π for which $\lambda(\pi)=2m+1$. Since the smallest part is $\geq m+1$, we may split π into a rectangle of dimension $n\cdot\times(m+1)$ where $\nu(\pi)=n$, and a Ferrers graph π_r to the right of this rectangle having $\lambda(\pi_r)=m$.

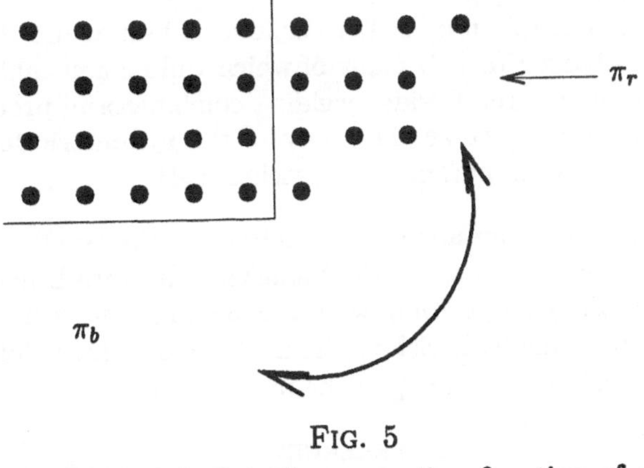

FIG. 5

The right hand side of (7.1) is the generating function of such Ferrers graphs. Next, move π_r and place it below the rectangle to get a Ferrers graph π' which is the conjugate of the type enumerated by the left side of (7.1). The procedure is reversible and this yields a bijective proof of (7.1).

What we would like to do next is to enumerate the number of different parts of π_r in Fig. 5. This the same as enumerating the number of different parts of π which are $> (m+1)$. After we move π_r and place it below the rectangle, we enumerate $\nu_d(\pi_b) = \nu_d(\pi_r)$ by considering the conjugate graph. This is the same as enumerating the number of different parts of π' which are $> n = \nu(\pi)$. Since ν_d is invariant under conjugation, this procedure leads to the following extension of (7.1):

$$(7.2) \qquad \sum_{n=0}^{\infty} \frac{(aq^{n+1})_n q^n}{(q^{n+1})_n} = \frac{1}{1-q} + \sum_{m=1}^{\infty} \frac{(1-a)(aq^{m+2})_{m-1} q^{2m+1}}{(q^{m+1})_{m+1}}.$$

Setting $a = 0$ in (7.2) yields (7.1).

8. Concluding remarks

The results proved in this paper are some of the important consequences of the invariance of ν_d under conjugation. There are many other partition theorems and q-series results which can be obtained using this invariance and a more complete discussion will be presented elsewhere [1]. For instance, just as there is a variant of Cauchy's identity (4.7) related to the classical Cauchy identity (4.1), there is also a Heine variant which we have not discussed here that can be deduced from (6.4). In fact (6.4) can be used to obtain interesting variants for several classical results such as the q-Gauss identity, the q-Kummer identity and Lebesgue's identity. In §4 it was observed that the symmetry (4.5) which immediately follows from conjugation, is not obvious from the series (4.3). A different expansion for $f(a, b, c; q)$ which renders this symmetry explicit can be derived using Durfee squares and conjugation. This leads to a variant of the Rogers-Fine identity (eqn. 14.1 in [5]). The proof of the Rogers-Fine identity in [5] utilizes transformation properties of the function $F(a, b; t)$, but subsequently, Andrews [3] gave a combinatorial proof. A cleaner combinatorial proof can be given by utilizing the invariance of ν_d under conjugation. All these will be presented in [1].

The next major step in this project is to conduct a systematic study of several results of Fine [5] involving $F(a, b; t)$ many of which will be amenable via the combinatorial ideas presented here. Besides yielding combinatorial proofs, this would also lead to generalizations just as the study of the symmetric form of Heine's transformation yielded the six parameter extension (6.4).

Acknowledgement: Some of the main ideas underlying the results presented here came up during my sabbatical visit to the Pennsylvania State University in 1992-93. I would like to thank George Andrews for arranging that visit and for several stimulating discussions. Finally, I wish to thank the referee for helpful suggestions and Bruce Berndt for a careful reading of the manuscript.

References

[1] K. Alladi, *Invariants under partition conjugation and q−series identities* (in preparation).
[2] G. E. Andrews, *Enumerative proofs of certain q−identities*, Glasgow Math. J., **8** (1967), 33–40.
[3] G. E. Andrews, *Two theorems of Gauss and allied identities proved arithmetically*, Pacific J. Math., **41** (1972), 563–578.

[4] G. E. Andrews, *The theory of partitions*, Encyclopedia of Mathematics and its Applications, **2**, Addison–Wesley, Reading (1976).
[5] N. J. Fine, *Basic hypergeometric and applications*, Math. Surveys and Monographs **27**, Amer. Math. Soc., Providence (1988).
[6] G. Gasper and M. Rahman, *Basic hypergeometric series*, Encyclopedia of Mathematics, **35**, Cambridge University Press, New York (1990).

E-mail address: `alladi@math.ufl.edu`

DEPARTMENT OF MATHEMATICS, UNIVERSITY OF FLORIDA, GAINESVILLE, FLORIDA 32611

CONGRUENCE PROPERTIES OF VALUES OF L-FUNCTIONS AND APPLICATIONS

J. H. BRUINIER, K. JAMES, W. KOHNEN,
K. ONO, C. SKINNER AND V. VATSAL

Dedicated to the memory of S. Chowla

1. INTRODUCTION

Ever since Dirichlet's introduction of the analytic class number formula, special values of L-functions have been the subject of much study and speculation. In this paper we survey some recent results about such values that were presented at this conference. Our attention is essentially restricted to the central values of L-functions associated to certain (holomorphic) newforms. These results have many applications to class numbers of imaginary quadratic fields, Selmer groups of elliptic curves, and K-groups of real quadratic fields, a few of which are included.

To describe the problem we will address, we need to introduce some notation. Let $F(z) = \sum_{n=1}^{\infty} a(n)q^n \in S_{2k}(M)$ ($q = e^{2\pi i z}$ as usual) be a newform of weight $2k$ on $\Gamma_0(M)$ with trivial Nebentypus character, and for $\mathrm{Re}(s) \gg 0$ let $L(F,s) = \sum_{n=1}^{\infty} a(n)n^{-s}$ be its L-function. Let D denote a fundamental discriminant of a quadratic field that is coprime to M; then χ_D shall denote the Kronecker character for the field $\mathbb{Q}(\sqrt{D})$.

The D-quadratic twist of F, denoted F_D, is given by

$$F_D(z) = \sum_{n=1}^{\infty} \chi_D(n)a(n)q^n,$$

and for $\mathrm{Re}(s) \gg 0$ its L-function is given by $L(F_D, s) = \sum_{n=1}^{\infty} \chi_D(n)a(n)n^{-s}$. These L-functions have analytic continuations to $\mathbb{C}$ and satisfy well known functional equations. If $\Lambda(F,s) = (2\pi)^{-s}\Gamma(s)M^{s/2}L(F,s)$, then

$$\Lambda(F,s) = \epsilon \cdot \Lambda(F, 2k-s),$$

where $\epsilon = \pm 1$ is the so-called sign of the functional equation, and the quadratic twists satisfy

$$\Lambda(F_D, s) = \epsilon \cdot \chi_D(-M)\Lambda(F_D, 2k-s).$$

1991 *Mathematics Subject Classification.* Primary 11F67 ; Secondary 11F37.

Key words and phrases. modular forms, critical values of L-functions.

The fourth author is supported by NSF grant DMS-9508976 and NSA grant MSPR-97Y012 and the fifth author is supported by NSF grant DMS-9304580 and by an Ostrowski Fellowship.

S.D. Ahlgren et al. (eds.), Topics in Number Theory, 115–125.

The value $L(F_D, k)$ is the *central value* of $L(F_D, s)$. Our motivating problem is to describe the behaviour of the family of values $L(F_D, k)$, as a function of D. Notice that if $\chi_D(-M)\epsilon = -1$, then $L(F_D, k) = 0$. Therefore at least 'half' of the $L(F_D, k)$ are trivially zero. As we shall see, the 'nontrivial zeros' (as one varies D) are quite mysterious. If F is a weight 2 newform associated to an elliptic curve E, then there are infinitely many non-trivial zeros, but in the case of Ramanujan's Delta-function $F = \Delta(z) \in S_{12}(1)$ there are no known non-trivial zeros.

For $X > 0$ let

$$N_F(X) = \#\{D \mid \mid D \mid < X \text{ and } L(F_D, k) \neq 0\}.$$

It is widely believed that

$$N_F(X) \gg_F X. \tag{1}$$

The following conjecture, due to Goldfeld, is more precise.

Goldfeld's Conjecture [G]. *If $F(z) \in S_{2k}(M)$ is a newform, then*

$$\sum_{|D|<X \text{ and } (D,M)=1} \operatorname{ord}_{s=k} L(F_D, s) \sim \frac{1}{2}\#\{D \mid \mid D \mid < X \text{ and } (D, M) = 1\}. \tag{2}$$

(Note: This conjecture was posed for weight 2 newforms associated to modular elliptic curves.)

Goldfeld's Conjecture is an analytic assertion, and it has been extensively studied as such, often with the help of sophisticated analytic techniques. Thanks to the work of Katz, Sarnak, Iwaniec, Kowalski, Michel, R. Murty and K. Murty, and others, much is known about the general phenomenon of the nonvanishing of values of L-functions and their derivatives. The results described in this paper follow from an essentially *algebraic* approach, based on the fact that the central value is a *critical* value (in the sense of Deligne). More concretely, this means that there exist nonzero complex numbers $\Omega_F^{\pm}$ known as *periods* for F, such that the quotient $L(F_D, k)/\Omega_F^{\mathrm{sign}(D)}$ is an algebraic integer, loosely referred to as the *algebraic part* of $L(F_D, k)$. Non-vanishing of $L(F_D, k)$ is equivalent to non-vanishing of the algebraic part, and the latter may be studied by using algebraic techniques. The key to the results in this paper is the observation that, to show nonvanishing of the central value, it suffices to show that the algebraic part is nonzero modulo p, for some prime p.

2. Statement of Results

The first result we wish to describe gives a strong estimate for the number of quadratic twists of F whose L-functions do not vanish at $s = k$. This result was proven by two of the present authors (see [O-S1]) by using the theory of 2-adic Galois representations and a congruence modulo the prime 2.

Theorem 1. **[O-S1, Cor. 3].** *If $F(z) \in S_{2k}(M)$ is a newform, then*

$$N_F(X) \gg_F \frac{X}{\log X}.$$

Recall that if $E/\mathbb{Q}$ is an elliptic curve given by

$$E: \quad y^2 = x^3 + Ax + B,$$

then E_D, its D-quadratic twist, is the curve given by

$$E_D: \quad y^2 = x^3 + AD^2x + BD^3.$$

Let $L(E_D, s)$ be the Hasse-Weil L-function for E_D. For modular E, Kolyvagin [Ko] proved that if $L(E_D, 1) \neq 0$, then E_D has rank zero. Theorem 1 together with Kolyvagin's theorem implies:

Corollary 1. *If $E/\mathbb{Q}$ is a modular elliptic curve, then the number of $|D| \leq X$ for which E_D has rank zero is $\gg_E X/\log X$.*

While Theorem 1 is very strong in that it applies to a general F, it falls short of the 'positive proportion' estimates predicted by Goldfeld. However, we will now describe a series of results showing that it is possible to do better in a class of special cases, namely the class of forms F for which there exist special congruence relations between $L(F_D, k)$ and class numbers of quadratic fields. When such relations exist modulo 3, results of Davenport and Heilbronn [D-H], suitably modified by Horie and Nakagawa [N-H], may be employed to prove the estimate (1). This approach was first carried out by James [Ja] for several weight 2 newforms associated to modular elliptic curves.

Using the same ideas, Kohnen [K] proved the following theorem for eigenforms with respect to the full modular group $SL_2(\mathbb{Z})$.

Theorem 2 **[K, Cor. 1].** *Let $k \geq 6$ be even. If $\epsilon > 0$ and $X \gg_\epsilon 0$, then there exists a Hecke eigenform $F(z) \in S_{2k}(1)$ for which*

$$N_F(X) \geq \left(\frac{9-\epsilon}{16 g_k \pi^2}\right) X$$

where $g_k = \dim(S_{2k}(1))$.

Corollary 2. *Let $\Delta(z) \in S_{12}(1)$ be Ramanujan's Delta-function. If $\epsilon > 0$ and $X \gg_\epsilon 0$, then*

$$N_\Delta(X) \geq \left(\frac{9-\epsilon}{16\pi^2}\right) X.$$

This technique can also be exploited in the context of certain elliptic curves with rational torsion points. But before stating the general theorem, we need to introduce some notation. Let E be a modular elliptic curve over $\mathbb{Q}$. Assume that E has a rational point of odd prime order p (so $p = 3$, 5, or 7), that the level M of E is squarefree, and that E has good reduction at p. Let q be any prime with $(q, M) = 1$ and $q \equiv 1 \pmod 9$ if $p = 3$ and $q \equiv 1 \pmod p$ if $p = 5$ or 7. Let M_1 denote the product of primes $\ell | M$ such that E has nonsplit multiplicative reduction, and let $M_2 = qM/M_1$.

Theorem 3 **[V, Th. 0.3].** *Let the elliptic curve E be as above. Then there exists a period Ω^- for E such that we have*

$$\begin{aligned}(1-\chi_D(q)/q)\cdot\tau(\chi_D)\frac{L(E_D,1)}{(-2\pi i)\Omega^-}\\ \equiv \frac{1}{2}\prod_{\ell\mid M_1}(1-\chi_D(\ell)/\ell)\prod_{\ell\mid M_2}(1-\chi_D(\ell))\cdot L(\chi_D,0)^2 \pmod p,\end{aligned}$$

for any $D<0$ prime to Mq. Here $\tau(\chi_D)$ denotes the usual Gauss sum associated to χ_D.

Observe now that $L(\chi_D,0)$ is essentially the class number $h(D)$ of $\mathbb{Q}(\sqrt{D})$. Thus if D is such that $h(D)$ and the various Euler factors are nonzero modulo p, then we may conclude that the value $L(E_D,1)$ is nonzero. The aforementioned results of Davenport-Heilbronn and Horie-Nakagawa then yield the following result:

Corollary 3. *If E is as in Theorem 3 and $p=3$, then*

$$\#\{-X<D<0 \mid L(E_D,1)\neq 0)\} \gg_E X.$$

Theorem 3 was already predicted by the work of Frey [F], who shows that the order of p-Selmer groups of certain curves E_D are trivial whenever $h(D)$ is prime to p. Thus our theorem may be viewed as an analytic counterpart to Frey's theorem, hence as verification of a weak form of the Birch and Swinnerton-Dyer Conjecture 'mod p' for rank zero quadratic twists E_D.

Work of Waldspurger [W1, W2] shows that the values $L(F_D,k)$ are essentially Fourier coefficients of modular forms of half-integral weight $k+\frac{1}{2}$. Theorems 1 and 2 are proved by studying the Fourier coefficients of such half-integral weight modular forms modulo p. There have been a number of investigations into the indivisibility of such coefficients in the works of Jochnowitz [J], Ono and Skinner [O-S2], and most recently Bruinier [B]. Our final theorem is a statement of the main result in [B]. First let us introduce some notation. Let k be an integer as before, M a positive integer divisible by 4 and χ a Dirichlet character modulo M. For convenience put $\chi^*=(\frac{-1}{\cdot})^k\chi$. If p is a prime, then let v_p denote a continuation of the usual p-adic valuation on $\mathbb{Q}$ to a fixed algebraic closure. Write $M_{k+1/2}(M,\chi)$ for the space of modular forms of weight $k+1/2$ with respect to $\Gamma_0(M)$ and Nebentypus character χ (in the sense of [Sh]).

Theorem 4 **[B, Th. 2].** *Let $f(z)=\sum_{n=0}^{\infty}a(n)q^n\in M_{k+\frac{1}{2}}(M,\chi)$ be an eigenform of all the Hecke operators $T(\ell^2)$ with corresponding eigenvalues λ_ℓ. If the coefficients $a(n)$ are algebraic integers, $p\nmid M$ is prime, and there is a positive integer m^* with $v_p(a(m^*))=0$, then define $w(f;p,m^*)$ by*

$$w(f;p,m^*):=\min_{\ell} v_p\left(\lambda_\ell-\left(\frac{m^*}{\ell}\right)\chi^*(\ell)(\ell^k+\ell^{k-1})\right),$$

Here the minimum is taken over all primes ℓ with $(\ell, Npm^*) = 1$ and $\ell \not\equiv 1 \pmod{p}$. Then there exist infinitely many square-free integers d with $v_p(a(d)) \leq w(f; p, m^*)$.

There are many immediate consequences of results like Theorem 4. Here we list a few exceptional examples. The deduction of these corollaries employs the aforementioned results of Waldspurger, which relate the algebraic parts of the values $L(F_D, k)$ to the Fourier coefficients at square-free integers of certain half-integral weight modular forms.

Corollary 4. *If $E/\mathbb{Q}$ has complex multiplication, then for every prime $p \gg_E 0$ there are infinitely many D for which*

$$rk(E_D) = 0 \quad \text{and} \quad p \nmid \#\text{Ш}(E_D).$$

Corollary 5. *Let $E/\mathbb{Q}$ be a modular elliptic curve for which $L(E, s)$ has a simple zero at $s = 1$. For all primes p outside a finite set which is effectively determinable* (see [O-S2])

$$ord_p(|\text{Ш}(E)|) \leq ord_p(\text{Sha}(E)),$$

where Sha(E) *denotes the order of* Ш(E) *as predicted by the Birch and Swinnerton-Dyer Conjecture.*

We also obtain a generalization of results due to Horie on the existence of certain infinite families of imaginary quadratic fields [Ho].

Corollary 6 [B, Theorem 7]. *Let $p_1, \ldots, p_r$ be distinct odd primes and $\varepsilon_1, \ldots, \varepsilon_r \in \{-1, 0, +1\}$. Let p be a prime ≥ 5 such that p does not divide $p_j(p_j - 1)(p_j + 1)$ for $j = 1, \ldots, r$. Then there are infinitely many fundamental discriminants $D < 0$ for which $h(D)$ is not divisible by p and $(\frac{D}{p_j}) = \varepsilon_j$ for $j = 1, \ldots, r$.*

Applying Theorem 4 to the Cohen-Eisenstein series, we obtain indivisibility results for certain values of Dirichlet L-series.

Corollary 7 [B, Theorem 6]. *Let k be a positive even integer and p a prime for which $p - 1 \nmid 2k$ and $v_p(\zeta(1-k)) = 0$. Then there exist infinitely many fundamental discriminants $D > 0$ with $v_p(L(1-k, \chi_D)) = 0$.*

Using the work of Mazur and Wiles [M-W] one immediately obtains the following from Corollary 7.

Corollary 8 [B, Cor. 2]. *Let p be a prime ≥ 7. Then there exist infinitely many real quadratic fields F such that the K-group $K_2 A_F$ of the ring of integers A_F of F contains no element of order p.*

In the remainder of this paper, we will briefly describe the proofs of the theorems stated above.

3. Discussion of Theorem 1

The proof of Theorem 1 (see [O-S1] for details) is based on the simple observation that if $\theta(z) = 1 + 2\sum_{n=1}^{\infty} q^{n^2}$ is the standard theta function, then $1 \equiv \theta(Nz) \pmod 2$ for any positive integer N. This is exploited as follows. The main results of [W1, W2] ensure the existence of a weight $k + \frac{1}{2}$ cusp form $f(z) = \sum_{n=1}^{\infty} c(n)q^n$ of some level $4M'$ and $\delta_F \in \{\pm\}$ such that if $(D, 4MM') = 1$ and if $\delta_F D > 0$, then $c(|D|)^2$ is essentially the algebraic part of $L(F_D, k)$. More precisely, let $P(F)$ denote the set of D's just described. Then for all $D \in P(F)$

$$c(|D|) \neq 0 \Rightarrow L(F_D, k) \neq 0.$$

The form $G(z) = f(z)\cdot\theta(Nz) = \sum_{n=1}^{\infty} b(n)q^n$ has integral weight $k+1$ and satisfies $b(n) \equiv c(n) \pmod 2$. Thus to show that $L(F_D, k) \neq 0$ for some $D \in P(F)$, it sufficies to show that $b(|D|) \not\equiv 0 \pmod 2$. For simplicity, suppose that $c(n) \in \mathbf{Z}$ for all n and that $G(z)$ is a newform. Suppose also that $2 \nmid c(|D_0|)$ for some $D_0 \in P(F)$. Write $D_0 = p_1 \cdots p_r$. Applying the Chebotarev Density Theorem to the mod 2 Galois representation associated to G, one finds that there are $\gg X/\log X$ sets of primes $\{q_1, ..., q_r\}$ such that $b(q_i) \equiv b(p_i) \pmod 2$, $q_1 \cdots q_r < X$, $\delta_F q_1 \cdots q_r \in P(F)$. As

$$\begin{aligned} b(q_1 \cdots q_r) &\equiv b(q_1) \cdots b(q_r) \\ &\equiv b(p_1) \cdots b(p_r) \equiv b(|D_0|) \pmod 2, \end{aligned}$$

the desired result follows. The proof in the general case is similar, but made more complicated by having to work in a general number field and by the fact that G is not usually a newform.

4. Discussion of Theorem 2

In this section we briefly sketch the proof of Theorem 2 (see [K] for details). Essentially, one uses a sufficiently explicit form of the Shimura correspondence and Waldspurger's theorem, due to Kohnen-Zagier, to find explicit relations between the twisted L-values and clas numbers of quadratic fields. One concludes the proof by using the Davenport-Heilbronn theorem, as mentioned previously.

Sketch of Proof of Theorem 2: Let k be even and $S^+_{k+1/2}$ be the space of cusp forms of weight $k + 1/2$ w.r.t. $\Gamma_0(4)$ having a Fourier expansion of the form $\sum_{n\geq 1} c(n)q^n$ with $c(n) = 0$ unless $n \equiv 0, 1 \pmod 4$. The spaces $S^+_{k+1/2}$ and $S_{2k} = S_{2k}(1)$ are isomorphic as modules over the Hecke algebra. If $f(z) = \sum_{n\geq 1} a(n)q^n$ is a normalized Hecke eigenform in S_{2k} and $g = \sum_{n\geq 1} c(n)q^n$ is a Hecke eigenform in $S^+_{k+1/2}$ corresponding to it, then for every fundamental discriminant $D > 0$ one has

$$\frac{|\, c(D)\,|^2}{\langle g, g\rangle} = \frac{(k-1)!}{\pi^k} D^{k-1/2} \frac{L(f, D, k)}{\langle f, f\rangle},$$

where $\langle g,g\rangle$ and $\langle f,f\rangle$ are appropriately normalized Petersson scalar products (cf. [K-Z]; the above identity is a more precise version of Waldspurger's formula [W1] in the special case of the full modular group). For $k \geq 4$ let

$$G_k(z) = \frac{1}{2}\zeta(1-k) + \sum_{n\geq 1} \sigma_{k-1}(n)q^n$$

($\sigma_\nu(n) := \sum_{d|n} d^\nu$) be the Eisenstein series of weight k and level 1, and let $\theta(z) = \sum_{n\in\mathbb{Z}} q^{n^2}$ be the standard theta series of weight 1/2.

We put

$$\delta_k(z) := \frac{1}{4\pi i}((\frac{k}{2}-1)G_{k-2}(4z)\theta'(z) - G'_{k-2}(4z)\theta(z))$$

and write $\alpha_k(n)$ for the Fourier coefficients of δ_k (cf. [K-Z], p. 187). One has $\delta_k \in S^+_{k+1/2}$; in fact, up to normalization δ_k is the first Rankin-Cohen bracket of $G_{k-2}(4z)$ and $\theta(z)$.

For simplicity let's suppose that k is not congruent to 1 modulo 3. It is easy to see that it is sufficient to prove that

$$\#\{0 < D < X, D \equiv -1 \pmod 3, \alpha_k(D) \neq 0\} \geq \left(\frac{9-\epsilon}{16\pi^2}\right) X.$$

To do this one first shows the congruence

$$\alpha_k(D) \equiv u_k h(-3D) \pmod 3 \qquad (D \equiv -1 \pmod 3)$$

where $u_k = -1$ resp 1 for $k \equiv 0 \pmod 3$ resp. $k \equiv 2 \pmod 3$.

Except for some elementary calculations modulo 3, the above congruence follows from an identity of Siegel which relates a certain finite sum involving σ_1 to the second generalized Bernoulli number of χ_D, and a classical formula of Lerch relating this Bernoulli number to the class number $h(-3D)$ modulo 3.

The proof then is finished using the results of [D-H] and [N-H], in a similar way as in [Ja]. Let m and N be positive integers, N odd and such that if p is an odd prime dividing (m, N), then $p^2 \mid N$ and p^2 does not divide m. Denote by $N_2^-(X, m, N)$ ($X > 0$) the number of fundamental discriminants D with $-X < D < 0$ and $D \equiv m \pmod N$.

The main result of [D-H, N-H] implies that

$$\sum_{-x<D<0, D\equiv m \pmod N, (h(D),3)=1} 1 \geq (\frac{1}{2}-\epsilon)N_2^-(X,m,N) \quad (X \gg_\epsilon 0).$$

Theorem 2 follows easily after applying the latter formula with $m = 3$ and $N = 9$ and the fact that $N_2^-(3X, 3, 9) \sim \frac{9}{8\pi^2}X$ for $X \to \infty$ (cf. e.g. [N-H], Prop. 2).

Q.E.D.

5. Discussion of Theorem 3.

We now sketch the proof of Theorem 3 (see [V] for details). The idea of the proof is simple: we will show that there exists a congruence modulo p between E and a suitable Eisenstein series G, and then relate the L-values of E to those of G. The latter are products of Dirichlet L-functions, and the special values are Bernoulli numbers whose relationship to class numbers is well-documented.

Proof of Theorem 3: Let ρ_0 denote the representation of $\mathrm{Gal}(\overline{\mathbb{Q}}/\mathbb{Q})$ on the p-division points $E[p]$ of E. Our hypotheses imply that there is an exact sequence

$$0 \to \mathbb{Z}/p\mathbb{Z} \to E[p] \to \mu_p \to 0.$$

Let $f(z) = \sum a_n(f)q^n$ be the newform associated to E; then the exact sequence above implies that $a_q(f) \equiv q+1 \pmod p$, for each prime q not dividing pM. Now let G be the non-holomorphic Eisenstein series of weight 2 and level 1. Thus we have

$$G(z) - \frac{1}{8\pi y} = \frac{-1}{24} + \sum \sigma(n)q^n,$$

where $\sigma(n) = \sum_{d|n} d$. Therefore $\sigma(n) \equiv a_n(f) \pmod p$ for all n with $(n, pM) = 1$. Observe also that we have the equality

$$\tau(\chi_D)\frac{L(1, G_D)}{(-2\pi i)} = \frac{1}{2}L(\chi_D, 0)^2.$$

Now theorem (3.3) of [V] shows how to modify f and G to obtain forms $f^* = \sum a_n(f)^* q^n$ and $G^* = \sum a_n(g)^* q^n$ of level Mq such that the congruence $a_n(f)^* \equiv a_n(g)^* \pmod p$ is valid at *all* integers n. Furthermore the Eisenstein series G^* has the property that the constant term in the Fourier expansion at every cusp is a rational integer divisible by p.

Put $\Gamma = \Gamma_1(Mq)$ and let $\delta \in H^1(\Gamma, \mathbb{Z})$ be the cocycle obtained by integrating G^*. Then δ vanishes modulo p on parabolic elements and we obtain a parabolic cocycle $\overline{\delta} \in H^1(\Gamma, \mathbb{Z}/p)$. Furthermore, it may be shown that δ lies in the minus eigenspace $H^1(\Gamma, \mathbb{Z})^-$ for the action of complex conjugation.

Let $\mathbb{T}$ be the Hecke ring generated over $\mathbb{Z}_p$ in the space of cuspforms for $\Gamma = \Gamma_1(Mq)$. Let m be the the maximal ideal determined by f^*. It can be shown that there is an isomorphism $\theta : S_2(\Gamma, \mathbb{Z}_p)_m \cong H^1(\Gamma, \mathbb{Z}_p)_m^-$ (see Theorem 2.7 in [V]). Thus we may define a canonical cocycle $\delta^* = \theta(f^*) \in H^1(\Gamma, \mathbb{Z}_p)^-$. One checks that there is a period Ω^- such that $\Omega^- \delta^* = (\omega)^-$, where ω is the differential form on $X_1(Nq)$ associated to f^*. It is a consequence of a theorem of Washington that the image of δ^* in $H^1(\Gamma, \mathbb{Z}/p)$ coincides up to unit with the Eisenstein cocycle $\overline{\delta}$ defined previously. Our Theorem 3 now follows upon computing the twisted L-values of δ^* and $\overline{\delta}$, using the definition for the former, and the theory of Dedekind sums for the latter (see [St], Lemma 2.2).

Q.E.D.

Applying Theorem 3 along with the techniques developed in [Ja] to all elliptic curves having a torsion point of order 3 and whose conductor is less than or equal to 50, we have compiled the following table. For each curve E, we list a Weierstrass equation for E, the conductor N_E of E, and a lower bound δ_E for

$$\liminf_{x\to\infty}\left[\frac{\#\{D \mid |D| < X \text{ and } L(E_D,1)\neq 0\}}{2X}\right].$$

E	N_E	δ_E
$y^2 = x^3 + x^2 + 72x - 368$	14	$21/64\pi^2$
$y^2 = x^3 + 4x^2 - 144x - 944$	19	$19/80\pi^2$
$y^2 = x^3 + x^2 + 4x + 4$	20	$15/72\pi^2$
$y^2 = x^3 + x^2 - 72x - 496$	26	$39/112\pi^2$
$y^2 = x^3 - 432$	27	$3/8\pi^2$
$y^2 = x^3 + x^2 + 24x + 144$	30	$15/128\pi^2$
$y^2 = x^3 + x^2 - 48x + 64$	34	$17/48\pi^2$
$y^2 = x^3 + 4x^2 + 144x + 80$	35	$35/192\pi^2$
$y^2 = x^3 + 1$	36	$3/4\pi^2$
$y^2 = x^3 + 4x^2 - 368x - 3184$	37	$37/114\pi^2$
$y^2 = x^3 + x^2 + 152x + 5776$	38	$19/160\pi^2$
$y^2 = x^3 + x^2 + 3x - 1$	44	$11/48\pi^2$
$y^2 = x^3 + 5x^2 - 200x - 14000$	50	$5/8\pi^2$

5. Discussion of Theorem 4

Here we describe the main ideas of the proof of Theorem 4 (see [B] for details). A related result and some more applications to elliptic curves are obtained in [O-S2] via the theory of p-adic Galois representations.

Here f denotes an element of $M_{k+1/2}(M,\chi)$ with algebraic integer Fourier coefficients $a(n)$. Define $v_p(f) = \inf_n(v_p(a(n)))$ and denote the usual Fricke involution by W_M. Using the q-expansion principle one may deduce that $f \mid W_M$ also has algebraic Fourier coefficients and moreover that $v_p(f) = v_p(f \mid W_M)$ for every prime p not dividing M ([B], Lemma 1).

Now let ℓ be a prime not dividing M and suppose that f is an eigenform of the Hecke operator $T(\ell^2)$. Then taking into account the lemma above and the properties of various operators defined on modular forms (in particular the commutation relation of a quadratic twist and W_M), it can be shown that a certain set of congruences modulo p for the Fourier coefficients implies a congruence for the Hecke eigenvalue λ_ℓ. Since this result might be of independent interest, let us state part of it in a more precise form.

Theorem 4′ [B, Theorem 1]. *Let ℓ be a prime not dividing M, $\varepsilon \in \{\pm 1\}$, and $\nu > 0$. Further let p be a prime with $(p, N\ell(\ell-1)) = 1$ and $v_p(f) = 0$. Suppose that f is an eigenform of $T(\ell^2)$ with corresponding eigenvalue λ_ℓ. If $v_p(a(n)) \geq \nu$ for all n with $\left(\frac{n}{\ell}\right) = -\varepsilon$, then the congruence $v_p\left(\lambda_\ell - \varepsilon\chi^*(\ell)(\ell^k + \ell^{k-1})\right) \geq \nu$ holds.*

As a *corollary* to Theorem 4' one infers that there always is an integer m_ε with $\left(\frac{m_\varepsilon}{\ell}\right) = -\varepsilon$ and $v_p(a(m_\varepsilon)) \leq v_p\left(\lambda_\ell - \varepsilon\chi^*(\ell)(\ell^k + \ell^{k-1})\right)$.

Proof of Theorem 4: We may choose a prime ℓ with $(\ell, Mpm^*) = 1$, $\ell \not\equiv 1 \pmod p$ and

$$v_p\left(\lambda_\ell - \left(\frac{m^*}{\ell}\right)\chi^*(\ell)(\ell^k + \ell^{k-1})\right) = w(f; p, m^*).$$

Then according to the above corollary there is an m with $(\frac{m}{\ell}) = -(\frac{m^*}{\ell})$ and $v_p(a(m)) \leq w(f; p, m^*)$. In fact, by an inductive argument ([B], Lemma 3) it can be seen that there are infinitely many such m with mutually distinct square-free part. Now, using the multiplicative properties of the Fourier coefficients the assertion can be deduced.

Q.E.D.

References

[B] J. H. Bruinier, *Non-vanishing modulo p of Fourier coefficients of half-integral weight modular forms*, (preprint).

[D–H] H. Davenport and H. Heilbronn, *On the density of discriminants of cubic fields.II*, Proc. Roy. Soc. London **ser. A 322** (1971), 405-420.

[F–H] S. Friedberg and J. Hoffstein, *Nonvanishing theorems for automophic L-functions on GL(2)*, Ann. Math. **142** (1995), 385-423.

[F] G. Frey, *On the Selmer group of twists of elliptic curves with ℚ-rational torsion points*, Can. J. Math. **XL** (1988), 649-665.

[G] D. Goldfeld, *Conjectures on elliptic curves over quadratic fields*, Number Theory, Carbondale, Springer Lect. Notes **751** (1979), 108-118.

[Ho] K. Horie, *Trace formulae and imaginary quadratic fields*, Math. Ann. **288** (1990), 605-612.

[Ja] K. James, *L-series with non-zero central critical value*, J. Amer. Math. Soc. **11** (1998), 635-641.

[J] N. Jochnowitz, *Congruences between modular forms of half-integral weights and implications for class numbers and elliptic curves*, (preprint).

[K] W. Kohnen, *On the proportion of quadratic character twists of L-functions attached to cusp forms not vanishing at the central point*, J. reine angew. math. (to appear).

[K–Z] W. Kohnen and D. Zagier, *Values of L-series of modular forms at the center of the critical strip*, Invent. Math. **64** (1981), 173-198.

[M–W] B. Mazur and A. Wiles, *Class fields of abelian extensions of ℚ*, Invent. Math. **76** (1984), 179-330.

[N–H] J. Nakagawa and K. Horie, *Elliptic curves with no rational points*, Proc. AMS **104, no.1** (1988), 20-24.

[O–S1] K. Ono and C. Skinner, *Nonvanishing of quadratic twists of modular L-functions*, Invent. Math., (to appear).

[O–S2] K. Ono and C. Skinner, *Fourier coefficients of half-integral weight modular forms modulo ℓ*, Annals of Math. **147** (1998), 451-468.

[Sh] G. Shimura, *On modular forms of half integral weight*, Annals of Math. **97**, 440-481.

[St] G. Stevens, *The cuspidal group and special values of L-functions*, Trans. A.M.S. **291**, 519-550.

[V] V. Vatsal, *Canonical periods and congruence formulae*, Duke Math. J. (to appear).

[V2] V. Vatsal, *Rank-one twists of a cetain elliptic curve*, Math. Annalen (to appear).

[W1] J.-L. Waldspurger, *Sur les coefficients de Fourier des formes modulaires de poids demi-entier*, J. Math. Pures et Appl. **60** (1981), 375-484.

[W2] J.-L. Waldspurger, *Correspondances de Shimura et quaternions*, Forum Math. **3** (1991), 219-307.

Mathematisches Institut, Universität Heidelberg, INF 288, D-69120 Heidelberg, Germany

E-mail address: bruinier@mathi.uni-heidelberg.de

Department of Mathematics, The Pennsylvania State University, University Park, Pa. 16802
E-mail address: klj@math.psu.edu

Mathematisches Institut, Universität Heidelberg, INF 288, D-69120 Heidelberg, Germany
E-mail address: winfried@mathi.uni-heidelberg.de

Department of Mathematics, The Pennsylvania State University, University Park, Pa. 16802
E-mail address: ono@math.psu.edu

School of Mathematics, Institute for Advanced Study, Princeton, New Jersey 08540
E-mail address: cskinner@math.ias.edu

Department of Mathematics, University of Toronto, Toronto, Canada M5S 1A1
E-mail address: vatsal@math.toronto.edu

ON TSCHIRNHAUS TRANSFORMATIONS

JOE BUHLER AND ZINOVY REICHSTEIN

ABSTRACT. We revisit the classical problem of simplifying polynomials by means of Tschirnhaus transformations. We consider Tschirnhaus transformations involving (i) no auxiliary radicals, (ii) arbitrary radicals, (iii) odd degree radicals, and (iv) odd degree radicals and the square root of the discriminant. In [BR] we showed that by using substitutions of type (i) one cannot reduce the general polynomial of degree n to a form with less than $[n/2]$ independent coefficients. In this paper we give a new proof of this result and also extend it to transformations of types (iii) and (iv). In the last section we present alternative proofs, based on the cohomological approach shown to us by J.-P. Serre.

1. INTRODUCTION

By the end of the nineteenth century a number of techniques were known for simplifying polynomials by using Tschirnhaus transformations. "Reducing the number of parameters" was one of the central motivations of Hilbert's 13th problem, which asks whether the root of a general seventh degree polynomial can be expressed in terms of functions of fewer than three variables. Our goal here is to revisit the topic of parameter reduction via Tschirnhaus transformations from a modern point of view.

We fix the notation of this paragraph throughout. Let k be a base field, of characteristic zero. We assume that all fields mentioned in this paper are function fields over k and that all varieties and maps are defined over k. Let n be a positive integer and $a_1, \cdots, a_n$ be algebraically independent over k. The polynomial

$$f(X) = X^n + a_1 X^{n-1} + \cdots + a_n$$

is a "general" monic polynomial of degree n; it can be specialized to any polynomial of degree n with coefficients lying in a field containing k. Let $K = k(a_1, \cdots, a_n)$, and let

$$L = K[X]/f \tag{1}$$

be the root field of f over K, so that L/K is, in some sense, a general extension of degree n. We let $x = X \pmod f$ so that $L = K(x)$. Finally, if K' is any extension of K, we let $L' = K'[X]/f = K'(x)$.

The second author was partially supported by NSA grant MDA904-9610022.

S.D. Ahlgren et al. (eds.), Topics in Number Theory, 127–142.

Now we consider the problem of finding a Tschirnhaus transformation that simplifies f. A Tschirnhaus transformation is determined by an element y of L

$$y = h(x) = b_0 + b_1 x + \cdots + b_{n-1}x^{n-1}\,, \tag{2}$$

where the b_i are in K. Denote the minimal polynomial g of y over K by

$$g(Y) = Y^n + c_1 Y^{n-1} + \cdots c_n\,;$$

this polynomial can be explicitly computed as the resultant of $h(X) - Y$ and $f(x)$, with respect to X. Let $K_y := k(c_1, \cdots, c_n)$ be the field generated by the coefficients of g. Our goal is to minimize the "number of parameters" in g, which we interpret as minimizing the transcendence degree of K_y. This motivates us to define

$$d(n) = \min\{\operatorname{trdeg} K_y : K(y) = L\}\,.$$

We emphasize that many of our definitions depend on the ground field k, so that $d_k(n)$ would be a more proper notation that $d(n)$, but we choose to suppress this throughout for the sake of simplicity. We also remark that if g is obtained as above by a Tschirnhaus transformation from the general polynomial $f(X)$ then g is generic in the sense that any extension of fields E/F of degree n (of fields containing k) can be obtained as a root field of a specialization of g (see [BR]).

One way to minimize the transcendence degree is to make coefficients zero. Hermite [He] showed in 1861 that the general quintic could be reduced, in the above sense, to the form

$$X^5 + aX^3 + bX + c.$$

Since it is easy to make the last two coefficients equal by scaling X, we see that $d(5) \le 2$. By similar techniques, Joubert [J] and Richmond [Ri] showed that a general sextic can be reduced to the form

$$X^6 + aX^4 + bX^2 + cX + c$$

so $d(6) \le 3$. For an elegant treatment of the these results that uses arithmetic properties of cubic surfaces, see [C].

A theorem of Klein ([Kl]) asserts that it impossible to reduce the general quintic to one-parameter form without using auxiliary radicals, thus proving that $d(5) \ge 2$. The following result generalizes this inequality.

Theorem 1.1. ([BR])
(a) $d(n+1) \ge d(n)$.
(b) $d(n) \ge [n/2]$.

Part (b) says that at least $[n/2]$ of the coefficients of any Tschirnhaus transformation $g(Y)$ of a general polynomial $f(X)$ are algebraically independent over k. It is not clear how close $[n/2]$ is to the true value of $d(n)$; the best upper bound we know ([BR]) is $d(n) \le n-3$ for all $n \ge 5$. In particular, $d(5) = 2$ and $d(6) = 3$, so the above-mentioned Hermite form of the quintic and Joubert-Richmont form of the sextic do, indeed, have the smallest possible numbers of independent parameters. The first unknown value of $d(n)$ is $d(7)$ which could be equal to either 3 or 4.

Much of the 19th century work was motivated by the problem of finding roots of polynomials; consequently, the above reductions were not nearly as prominent as other normal forms, obtained by allowing auxiliary radicals. For instance, Bring and then Jerrard proved that the general quintic can be reduced to the form

$$X^5 + X + a\,.$$

One-parameter forms of the quintic were used by Klein and others to reduce the solution of the quintic to the evaluation of special functions. More recently Doyle and McMullen [DM] used this technique in their solution of the quintic by iterations.

To formalize the notion of auxiliary radicals, we consider Tschirnhaus transformations of the form

$$y' = b_0 + b_1 x + \cdots + b_{n-1} x^{n-1}\,,$$

where the b_i lie in a solvable extension K' of K. Then we let c_i' denote the coefficients of the minimal polynomial of y' over K', set $K_y' = k(c_1', \cdots, c_n')$. We are only interested in y that generate the full extension in the sense that $K'(y) = L' = K'(x)$. Thus we define

$$d^{solv}(n) = \min\{\operatorname{trdeg} K_y' \mid K'(y) = K'(x)\}$$

where K' ranges over all solvable extensions of K. Obviously, $d^{solv}(n) \leq d(n)$. Since equations of degree ≤ 4 can be solved in radicals, we shall assume from now on that $n \geq 5$.

It is easy to show that $d^{solv}(n) \leq n - 4$ (see, e.g., [H2]); however, we know no interesting lower bounds on $d^{solv}(n)$. In an attempt to better understand how close $d^{solv}(n)$ is to $d(n)$, we will consider two related functions that lie between them:

$$d^{odd}(n) = \min\{\operatorname{trdeg}_k K_y'' \mid K''(y) = K''(x)\}$$

where K'' ranges over all solvable extensions of K of odd degree, and

$$d^{odd}_{\Delta}(n) = \min\{\operatorname{trdeg}_k K_y''' \mid K'''(y) = K'''(x)\}$$

where $\Delta \in K$ is the discriminant of $f(x)$ and K''' ranges over all solvable extensions of $K(\sqrt{\Delta})$ of odd degree. The number $d^{odd}(n)$ tells us to what extent we can simplify $f(x)$ by performing a Tschirnhaus transformations $x \longrightarrow y$, where y is in (2) and each b_i is a radical expression in $a_1, \ldots, a_n$ (with coefficients in k) involving only odd degree radicals. The number $d^{odd}_{\Delta}(n)$ answers a similar question when the α_i involve odd degree radicals and the square root of the discriminant. It is clear from the definitions that

$$d^{solv}(n) \leq d^{odd}_{\Delta}(n) \leq d^{odd}(n) \leq d(n)\,.$$

The main new results of this paper are Theorem 6.1, which says that $d^{solv}(n)$, $d^{odd}(n)$, and $d^{odd}_{\Delta}(n)$ are non-decreasing functions of n, and Theorem 7.1, which says that $d^{odd}(n) \geq [n/2]$ and $d^{odd}_{\Delta}(n) \geq 2[n/4]$. Note that Theorem 7.1 implies, in particular, that the above-mentioned Bring-Jerrard form (or any other 1-parameter form) of the quintic cannot be attained by a substitution involving only odd degree radicals and $\sqrt{\Delta}$.

The rest of this paper is organized as follows. In an attempt to understand $d(n)$ in more general terms, we will introduce the notion of "essential dimension" in the second section. Then we consider the relationship of these ideas to Hilbert's 13th problem. In section 4 we give a new proof of the lower bound $d(n) \geq [n/2]$ that relies on Tsen-Lang theory. In section 5, 6, and 7 we study the functions $d^{solv}(n)$,

$d^{odd}(n)$, and $d^{odd}_{\Delta}(n)$. In the final section we outline Serre's proof of the inequality $d(n) \geq [n/2]$, and then show how his cohomological techniques can be used to give alternative proofs of the lower bounds on $d^{odd}(n)$ and $d^{odd}_{\Delta}(n)$.

Acknowledgements: We would like to thank J.-P. Serre and B. Kahn for helpful discussions.

2. Essential Dimension

Let E/F be a field extension of degree n such that F contains k. We shall say that E is defined over a subfield F_0 of F if there is an extension E_0/F_0 of degree n such that $E_0F = E$; thus, $E = E_0 \otimes_{F_0} F$. (As always, F_0 and all other fields are function fields over k.) In other words, E is defined over F_0 if there exists an element $y \in E$ whose minimal polynomial over F is of degree n and has all of its coefficients in F_0. The ***essential dimension*** of the extension E/F, denoted $\mathrm{ed}(E/F)$, is defined to be the minimum value of $\mathrm{trdeg}_k F_0$, where E is defined over F_0.

Lemma 2.1. *Let $K = k(a_1, \ldots, a_n)$, $f = X^n + a_1X^{n-1} + \cdots + a_n$, and $L = K[X]/f$ be as in (1). If K'/K is a solvable extension, denote the field $K'[X]/f$ by L'. Then*
(a) $d(n) = \mathrm{ed}_k(L/K)$.
(b) $d^{solv}(n) = \min\{\mathrm{ed}_k(L'/K') \mid K'/K \text{ solvable}, \sqrt{\Delta} \in K'\}$,
(c) $d^{odd}(n) = \min\{\mathrm{ed}_k(L'/K') \mid K'/K \text{ solvable of odd degree}\}$,
(d) $d^{odd}_{\Delta}(n) = \min\{\mathrm{ed}_k(L'/K') \mid K'/K(\sqrt{\Delta}) \text{ solvable of odd degree}\}$.

Proof. The lemma follows from the definitions and discussion in the previous section; note that in part (a) we can, indeed, let K' range only over those solvable extensions of K which contains $\sqrt{\Delta}$, since making K' bigger (e.g., by adjoining $\sqrt{\Delta}$) cannot increase $\mathrm{ed}_k(L'/K')$. □

Two facts about essential dimension of fields will be useful when we consider a geometric interpretation of this concept. The first says that the essential dimension is unchanged if a further algebraically independent variable is adjoined. The second says that essential dimension is unchanged when one passes to the normal closure; therefore it suffices to consider the essential dimension of normal extensions of fields.

Lemma 2.2. *Let E/F be a finite field extension and let t be an independent variable over F. Then* $\mathrm{ed}(E(t)/F(t)) = \mathrm{ed}(E/F)$.

Proof. The inequality $\mathrm{ed}(E(t)/F(t)) \leq \mathrm{ed}(E/F)$ follows directly from the definition, and the opposite inequality follows from [Ro, Lemma 1]. □

Lemma 2.3. (*Lemma 2.3 in* [BR]) *Let E/F be a field extension and let $E^{\#}$ be the normal closure of E over F. Then*

$$\mathrm{ed}(E/F) = \mathrm{ed}(E^{\#}/F).$$

If G is a finite group and X is an irreducible algebraic variety on which G acts faithfully, then a ***compression*** of this action is a faithful G-variety Y together with a dominant rational map $X \longrightarrow Y$ of faithful G-varieties. The ***essential*** *dimension* of X, denoted $\mathrm{ed}(X)$, is the minimal dimension of Y, as Y ranges over all compressions of X.

The essential dimension of a G-variety is determined by the action of G on its function field. More precisely

$$\mathrm{ed}(X) = \mathrm{ed}(k(X)/k(X)^G) ;$$

see [BR, Lemma 2.2]. Combining this with the previous lemma shows that if $G = S_n$ and $V = k^n$ is made into an S_n-variety by using the natural action of the symmetric group on n-tuples, then

$$d(n) = \mathrm{ed}(V) . \tag{3}$$

Next we prove the somewhat surprising fact that $\mathrm{ed}(V)$ is independent of V, as V ranges over all faithful linear representations of G. As a first step, we consider the following lemma from [Re1].

Lemma 2.4. *Let X be a faithful G-variety, V be a linear G-variety of dimension d, and V_0 be a d-dimensional vector space, which we shall view as a G-variety with trivial G-action. Then $X \times V$ and $X \times V_0$ are birationally isomorphic (as G-varieties).*

Proof. This is a variant of the "no-name" lemma. To derive it from the "usual" form (see [EM, Prop. 1.1]), denote $k(X \times V)$ by $E = k(X)(t_1, \dots , t_d)$. It is easy to see that the embeddings $E^G \hookrightarrow E$ and $k(X) \hookrightarrow E$ induce a G-equivariant isomorphism of fields $k(X) \otimes_{k(X)^G} E^G \simeq E$. By the usual no-name lemma $E^G = k(X)^G(s_1, \dots , s_d)$ for some $s_1, \dots , s_d \in E^d$. Thus we have G-equivariant isomorphisms

$$E \simeq k(X) \otimes_k (X)^G E^G \simeq k(X)(s_1, \dots , s_d) \simeq k(X \times V_0) ,$$

and the lemma follows. □

As an easy corollary, we see that $\mathrm{ed}(X \times V) = \mathrm{ed}(X)$, where X is a faithful G-variety and V is a faithful linear G-variety. Indeed, by the lemma, $\mathrm{ed}(X \times V) = \mathrm{ed}(X \times V_0)$ where V_0 is a vector space of the same dimension as V on which V acts trivially. The function field of $X \times V_0$ is obtained from $k(X)$ by adjoining independent variables, so

$$\mathrm{ed}(X \times V) = \mathrm{ed}(X \times V_0) = \mathrm{ed}(X) ,$$

by Lemma 2.2. If V and V' are faithful linear representations of G then this fact says that

$$\mathrm{ed}(V) = \mathrm{ed}(V \times V') = \mathrm{ed}(V').$$

Thus the essential dimension of a linear G-variety depends only on the finite group G, and not on the choice of V; we let $\mathrm{ed}(G)$ denote this dimension, and refer to this as the essential dimension of G.

In particular $d(n) = \mathrm{ed}(S_n)$; see (3). We remark that if H is a subgroup of G then $\mathrm{ed}(H) \leq \mathrm{ed}(G)$, from the definition. It follows, since S_n is a subgroup of S_{n+1}, that $d(n) \leq d(n+1)$, as claimed in Theorem 1.1(a).

We also remark that the concept of essential dimension can be extended to algebraic groups (see [Re1]), and that most of the basic results here extend to that context.

3. Hilbert's 13-th problem

Hilbert phrased his 13th problem in broad terms. He viewed a root x of $f(x) = x^n + a_1x^{n-1} + \ldots + a_n$ as an algebraic function of $a_1, \ldots, a_n$; the 13th problem specifically asks if this function can be expressed as compositions of continuous functions of 2 variables. Kolmogorov and Arnold answered this question by showing, roughly speaking, that every continuous function of any number of variables can be expressed in terms of continuous functions of one variables and the binary function "addition" (see [Lo] for this, as well as several further developments in this direction).

However, Hilbert also referred to expressions for x in terms of algebraic functions, and it seems that this was a central motivation for the problem. Some support for this can be seen from amount of space devoted to this version of the question in [H2], which was apparently Hilbert's last published paper. A fuller discussion of the history of algebraic versions of the problem can be found in [AS] and [Di]. Further algebraic interpretations and interesting results can be found in a recent paper of Abhyankar ([A]).

One algebraic version of the question is as follows (this is equivalent to those in [AS] and [Di]). Let $n = 7$, so that $L = K[X]/f$ is the root field of the general septic polynomial; see (1). Is L contained in the top field K_r of a tower

$$K = K_0 \subset K_1 \subset \cdots \subset K_r$$

where each K_{i+1}/K_i has essential dimension at most three? More generally, for arbitrary n, we let $s(n)$ be the minimal integer such that L is contained in a tower of extensions of essential dimension at most $s(n)$.

Hilbert proves in [H2] that $s(n) \leq n - 5$ for $n \geq 9$. More generally, $s(n)$ can be bounded by $n - t(n)$ where $t(n)$ is a very slowly growing function of n (see the discussion in in [Di]). It is very unclear whether this is even a rough approximation to the correct value of $s(n)$.

We do note that $s(n) \leq d^{solv}(n)$. Indeed, solvable extensions are of essential dimension one. By the definition of $d^{solv}(n)$ there is a sequence of solvable extensions followed by an extension of essential dimension $d^{solv}(n)$ that contains L. Thus $s(n) \leq d^{solv}(n)$.

4. A lower bound on $d(n)$.

Since S_n has a subgroup which is isomorphic to $(\mathbf{Z}/2\mathbf{Z})^{[n/2]}$, we have

$$d(n) = \mathrm{ed}(S_n) \geq \mathrm{ed}(\mathbf{Z}/2\mathbf{Z})^{[n/2]}.$$

Thus the lower bound on $d(n)$ given in Theorem 1.1(b) is a consequence of the following result.

Theorem 4.1. $\mathrm{ed}((\mathbf{Z}/2\mathbf{Z})^m) = m$.

The proof in [BR, Corollary 5.5] uses induction on m; the inductive step involves passing to the residue field for an appropriate valuation. Another argument, due to J.-P. Serre, can be found in Section 8. Here we give a short proof using Tsen-Lang theory.

Proof. The group $G_r = (\mathbf{Z}/2\mathbf{Z})^r$ has a faithful r-dimensional representation $V = k^r$ given by $\zeta x_i \longrightarrow (-1)^{\zeta_i} x_i$ where $\zeta = (\zeta_1, \dots, \zeta_r) \in G_r$. Thus $\mathrm{ed}(G_r) \leq m$. To prove the opposite inequality, we may assume without loss of generality that the base field k is algebraically closed; in fact, it is clear from the definition that the essential dimension can only decrease when the base field k is enlarged.

Let $E = k(V) = k(x_1, \cdots, x_r)$ be the function field of V and let $F = E^{G_r} = k(a_1, \dots, a_r)$, where $a_i = x_i^2$. Then $\mathrm{ed}(G_r) = \mathrm{ed}(V) = \mathrm{ed}(E/F)$. We want to show that the extension E/F is not defined over any field F_0 with $\mathrm{trdeg}_k(F_0) < r$. In order to do so, we consider the trace form $q_{E/F}(x) = \mathrm{Tr}_{E/F}(x^2)$. This is a quadratic form $E \longrightarrow F$, where E is viewed as a 2^r-dimensional vector space over F. We claim that this form is anisotropic (i.e., has no zeroes) over F. Indeed, set $x^I = x_1^{i_1} \dots x_r^{i_r}$. The elements x^I form an F-basis of E, where $i_1, \dots, i_r \in \{0, 1\}$. Writing $x = \sum t_I x^I$ we see that

$$q_{E/F}(x) = 2^r \sum_{I \in \{0,1\}^r} a^I t_I^2 ,$$

where $a^I = a_1^{i_1} \dots a_r^{i_r}$. In other words, $q_{E/F}$ is a constant multiple of the r-fold Pfister form $<< a_1, \dots, a_r >>$, which is known to be anisotropic (see, e.g., [P, Ch. 8, Ex. 1.2(6)]). This proves our claim.

We are now ready to finish the proof of the theorem. Suppose the extension E/F is defined over a field F_0 with $\mathrm{trdeg}_k(F_0) \leq r - 1$. Since we are assuming that k is algebraically closed, and $q_{E/F}$ is a quadratic form in 2^r variables, the Tsen-Lang theorem asserts that $q_{E/F}$ has a nontrivial zero, i.e., is isotropic; see [P, Ch. 5, Cor. 1.5]. This contradiction shows that $\mathrm{ed}(E/F) = r$, thus completing the proof of Theorem 4.1. □

5. Relative essential dimension

In order to study $d^{solv}(n)$, $d^{odd}(n)$, and $d^{odd}_{\Delta}(n)$, we need to relativize the notion of essential dimension.

Let G and H be finite groups, let m be an integer, and let X be a faithful G-variety. We shall say that a G-equivariant map $\pi : Y \longrightarrow X$ is an (m, H)-cover if π corresponds to a finite field extension $k(Y)/k(X)$ whose degree divides m and whose galois group $\mathrm{Gal}(k(Y), k(X))$ is isomorphic to a subgroup of H. Note that we do not require $k(Y)$ to be Galois over $k(X)$; here, as usual, $\mathrm{Gal}(k(Y), k(X))$ denotes $\mathrm{Aut}_{k(X)}(k(Y)^{\#})$, where $k(Y)^{\#}$ is the normal closure of $k(Y)$ over $k(X)$.

Definition 5.1. Let X be a faithful G-variety. The relative essential dimension $\mathrm{ed}^{m,H}(X)$ is the minimal value of $\mathrm{ed}(Y)$, as

$$\pi : Y \longrightarrow X$$

ranges over all (m, H)-covers of X.

Observe that the identity map $id_X : X \longrightarrow X$ is an (m, H)-cover for any m and H. Thus the minimum in the above definition is taken over a non-empty set and $\mathrm{ed}^{m,H}(X) \leq \mathrm{ed}(X)$ for any m and H. On the other hand, for some m and H there are no (m, H)-covers other than id_X, and, consequently, the above inequality becomes an equality by default. For example, this happens when m and $|H|$ are relatively prime.

We record an algebraic form of Definition 5.1 for future reference.

Lemma 5.2. *Let X be a G-variety with function field $E = k(X)$ and let $F = k(X)^G$. Then $\mathrm{ed}_H(X) = \min\{\mathrm{ed}(E \otimes_F F'/F')\}$, where F' ranges over all extensions F'/F such that (i) $[F' : F']$ divides m, (ii) $E \otimes_F F'$ is a field and (iii) $\mathrm{Gal}(E'/E)$ is isomorphic to a subgroup of H.*

Proof. Recall that all fields considered in this paper are function fields over k. Thus the extensions F'/F satisfying (i) and (ii) are in 1-1 correspondence with the (m, H)-covers $Y \longrightarrow X$ via $F' = k(Y)^G$ and Y is a variety with function field $E \otimes_F F'$. □

We now prove the relative form of Lemma 2.2.

Lemma 5.3. *Let X be a faithful G-variety. Then*

$$\mathrm{ed}^{m,H}(X \times \mathbf{A}^1) = \mathrm{ed}^{m,H}(X)$$

for any m and H. Here $\mathbf{A}^1$ is an affine line with trivial G-action.

Proof. Suppose $\pi : Y \longrightarrow X$ is an (m, H)-cover. Then

$$\pi \times id : Y \times \mathbf{A}^1 \longrightarrow X \times \mathbf{A}^1$$

is also an (m, H)-cover. Since $\mathrm{ed}(Y \times \mathbf{A}^1) = \mathrm{ed}(Y)$, we conclude that $\mathrm{ed}^{m,H} H(X \times \mathbf{A}^1) \leq \mathrm{ed}^{m,H}(X)$.

The opposite inequality is a consequence of the Hilbert Irreducibility Theorem. Indeed, suppose $\mathrm{ed}^{m,H}(X \times \mathbf{A}^1) = d$. Let $\pi : Y \longrightarrow X \times \mathbf{A}^1$ be an H-cover such that there exists a compression $\alpha : Y \longrightarrow Z$ with $\dim(Z) = d$. After replacing Y and Z by appropriate G-invariant open subsets, we may assume without loss of generality that α is a regular map and that G acts freely on Z (and, hence, on Y).

Given an element $f \in k(X)^G$, let X_f be the graph of the rational map $f : X \longrightarrow \mathbf{A}^1$. Note that X_f is a G-invariant subvariety of $X \times \mathbf{A}^1$; moreover, X and X_f are birationally isomorphic as G-varieties. Let $Y_f = \pi^{-1}(X_f)$. Then by the Hilbert Irreducibility Theorem we can choose f so that Y_f is irreducible and

$$\pi : Y_f \longrightarrow X_f$$

is an (m, H)-cover; see [FJ, Cor. 11.7]. Since $\alpha_{|Y_f}$ compresses Y_f to a subvariety of Z, we have $\mathrm{ed}(Y_f) \leq d$ and hence, $\mathrm{ed}^{m,H}(X) = \mathrm{ed}^{m,H}(X_f) \leq \mathrm{ed}(Y) \leq d$, as claimed. □

We can now define the relative essential dimension $\mathrm{ed}^{m,H}(G)$ by mimicking the definition of $\mathrm{ed}(G)$ given in section 2. That is, we define $\mathrm{ed}^{m,H}(G)$ as $\mathrm{ed}^{m,H}(V)$, where V is a faithful representation of G. To make sure this number is well defined, we need to show that

$$\mathrm{ed}^{m,H}(V_1) = \mathrm{ed}^{m,H}(V_2)$$

for any two faithful representations V_1 and V_2 of G. Indeed, combining the above lemma with Lemma 2.4, we see that $\mathrm{ed}^{m,H}(X \times V) = \mathrm{ed}^{m,H}(X)$, where X is a faithful G-variety and V is a G-representation. Substituting $X = V_1$ and $V = V_2$, we obtain $\mathrm{ed}^{m,H}(V_1 \times V_2) = \mathrm{ed}^{m,H}(V_1)$. Similarly, $\mathrm{ed}^{m,H}(V_1 \times V_2) = \mathrm{ed}^{m,H}(V_2)$, which proves the desired equality.

Corollary 5.4. *If G_1 is a subgroup of G then $\mathrm{ed}^{m,H}(G_1) \leq \mathrm{ed}^{m,H}(G)$ for any group H and any integer m.*

Proof. Immediate from the definition. □

6. The d-functions are non-decreasing

In this section we prove the following theorem.

Theorem 6.1. *(a) $d^{solv}(n+1) \geq d^{solv}(n)$.*
(b) $d^{odd}(n+1) \geq d^{odd}(n)$.
(c) $d^{odd}_{\Delta}(n+1) \geq d^{odd}_{\Delta}(n)$.

The idea of the proof is to interpret the functions $d^{solv}(n)$, $d^{odd}(n)$, and $d^{odd}_{\Delta}(n)$ in terms of essential dimension.

Lemma 6.2. *(a) $d^{solv}(n) = \min\{\mathrm{ed}^{m,H}(A_n) \mid H \text{ solvable}, m \geq 1\}$.*
(b) $d^{odd}(n) = \min\{\mathrm{ed}^{m,H}(S_n) \mid H \text{ solvable}, m \text{ odd}\}$.
(c) $d^{odd}_{\Delta}(n) = \min\{\mathrm{ed}^{m,H}(A_n) \mid H \text{ solvable}, m \text{ odd}\}$.

Proof. Let $V = k^n$, viewed as an S_n-variety with the natural action. Denote the function field of V by $E = k(V) = k(x_1, \ldots, x_n)$ and let $K = E^{S_n} = k(a_1, \ldots, a_n)$, where $f(x) = x^n + a_1x^{n-1} + \ldots + a_n$ is the minimal polynomial of x_1. Note that this is the same polynomial f and the same field K we considered in Section 1 and that the field $L = K[x]/f$ can now be identified with $K(x_1)$. Also note that $E^{A_n} = K(\sqrt{\Delta})$, where Δ is the discriminant of f.

Part (a) now follows from the characterization of $d^{solv}(n)$ in terms of essential dimension given by Lemma 2.1(b). Indeed, recall that by definition $\mathrm{ed}^{m,H}(A_n) = \mathrm{ed}^{m,H}(V)$, where V is viewed as a linear A_n-variety. Moreover, by Lemma 5.2 $\mathrm{ed}^{m,H}(V)$ is the minimum of $\mathrm{ed}(E'/K')$, where K' ranges over all extensions of $K(\sqrt{\Delta})$ such that (i) $[K' : K(\sqrt{\Delta})]$ divides m, (ii) $E' = E \otimes_{K(\sqrt{\Delta})} K'$ is a field, and (iii) $\mathrm{Gal}(K', K(\sqrt{\Delta}))$ is isomorphic to a subgroup of H. Note that if H is a solvable group then (ii) is automatic, since f remains irreducible over K'. (Recall that we are assuming throughout that $n \geq 5$.) Moreover, E' equals the splitting field of $L' = K'[x]/f$ over K' and thus, by Lemma 2.3, $\mathrm{ed}(E'/K') = \mathrm{ed}(L'/K')$. Letting m range over all positive integers and H over all solvable groups, we see that the right hand side of (a) is equal to $\min\{\mathrm{ed}(L'/K')\}$, as K' ranges over all solvable extensions of $K(\sqrt{\Delta})$. Part (a) now follows from Lemma 2.1(b). Parts (b) and (c) are proved in the same way. □

Proof of Theorem 6.1 Since $A_n \subset A_{n+1}$, Corollary 5.4 implies

$$\mathrm{ed}^{m,H}(A_n) \le \mathrm{ed}^{m,H}(A_{n+1})$$

for any group H. By the previous Lemma this translates into

$$d^{solv}(n+1) \ge d^{solv}(n)\,.$$

The other two parts of the Theorem follow similarly. □

7. Lower bounds on d^{odd} and d^{odd}_{Δ}

Our goal is to prove the following lower bounds on $d^{odd}(n)$ and $d^{odd}_{\Delta}(n)$.

Theorem 7.1. *(a)* $d^{odd}(n) \ge [n/2]$.

(b) $d^{odd}_{\Delta}(n) \ge 2[n/4]$.

We begin by generalizing Theorem 4.1 to the relative situation.

Theorem 7.2. *Let* $G_r = (\mathbf{Z}/2\mathbf{Z})^r$ *and let* m *be an odd integer. Then* $\mathrm{ed}^{m,H}(G_r) = r$ *for any group* H.

Proof. We mimic the proof of Theorem 4.1. Let V be the faithful r-dimensional representation of G_r given in that proof,

$$E = k(V) = k(x_1, \dots, x_r),$$

and

$$F = k(V)^{G_r} = k(a_1, \dots, a_r),$$

where $a_i = x_i^r$. Since G_r has a faithful
r-dimensional representation, $\mathrm{ed}^{(m,H)}(G_r) \le r$. It therefore remains to prove the opposite inequality. At this point we may assume without loss of generality that the base field k is algebraically closed; indeed, the essential dimension cannot decrease if the base field is enlarged. By Definition 5.1, it is sufficient to prove that $\mathrm{ed}^{m,H}(Y) \ge r$ for any (m,H)-cover $\pi : Y \longrightarrow V$. Denote $k(Y)$ by E' and $k(Y)^G$ by F'. Note that $m' = [E' : E]$ divides m and is, therefore, odd. We now consider the following diagram.

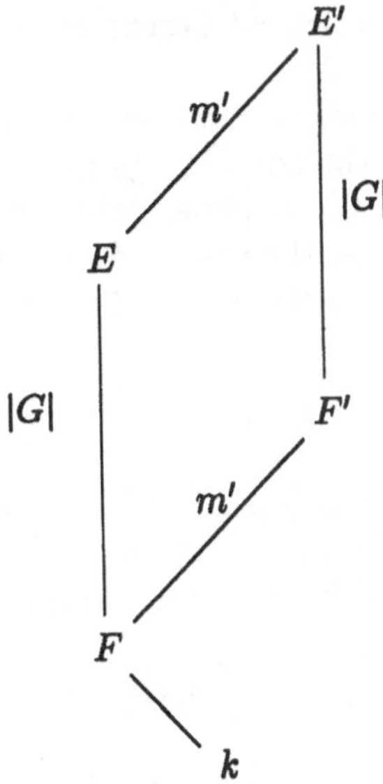

In section 4 we showed that $q_{E/F}(x) = \mathrm{Tr}_{E/F}(x^2)$ is anisotropic. Thus by a theorem of Springer $q_{E'/F'}(x) = \mathrm{Tr}_{E'/F'}(x^2)$ is also anisotropic (see, e.g., [P, Thm 1.12]). The rest of the proof of Theorem 4.1 goes through unchanged. That is, assume to the contrary, that the extension E'/F' is defined over a field F_0 with $\mathrm{trdeg}_k(F_0) \leq r-1$. Then $q_{E'/F'}$ is also defined over F_0. By the Tsen-Lang Theorem F_0 is a C_{r-1}-field. Thus every quadratic form in 2^r variables defined over F_0 is isotropic, a contradiction. □

Proof of Theorem 7.1 (a) Recall that S_n contains a copy of $G_r = (\mathbf{Z}/2\mathbf{Z})^r$ for $r = [n/2]$. Thus for any group H of odd order

$$\mathrm{ed}^{m,H}(S_n) \geq \mathrm{ed}^{m,H}(G_{[n/2]}) = [n/2] \, ;$$

see Corollary 5.4. Combining this with Lemma 6.2(b), we obtain the desired inequality $d^{odd}(n) \geq [n/2]$.

(b) Note that A_n contains $G_r = (\mathbf{Z}/2\mathbf{Z})^r$ for $r = 2[n/4]$. Indeed, A_n contains $(A_4)^{[n/4]}$ and each copy of A_4 contains $(\mathbf{Z}/2\mathbf{Z}) \times (\mathbf{Z}/2\mathbf{Z})$. Thus for any group H of odd order

$$\mathrm{ed}^{m,H}(A_n) \geq \mathrm{ed}^{m,H} G_{2[n/4]} = 2[n/4] \, .$$

Combining this with Lemma 6.2(c), we obtain $d^{odd}(n) \geq 2[n/4]$. □

8. A COHOMOLOGICAL APPROACH

Throughout this section k will be assumed to be algebraically closed. Suppose F/k is a field extension of transcendence degree $\leq d$. The proofs above of Theorems 1.1(b) and 7.1 were based on the fact that such a field F is a C_d-field; see e.g., [P, Sect. 5]. An alternative proof of Theorem 1.1(b), communicated to us by J.-P. Serre, is based on the fact that the cohomological dimension of F is at most

d; see [S, II.4]. In this section we present Serre's argument and use it to give an alternative proof of Theorem 7.1.

If F is a field, then we let $\overline{F}$ be its algebraic closure, and let $G_F = \mathrm{Gal}(\overline{F}, F)$ be its Galois group. We will denote the cohomology group $H^i(G_F, \mathbf{Z}/2\mathbf{Z})$ by $H^i(F)$; here $\mathbf{Z}/2\mathbf{Z}$ is viewed as a trivial G_F-module. The sum $H^*(F) = \oplus_i H^i(F)$ is a ring with the multiplication given by the cup product. Since our coefficient module is $\mathbf{Z}/2\mathbf{Z}$, this ring is, in fact, commutative. A field extension F'/F induces the restriction homomorphism

$$\mathrm{Res}_{F'/F} : H^i(F) \longrightarrow H^i(F')$$

for every i. By Kummer theory $H^1(F) \simeq F^*/F^{*2}$; we will denote the class corresponding to $x \pmod{x^2}$ by (x). If $q(x_1, \ldots, x_n) = a_1x_1^2 + \ldots + a_nx_n^2$ is a quadratic form over F then the r-th Stiefel-Whitney class $w_r(q) \in H^r(F)$ is given by

$$w_r(q) = \sum_{1 \le i_1 < \ldots < i_r \le n} (a_{i_1})(a_{i_2}) \ldots (a_{i_r}) . \tag{4}$$

By a theorem of Delzant $w_r(q)$ depends only on the equivalence class of the quadratic form $q = < a_1, \ldots, a_r >$ and not on the specific presentation of q as a sum of squares; see [De].

We shall be particularly interested in the quadratic forms which arise as trace forms of etale algebras. Recall that E is an etale algebra over F if E is a direct sum of field extensions of F of finite degree. We shall denote the trace form of such an algebra by $q_{E/F}$. That is, $q_{E/F}(x) = \mathrm{Tr}_{E/F}(x^2)$ for any $x \in E$.

Lemma 8.1. *Let k be an algebraically closed field, F be a field containing k, and E/F be a finite field extension with* $\mathrm{ed}(E/F) = d$. *Then $w_r(q_{E/F}) = 0$ for every $r > d$.*

Proof. Since $\mathrm{ed}(E/F) = d$, there exists a subfield F_0 of F and a degree n extension E_0/F_0 such that $\mathrm{trdeg}_k(F_0) = d$ and $E \simeq E_0 \otimes_{F_0} F$. Then $w_r(q_{E/F}) = \mathrm{Res}_{E/F}(w_r(q_{E_0/F_0}))$. Since the cohomological dimension of F_0 is $\le d$, we have $H^r(F_0) = (0)$ for all $r > d$. In particular, $w_r(q_{E_0/F_0}) = 0$, and the lemma follows. □

Lemma 8.2. *Let $F_m = k(b_1, \ldots, b_m)$, where $b_1, \ldots, b_m$ are independent variables over k. Then $(b_1) \ldots (b_m) \neq 0$ in $H^m(F_m)$.*

Proof. We proceed by induction on m. Since b_1 is not a square in F_1, the lemma is true for $m = 1$. To prove the induction step, let F be a field and let t be an independent variable over F. (We will later set $F = F_{m-1}$ and $t = b_m$.) Let

$$G = G_{F(t)} = \mathrm{Gal}(\overline{F(t)}, F(t)),\ N = \mathrm{Gal}(\overline{F(t)}, \overline{F}(t)),$$

and thus $G/N = \mathrm{Gal}(\overline{F}(t), F(t)) \simeq \mathrm{Gal}(\overline{F}, F) = G_F$. Consider the Hochschild-Serre sequence

$$0 \longrightarrow H^m(F) \longrightarrow H^m(F(t)) \xrightarrow{\alpha} H^{m-1}(G_F, H^1(\overline{F}(t))) \longrightarrow 0 ;$$

see [S, (1.4), p. 111]. The valuation $\overline{F}(t) \longrightarrow \mathbf{Z}$ given by the degree in t descends to

$$H^1(\overline{F}(t)) \simeq \overline{F}(t)^*/(\overline{F}(t)^*)^2 \longrightarrow \mathbf{Z}/2\mathbf{Z}\,,$$

which in turn gives rise to

$$\beta : H^{m-1}(G/N, H^1(\overline{F}(t)) \longrightarrow H^{m-1}(G/N, \mathbf{Z}/2\mathbf{Z}) = H^{m-1}(F)\,.$$

These maps are described in some detail in [S, II, Appendix 1 and II.4.2]. Tracing through the definitions, we see that $\gamma(a.(t)) = a$; where

$$\gamma = \beta \circ \alpha : H^m(F(t)) \longrightarrow H^{m-1}(F)$$

and $a.(t)$ denotes the cup product of a and (t). In particular, $a.(t)$ is nonzero if a is nonzero. Setting $F = F_{m-1}$, $t = b_m$, and $a = (b_1)\dots(b_{m-1})$, yields the desired induction step. □

Now let E be an etale algebra of dimension n over F and let Σ be the set of F-homomorphisms $E \longrightarrow \overline{F}$. Then Σ is a set of n elements. The group $G_F = \mathrm{Gal}(\overline{F}, F)$ acts on Σ; this action is given by a (continuous) homomorphism $\phi : G_F \longrightarrow S_n$. This map is uniquely determined by E up to an inner automorphisms of S_n. We shall refer to the image of ϕ as the Galois group of E over F and denote it by $\mathrm{Gal}(E, F)$. Note that if E is a finite field extension of F then this definition coincides with the usual one, i.e., $\mathrm{Gal}(E, F) = \mathrm{Aut}_F(E^\#)$, where $E^\#$ is the normal closure of E over F.

Proposition 8.3. *Let L/K be the general field extension, as in (1). Let F be a field extension of k, E be an n-dimensional an etale algebra over F, and $t = (t_1, \dots, t_n)$ be an n-tuple of algebraically independent variables over F. Then*

(a) there is an inclusion of fields $K \hookrightarrow F(t)$ such that $E(t) \simeq L \otimes_K F(t)$. Moreover,

(b) if $\mathrm{Gal}(E, F) \subset A_n$ then there exists an inclusion of fields $K(\sqrt{\Delta}) \hookrightarrow F(t)$ such that $E(t) \simeq L(\sqrt{\Delta}) \otimes_{K(\sqrt{\Delta})} F(t)$.

Proof. (a) It is sufficient to find an element y of $E(t) = E \otimes_F F(t)$ such that (i) $E = F[y]$ and (ii) the minimal polynomial $g(t) = t^n + b_1 t^{n-1} + \dots + b_n$ of y over $F(t)$ has algebraically independent coefficients. Indeed, if such an element can be found then the desired inclusion $K \hookrightarrow F$ is given by $a_i \longrightarrow b_i$. If $e_1, \dots, e_n$ is a basis for E as an F-vector space then $y = t_1 e_1 + \dots + t_n e_n$ satisfies (i) and (ii); this is proved in [Re2].

(b) Suppose $y \in F$ satisfies conditions (i) and (ii) above. Denote the discriminant of g by Δ_y. Then, since $\mathrm{Gal}(E, F) \subset A_n$, $\Delta_y = z^2$ for some $z \in F$. Now the desired embedding $K(\sqrt{\Delta}) \hookrightarrow F$ is given by $a_i \longrightarrow b_i$ and $\sqrt{\Delta} \longrightarrow z$. □

Theorem 8.4. *(a) Let $K = k(a_1, \dots, a_n)$, $f(X)$, and $L = K[x]/f$ be as in (1). Then*

(a) $w_{[n/2]}(q_{L/K}) \neq 0$.

(b) $w_{2[n/4]}(q_{L(\sqrt{\Delta})/K(\sqrt{\Delta})}) \neq 0$.

Proof. (a) Assume, to the contrary, that $w_{[n/2]}(q_{L/K}) = 0$. Let E be an n-dimensional etale algebra over F and let $t = (t_1, \ldots, t_n)$ be an n-tuple of independent variables over F. Then by Proposition 8.3(a), $w_{[n/2]}(q_{E(t)/F(t)}) = 0$, since it is the image of $w_{[n/2]}(q_{L/K}) = 0$ under the restriction map $\mathrm{Res}_{F(t)/K}$. On the other hand, since $\mathrm{Res}_{F(t)/F}$ is injective ([S, I.4.3]), $w_{[n/2]}(q_{E(t)/F(t)}) = 0$ implies $w_{[n/2]}(q_{E/F}) = 0$. Thus, in order to obtain a contradiction, it is sufficient to construct a field F containing k and an n-dimensional etale F-algebra E such that $w_{[n/2]}(q_{E/F}) \neq 0$.

First suppose $n = 2m$ is even. Let $b_1, \ldots, b_m$ be independent variables over F, $F = k(b_1, \ldots, b_m)$, and $E = E_1 \oplus \ldots \oplus E_m$, where $E_i = F(\sqrt{b_i})$. Then the elements $1_{E_1}, \sqrt{b_1}, 1_{E_2}, \sqrt{b_2}, \ldots, 1_{E_m}, \sqrt{b_m}$ form a basis for E over F. Writing out $q_{E/F}$ in the basis, we see that $q_{E/F} = < 2, 2b_1, \ldots, 2, 2b_m >$ and, hence, $w_{2m}(q_{E/F}) = (b_1)(b_2)\ldots(b_m)$. By Lemma 8.2 this class is non-zero in $H^m(F)$, as claimed.

If $n = 2m+1$ is odd, we repeat the same argument with

$$E_1 = F \oplus E.$$

Then $q_{E_1/F} = < 1 > \oplus q_{E/F}$ and $w_{2m}(q_{E_1/F}) = (b_1)(b_2)\cdots(b_m)$, which is again non-zero by Lemma 8.2.

(b) Arguing as in part (a) (but using Proposition 8.3(b) in place of 8.3(a)), we see that it is sufficient to construct an n-dimensional etale algebra E over a field F with $w_{2[n/4]}(q_{E/F}) \neq 0$ and $\mathrm{Gal}(E, F) \subset A_n$.

First consider the case $n = 4m$. Let $F = k(b_1, \ldots, b_{2m})$, as before, and let $E = E_1 \oplus \ldots \oplus E_m$, with $E_i = F(\sqrt{b_{2i-1}}, \sqrt{b_{2i}})$. Note that each E_i is Galois over F and $\mathrm{Gal}(E_j, F) \simeq \mathbf{Z}/2\mathbf{Z} \times \mathbf{Z}/2\mathbf{Z} \subset S_4$. Moreover, since E_i is a field, this subgroup is transitive. The group S_4 has only one such subgroup, namely, the normal subgroup of order 4; in particular, this group is contained in A_4. Consequently,

$$\mathrm{Gal}(E, F) \subset A_4 \times \ldots \times A_4 \ \ (m \ \text{times}) \subset A_n \ .$$

To show that $w_{2m}(q_{E/F}) \neq 0$, we write $q_{E/F}$ in the basis $z_1, \ldots, z_{4m}$, where

$$z_{4i-3} = 1_{E_i}, z_{4i-2} = \sqrt{b_{2i-1}}, z_{4i-1} = \sqrt{b_{2i-2}}, z_{4i} = \sqrt{b_{2i-1}b_{2i-2}}$$

for every $i = 1, \ldots, m$. In this basis

$$q = \oplus_{i=1}^{m} < 4, 4b_{2i-1}, 4b_{2i}, 4b_{2i-1}b_{2i} > \ .$$

Now formula (4) yields $w_{2m}(q_{E/F}) = (b_1)(b_2)\ldots(b_{2m})$. By Lemma 8.2 this class is non-zero in $H^{2m}(F)$, as claimed.

This proves part (b) in the case $n = 4m$. If $n = 4m + s$, where $s = 0, 1, 2, 3$ then we replace E by $F^s \oplus E$; the rest of the argument goes through unchanged. □

We are now ready to give alternative proofs of Theorems 1.1(b) and 7.1. Note that in both cases we may assume without loss of generality that the base field k is algebraically closed.

Cohomological proof of Theorem 1.1(b) (a) Assume the contrary: $d(n) < [n/2]$. That is, $\mathrm{ed}(L/K) < [n/2]$; see Lemma 2.1(a). By Lemma 8.1 this implies $w_{[n/2]}(q_{L/K}) = 0$ and thus contradicts Theorem 8.4(a). □

Cohomological proof of Theorem 7.1 (a) Assume to the contrary that $d^{odd}(n) < [n/2]$. Then by Lemma 2.1(c) there exists a solvable odd degree extension K'/K such that $\text{ed}(L'/K') < [n/2]$. Hence, $w_{[n/2]}(q_{L'/K'}) = 0$ in $H^{[n/2]}(K')$ by Lemma 8.1. On the other hand, since K'/K is an odd degree extension, the restriction map $\text{Res}_{F'/F} : H^i(F) \longrightarrow H^i(F')$ as injective for every i; see [S, Sect. 2.4]. Thus $w_{[n/2]}(q_{L/K})$, being the preimage of $w_{[n/2]}(q_{L'/K'})$ under this map, is also 0, contradicting Proposition 8.3(a).

(b) Assume the contrary: $d^{odd}(n) < 2[n/4]$. Then there exists a solvable odd degree extension $K'/K(\sqrt{\Delta})$ such that

$$\text{ed}(L'/K') < [n/2].$$

Hence, $w_{2[n/4]}(q_{L'/K'}) = 0$ in $H^{2[n/4]}(K')$. Reasoning as in part (a), we see that this implies $w_{2[n/4]}(q_{L(\sqrt{\Delta})/K(\sqrt{\Delta})}) = 0$, contradicting Proposition 8.3(b). □

REFERENCES

[A] S. Abhyankar, *Hilbert's Thirteenth Problem*, Société Mathématique de France, Seminaires et Congres, Mémoire **67**, 1997.

[AS] V. Arnold, and G. Shimura, *Superposition of algebraic functions*, Proc. Symposia in pure math., **28** (1976), AMS, Providence, 45-46.

[BR] J. Buhler, Z. Reichstein, *On the essential dimension of a finite group*, Compositio Math., **106** (1997), 159-179.

[C] D. Coray, *Cubic hypersurfaces and a result of Hermite*, Duke J. Math. **54** (1987), 657-670.

[De] A. Delzant, *Définition des classes de Stiefel-Whitney d'un module quadratique sur un corps de caractéristique différente de 2*, C. R. Acad. Sci. Paris **255** (1962), 1366-1368.

[Di] J. Dixmier, *Histoire du 13e probleme de Hilbert*, Cahiers Sém. Hist. Math. S/'er. 2, 3, 85-94 Univ. Paris VI, Paris, 1993.

[DM] P. Doyle and C. McMullen, *Solving the quintic by iteration*, Acta Math. **163** (1989), 151-180.

[EM] S. Edno and T. Miyata, *Invariants of finite abelian groups*, J. Math. Soc. Japan, **25** (1973), no. 1, 7-26.

[FJ] M. Fried and M. Jarden, *Field Arithmetic*, Springer-Verlag, Berlin, Heidelberg and New York, 1986.

[He] C. Hermite, *Sur l'invariant du dix-huitième ordre des formes du cinquième degré*, J. Crelle **59** (1861), 304–305.

[H1] D. Hilbert, Mathematical problems. Lecture delivered before the international congress of mathematicians at Paris in 1900, in *Mathematical developments arising from Hilbert Problems*, Proceedings of Symposia in Pure Mathematics **28**, American Mathematical Society, 1976.

[H2] D. Hilbert, *Über die Gleichung neunten Grades*, Mathematische Annalen **97** (1927), 243–250.

[J] P. Joubert, *Sur l'equation du sixième degré*, C-R. Acad. Sc. Paris **64** (1867), 1025–1029.

[Kl] F. Klein, *Vorlesungen über das Ikosaeder und die Auflösung der Gleichungen vom fünften Grade*, Teubner, Leipzig, 1884. English translation: *Lectures on the icosahedron and solution of equations of the fifth degree*, translated by G.G. Morrice, 2nd and rev. edition, New York, Dover Publications, 1956.

[Lo] G. G. Lorentz, The 13-th problem of Hilbert, Proc. Symposia in pure math., **28** (1976) AMS, Providence, 419-430.

[P] A. Pfister, *Quadratic Forms with Applications to Geometry and Topology*, Cambridge University Press, 1995.

[Re1] Z. Reichstein, *On the essential dimension of an algebraic group*, preprint.

[Re2] Z. Reichstein, *On a theorem of Hermite and Joubert*, submitted for publication; preprint on `www.orst.edu/ ~reichtsz/pub.html`.

[Ri] H. W. Richmond, *Note on the invariants of a binary sextic*, Quarterly J. of Math. **31** (1900), 57–59.

[Ro] P. Roquette, *Isomorphisms of generic splitting fields of simple algebras*, J. Reine Angew. Math. **214/215** (1964), 207–226.

[S] J-P. Serre, Galois Cohomology, Springer, 1997.

DEPARTMENT OF MATHEMATICS, REED COLLEGE, PORTLAND, OR 97202, USA

E-mail address: `jpb@reed.edu`

DEPARTMENT OF MATHEMATICS, OREGON STATE UNIVERSITY, CORVALLIS, OR 97331, USA

E-mail address: `zinovy@math.orst.edu`

HECKE OPERATORS AND THE NONVANISHING OF L–FUNCTIONS

J.B. CONREY
D.W. FARMER

Dedicated to the memory of S. Chowla.

ABSTRACT. We determine some conditions which imply the nonvanishing of $L_f(k/2)$ for f a cusp form of weight $k \equiv 0 \mod 4$ for the full modular group. These conditions are verified for weights $k \leq 500$.

1. INTRODUCTION AND STATEMENT OF RESULTS

Let $\mathcal{S}_k(1)$ be the space of holomorphic cusp forms of even integral weight k for the full modular group. This space has a basis of Hecke eigenforms

$$B = \{f_1, \dots, f_d\}, \qquad d = \begin{cases} [\frac{k}{12}] - 1 & \text{if } k \equiv 2 \bmod 12; \\ [\frac{k}{12}] & \text{otherwise,} \end{cases}$$

satisfying $T_n f_j = a_j(n) f_j$ for all n, where T_n is the nth Hecke operator. The f_j have Fourier expansions

$$f_j(z) = \sum_{n=1}^{\infty} a_j(n) e(nz)$$

where $e(z) = \exp(2\pi i z)$. We normalize f_j by setting $a_j(1) = 1$.

A Dirichlet series is associated to f_j by

$$L_{f_j}(s) = \sum_{n=1}^{\infty} \frac{a_j(n)}{n^s}.$$

In this paper we study the "critical value" $L_{f_j}(k/2)$. Since $L_{f_j}(s)$ has functional equation

$$(2\pi)^{-s}\Gamma(s)L_{f_j}(s) = (-1)^{k/2}(2\pi)^{s-k}\Gamma(k-s)L_{f_j}(k-s),$$

it follows that $L_{f_j}(k/2) = 0$ if $k \equiv 2 \bmod 4$. If $k \equiv 0 \bmod 4$, then $L_{f_j}(k/2) = 0$ implies that $L_{f_j}(s)$ has a zero of order at least 2 at $s = k/2$. In this paper we make progress toward showing that $L_{f_j}(k/2) \neq 0$ for $k \equiv 0 \bmod 4$.

Research of the first author supported in part by a grant from the NSF.

S.D. Ahlgren et al. (eds.), Topics in Number Theory, 143–150.

Our motivation for this work came from a recent conjecture of Duke and Imamoglu [DI] concerning Siegel modular forms. If n and k are even integers and f is a classical cusp form of weight $2k-n$, they conjecture that there is a Siegel cusp form F of degree n and weight k whose standard L-function factors as

$$L(F,s) = \xi(s) \prod_{j=1}^{n-1} L(f, s-j/2)L(f, s+j/2),$$

where all L-functions have been normalized to have functional equations sending $s \mapsto 1-s$. If $L_f(s)$ vanishes at the critical point then the above expression does not have a pole at $s=1$. Thus, the vanishing of $L_f(s)$ is related to whether or not the Siegel modular form F is a linear combination of theta series [Bo].

Our main result is

Theorem 1. *Suppose $k \equiv 0 \bmod 4$. If, for some n, the characteristic polynomial of T_n acting on $S_k(1)$ is irreducible over $\mathbb{Q}$, then $L_{f_j}(k/2) \neq 0$ for all j.*

Maeda has conjectured that the Hecke algebra of $SL_2(\mathbb{Z})$ is simple, which implies that the condition in Theorem 1 always holds. It appears to be possible that the characteristic polynomial is irreducible for all n. Theorem 1 is also stated by Kohnen and Zagier [KZ].

We have verified the irreducibility of the characteristic polynomial of T_2 acting on $S_k(1)$ for $k \leq 500$, yielding

Theorem 2. *$L_{f_j}(k/2) \neq 0$ for $k \leq 500$, $k \equiv 0 \bmod 4$.*

Questions concerning the characteristic polynomials of Hecke operators have received considerable attention lately. For example, if N is large then the characteristic polynomial of T_p acting on $S_2(\Gamma_0(N))$ will have lots of linear factors corresponding to elliptic curves, but one can look for irreducible factors of a large degree (see [K]). At the other extreme, one expects the characteristic polynomial of T_p acting on $S_k(\Gamma(1))$ to be irreducible, so Buzzard [B] began a study of its Galois group. We applied the method of Buzzard to show

Theorem 3. *For $k \leq 500$, $k \equiv 0 \bmod 4$, let $T_{2,k}(x) \in \mathbb{Z}[x]$ be the characteristic polynomial of T_2 acting on $S_k(1)$. Then the Galois group of the splitting field of $T_{2,k}(x)$ over $\mathbb{Q}$ is the symmetric group S_d.*

Corollary 4. *For $k \leq 500$, $k \equiv 0 \bmod 4$, let $\mathbb{K}_k$ be the field obtained by adjoining to $\mathbb{Q}$ all of the Fourier coefficients of all of the Hecke eigenforms f_j. Then $\mathbb{K}_k/\mathbb{Q}$ is a Galois extension of degree d with $Gal(\mathbb{K}_k/\mathbb{Q}) = S_d$.*

2. Background and Notation

Our methods involve the Hecke operators T_n. These can be defined as linear maps $T_n : S_k(1) \to S_k(1)$ given by

$$T_n\left(\sum_{m=1}^{\infty} a(m)e(mz)\right) = \sum_{m=1}^{\infty}\left(\sum_{d|(n,m)} d^{k-1}a\left(\frac{mn}{d^2}\right)\right) e(mz)$$

There is an inner product on $\mathcal{S}_k(1)$, called the Petersson inner product,

$$\langle f,g\rangle = \iint_{\mathcal{D}} f(z)\overline{g(z)}y^k d\mu(z),$$

where $\mathcal{D}$ is a fundamental domain for the full modular group, and $d\mu(z) = y^{-2}dxdy$. It can be shown that the Hecke operators mutually commute and are self-adjoint with respect to the Petersson inner product. Thus there is a basis of simultaneous eigenfunctions of the T_n, which we have already denoted by $B = \{f_1, \dots, f_d\}$.

The holomorphic Eisenstein series $E_k(z) \in M_k(1)$ is defined by

$$E_k(z) = \frac{1}{2\zeta(k)} \sum_{\gamma\in\Gamma_\infty\backslash\Gamma} j(\gamma,z)^{-k} = 1 - \frac{2k}{B_k}\sum_{m=1}^{\infty} \sigma_{k-1}(m)e(mz),$$

where $j(\gamma, z) = cz + d$ if γ is a 2×2 matrix with bottom row $(c \ \ d)$, the B_k are Bernoulli numbers, and

$$\sigma_\ell(n) = \sum_{d|n} d^\ell.$$

The Eisenstein series $E_k(z)$ is also an eigenfunction of the T_n, and $B \cup \{E_k\}$ is a basis for $M_k(1)$. For more information on modular forms for the full modular group, see Apostol's book [A].

We make extensive use of the fact that $\mathcal{S}_k(1)$ has a basis of functions possessing rational integer Fourier coefficients. Set $B_\mathbb{Q} = \{F_1, \dots, F_d\}$ where

$$F_j = \Delta^j E_4^{\frac{k}{4}-3j} = \sum_{m=j}^{\infty} A_j(m)e(mz), \qquad A_j(m) \in \mathbb{Z}, \tag{2.1}$$

say, where $\Delta \in \mathcal{S}_{12}(1)$ is the cusp form of weight 12. We denote by $T_{n,k}(x)$ the characteristic polynomial of T_n acting on $\mathcal{S}_k(1)$. The matrix representing T_n in the basis $B_\mathbb{Q}$ has rational entries, so $T_{n,k}(x) \in \mathbb{Z}[x]$.

We require some information about the field $\mathbb{K}_k$ obtained by adjoining to $\mathbb{Q}$ all of the Fourier coefficients $a_j(n)$ of all of the f_j. This is the same as the splitting field of all of the $T_{n,k}(x)$.

Lemma 1. $\mathbb{K}_k = \mathbb{Q}(\{a_j(n) \ : \ 1 \le j, n \le d\})$. *In particular,* $\mathbb{K}_k$ *is a finite Galois extension of* $\mathbb{Q}$.

Proof. Since $B_\mathbb{Q}$ is a basis, we can write

$$f_j = \sum_{1\le m\le d} \alpha_{j,m}F_m \tag{2.2}$$

The $\alpha_{j,m}$ can be found by solving the "upper triangular" system

$$a_j(n) = \sum_{1\le m\le d} \alpha_{j,m}A_m(n), \qquad 1 \le j,n \le d.$$

Thus, the $\alpha_{j,m}$ are in the field $\mathbb{Q}(\{a_j(n) : 1 \le j, n \le d\})$. But then by (2.2) all of the Fourier coefficients of the f_j are in that field, so $\mathbb{K}_k$ is as described.

The Galois group $Gal(\mathbb{K}_k/\mathbb{Q})$ acts on functions with Fourier coefficients in $\mathbb{K}_k$ by

$$\sigma\left(\sum_{m=1}^{\infty} a(m)e(mz)\right) = \left(\sum_{m=1}^{\infty} \sigma\left(a(m)\right)e(mz)\right).$$

Lemma 2. *The group $Gal(\mathbb{K}_k/\mathbb{Q})$ acts on $B = \{f_1, \dots, f_d\}$. If $T_{n,k}(x)$ is irreducible for some n, then the action is transitive. Furthermore, if $T_{n,k}(x)$ is irreducible, then $\mathbb{K}_k/\mathbb{Q}$ is the splitting field of $T_{n,k}(x)$, and $Gal(\mathbb{K}_k/\mathbb{Q}) = Gal(T_{n,k})$.*

Proof. $Gal(\mathbb{K}_k/\mathbb{Q})$ acts on B because the elements of $Gal(\mathbb{K}_k/\mathbb{Q})$ commute with each Hecke operator T_n, and the f_j are the simultaneous eigenfunctions of the T_n.

If $T_{n,k}(x)$ is irreducible, then the Galois group $Gal(\mathbb{K}_k/\mathbb{Q})$ acts transitively on $\{a_j(n) : 1 \le j \le d\}$, which is the set of nth Fourier coefficients of the f_j. This forces the action to be transitive on B. Furthermore, $Gal(\mathbb{K}_k/\mathbb{Q}) \subset Gal(T_{n,k})$ by considering the action of $Gal(\mathbb{K}_k/\mathbb{Q})$ on B and restricting to the set of nth Fourier coefficients. By definition the splitting field of $T_{n,k}(x)$ is contained in $\mathbb{K}_k$, so they must be equal.

In the next section we relate the transitive action of $Gal(\mathbb{K}_k/\mathbb{Q})$ on B to the nonvanishing of $L_{f_j}(k/2)$. In the following section we describe our calculations which show that $T_{2,k}(x)$ is irreducible for small k.

3. Proof of Theorem 1

Let

$$f(z) = \sum_{m=1}^{\infty} a(m)e(mz)$$
$$= f_j(z),$$

for some j. Set

$$F(s,\ell) = \sum_{n=1}^{\infty} \frac{a(n)\sigma_{\ell-1}(n)}{n^s}.$$

The next two propositions make use of ideas of Shimura [Sh], and can also be found in [Z].

Proposition 1. *We have $F(k-1, k/2) = 0$ if and only if*

$$L_f(k/2) = 0.$$

Proof. By comparing Euler products, we find that

$$F(s,\ell) = \frac{L_f(s)L_f(s+1-\ell)}{\zeta(2s+2-k-\ell)}.$$

(For a discussion of this method, see Bump's article [Bu].) In particular,

$$F(k-1,k/2)=\frac{L_f(k-1)L_f(k/2)}{\zeta(k/2)}.$$

To finish the proof, note that $k-1$ is strictly within the plane of absolute convergence for the Euler product for $L_f(s)$, so $L_f(k-1)\neq 0$.

Proposition 2. *We have* $\langle E_{k/2}^2, f\rangle = C_k\, L_f(k-1)L_f(k/2)$, *where*

$$C_k=\frac{(-1)^{\frac{k}{4}}}{\sqrt{2}}\left(\frac{2}{\pi}\right)^{\frac{k-1}{2}}\frac{\Gamma\left(\frac{k-1}{2}\right)}{\zeta(\frac{k}{2})^3}.$$

In particular, $\langle E_{k/2}^2, f\rangle = 0$ *if and only if* $L_f(k/2)=0$.

Proof. Let

$$I(s)=\int_0^\infty\int_0^1 \overline{f(z)}E_\ell(z)y^{s-1}\,dx\,dy.$$

By substituting the Fourier series for $E_\ell(z)$ and $f(z)$ into the equation for $I(s)$ and integrating term–by–term, we find that

$$I(s)=-\frac{2\ell}{B_\ell}(4\pi)^{-s}\Gamma(s)F(s,\ell).$$

But by the Rankin–Selberg unfolding method [Bu], we also have

$$I(s)=\iint_{\mathcal{D}}\overline{f(z)}E_\ell(z)\sum_{\gamma\in\Gamma_\infty\backslash\Gamma}\overline{j(\gamma,z)}^{k}j(\gamma,z)^{\ell}y(\gamma z)^{s+1}d\mu(z),$$

where $\mathcal{D}$ is a fundamental domain for the full modular group. The sum over γ equals $2\zeta(k-\ell)y(z)^kE_{k-\ell}(s+1-k,z)$, which involves the non-holomorphic Eisenstein series defined by

$$E_k(s,z)=\frac{1}{2\zeta(k)}\sum_{\gamma\in\Gamma\infty\backslash\Gamma}j(\gamma,z)^{-k}y(\gamma z)^s,$$

where $y(u+iv)=v$. In the notation of the Petersson inner product, we have

$$-\frac{2\ell}{B_\ell}(4\pi)^{-s}\Gamma(s)F(s,\ell)=2\zeta(k-\ell)\langle E_\ell(\cdot)E_{k-\ell}(s+1-k,\cdot),f(\cdot)\rangle.$$

If we take $s=k-1$ and $\ell=k/2$, and use the fact that $E_{k/2}(0,z)=E_{k/2}(z)$, we obtain the equation in the proposition, after expressing $B_{k/2}$ in terms of $\zeta(k/2)$ and using the duplication formula for the Γ–function.

Proof of Theorem 1. A basis for the space of modular forms of weight k is

$$\{f_1,f_2,\ldots,f_d,E_k\}.$$

Since $E_{k/2}^2$ is a modular form of weight k, there is a unique way to write

$$E_{k/2}^2 = E_k + \sum_{1 \le j \le d} c_j f_j \tag{3.1}$$

for some coefficients $c_j \in \mathbb{K}_k$. In particular,

$$\langle E_{k/2}^2, f_j \rangle = c_j \langle f_j, f_j \rangle,$$

so by Proposition 2, $L_{f_j}(k/2) = 0$ if and only if $c_j = 0$.

Applying $\sigma \in Gal(\mathbb{K}_k/\mathbb{Q})$ to (3.1) gives

$$E_{k/2}^2 = E_k + \sum_{1 \le j \le d} \sigma(c_j) f_{\overline{\sigma}(j)},$$

where $\overline{\sigma}$ is the permutation of $\{1, 2, \ldots, d\}$ induced by the action of σ on $B = \{f_1, \ldots, f_d\}$. Since the expression (3.1) is unique, it follows that $\sigma(c_j) = c_{\overline{\sigma}(j)}$.

Suppose $Gal(\mathbb{K}_k/\mathbb{Q})$ acts transitively on B, and c_j=0 for some j. Then by the previous observation we find that $c_j = 0$ for all j. This implies $E_{k/2}^2 = E_k$. This is a contradiction if $k > 8$, as can be seen by looking at the second Fourier coefficient of E_k and $E_{k/2}^2$ and using elementary properties of the Bernoulli numbers. So by Lemma 2, if $T_{n,k}(x)$ is irreducible for some n, then $c_j \neq 0$. This proves Theorem 1.

Note: In the above proof we required the fact that $E_{k/2}^2 = E_k$ does not hold for $k > 8$. E. Ghate [G] has used the ideas in this paper to give an interesting proof that the only identities among Eisenstein series are the classical ones: $E_4^2 = E_8$, $E_4 E_6 = E_{10}$, and $E_6 E_8 = E_{14}$.

4. Calculation of the Hecke operators

In this section we describe some calculations involving the Hecke operator T_2 acting on $S_k(1)$. Our goal is to show that the the characteristic polynomial $T_{2,k}(x)$ of T_2 is irreducible and has Galois group S_d, where $d = [k/12] = dim(S_k(1))$.

Direct calculation of the polynomial $T_{2,k}(x)$ is difficult for large k because of the huge size of the coefficients ($T_{2,k}(x)$ is a polynomial of degree d, with roots of size 2^{6d}). To verify that $T_{2,k}(x)$ is irreducible we calculate $T_{2,k}(x)$ mod p for small p and check if the reduced polynomial is irreducible mod p. Having established that the polynomial is irreducible, we use the following extension of the method of Buzzard [B] to show that it has Galois group S_d.

Lemma 3. *Suppose G is a transitive subgroup of the symmetric group S_d. If G contains both a transposition and a p–cycle for some prime $p > d/2$, then $G = S_d$.*

Proof. Put a relation $\sim$ on $S = \{1, 2, \ldots, d\}$ by $a \sim b$ if $a = b$ or if the transposition $(a\ b) \in G$. This is easily seen to be an equivalence relation. Since G is transitive, each equivalence class has the same number of elements, say n, and it follows that $n|d$. Note that $n > 1$ because G contains at least one transposition.

Let $T = \{e_1, e_2, \ldots, e_p\}$ be the subset of S permuted by the assumed p–cycle in G, and let G_T be the subgroup of G fixing the elements not in T. Define an equivalence relation on T by $a \approx b$ if $a = b$ or if the transposition $(a\ b) \in G_T$. As above, each equivalence class has the same number of elements, say m, and $m|p$. Since $n > 1$, we have $m > 1$, so $m = p$ because p is prime. But $n \geq m$ because $G_T \subset G$. Thus $n > d/2$, so $n = d$. This implies $G = S_d$.

Lemma 4. *Let $f(x) \in \mathbb{Z}[x]$ be a monic irreducible polynomial of degree d, with splitting field $\mathbb{K}/\mathbb{Q}$. Suppose there is a prime q such that the* mod q *reduction $f_q \in \mathbb{F}_q[x]$ satisfies*

$$f_q = g_0 g_1 \cdots g_j$$

for distinct irreducible $g_i \in \mathbb{F}_q[x]$ with

$$deg(g_0) = 2$$
$$deg(g_i) \quad odd\ for\ i \geq 1.$$

Suppose there also is a prime r with the mod r *reduction $f_r \in \mathbb{F}_r[x]$ satisfying*

$$f_r = h_0 h_1 \cdots h_l$$

for distinct irreducible $h_i \in \mathbb{F}_r[x]$ with $deg(h_0) = p$, with $p > d/2$ prime. Then $Gal(\mathbb{K}/\mathbb{Q})$ is the full symmetric group S_d.

Proof. Identify $G = Gal(\mathbb{K}/\mathbb{Q})$ as a subgroup of S_d. We have assumed that $f(x)$ has distinct roots mod q and mod r. So if $\mathcal{Q}$ and $\mathcal{R}$ are primes of $\mathbb{K}$ above q and r, respectively, then we have $\mathrm{Frob}_{\mathcal{Q}}, \mathrm{Frob}_{\mathcal{R}} \in G$. Appropriate powers $\mathrm{Frob}_{\mathcal{Q}}^{t_1}$ and $\mathrm{Frob}_{\mathcal{R}}^{t_2}$ are a transposition and a p–cycle, respectively. Lemma 3 applies, so $G = S_d$.

The following is our procedure to check that $T_{2,k}(x)$ is irreducible and has Galois group S_d.

Step 1. Generate the basis $B_{\mathbf{Q}} = \{F_1, \ldots, F_d\}$ given in (2.1). We use the first 200 terms in the Fourier expansion.

Step 2. Form the matrices $\mathcal{A} = (a_{j,l})$ with $a_{j,l} = A_j(2l) + 2^{k-1}A_j(l/2)$ and $\mathcal{B} = (A_j(l))$, where $A_j(l)$ is given in (2.1).

Note that $\mathcal{A}\mathcal{B}^{-1}$ is the matrix of T_2 in the basis $B_{\mathbf{Q}}$, and $\det(\mathcal{A} - x\mathcal{B}) \in \mathbb{Z}[x]$ is the characteristic polynomial of T_2.

Step 3. Reduce $\mathcal{A} - x\mathcal{B}$ mod p, compute the determinant of the reduced matrix, and factor that polynomial mod p.

Step 3 is repeated until we find a prime p for which the polynomial is irreducible, and primes q and r as in Lemma 4.

The above procedure was implemented in Mathematica on a Sun SparcCenter 1000 with 4 processors. The program took 867 CPU minutes for the range $24 \leq k \leq 400$, and an additional 2103 minutes for the range $404 \leq k \leq 500$. It was found that for all k in those ranges the characteristic polynomial $T_{2,k}(x)$ is irreducible and has Galois group S_d.

Combining the computer calculation with Theorem 1 proves Theorem 2. Combining the computer calculation with Lemma 4 proves Theorem 3. Corollary 4 follows because, by Lemma 2, if $T_{2,k}(x)$ is irreducible then $\mathbb{K}_k = \mathbb{Q}(\{a_j(2)\ :\ 1 \leq j \leq d\})$.

The results of the calculation, for small k, are given in the table below. The table shows the smallest primes p, q, r for which the polynomial $T_{2,k}(x)$ is irreducible mod p, and q, r are as in Lemma 4. (If d is prime, then we put $r = p$).

k	p	q	r
24	23	23	23
28	29	29	29
32	29	29	29
36	37	29	37
40	37	41	37
44	61	41	61
48	53	23	43
52	67	29	53
56	71	29	47
60	83	37	83
64	73	29	73
68	97	23	97
72	107	23	73
76	137	47	73
80	103	29	89

k	p	q	r
84	307	37	307
88	349	31	349
92	107	41	107
96	149	23	101
100	179	41	89
104	127	37	89
108	223	29	83
112	127	23	107
116	263	53	107
120	293	23	113
124	163	29	107
128	149	59	127
132	479	37	479
136	139	29	139
140	149	41	149

References

[A] *T. Apostol*, Modular functions and Dirichlet series in number theory, Springer–Verlag GTM 41, 1990.

[Bo] *S. Böcherer*, Siegel modular forms and theta series, Theta Functions, Bowdoin 1987, AMS Proceedings of Symposia in Pure Mthematics, vol. 49, 3–17.

[Bu] *D. Bump*, The Rankin-Selberg method: a survey, in Number theory, trace formulas and discrete groups (Oslo, 1987), 49–109, Academic Press, Boston, MA, 1989.

[B] *K. Buzzard*, On the eigenvalues of the Hecke operator T_2, J. Number Theory **57** (1996), no. 1, 130–132.

[DIm] *W. Duke* and *O. Imamoglu*, personal communication.

[G] *E. Ghate*, On monomial relations between Eisenstein series, preprint.

[K] *S. Kamienny*, On the eigenvalues of Hecke operators, Comm. Algebra **23** (1995), no. 3, 995–998.

[KZ] *W. Kohnen* and *D. Zagier*, Values of *L*-series of modular forms at the center of the critical strip, Invent. Math. **64** (1981), no. 2, 175–198.

[Sh] *G. Shimura*, The special values of the zeta functions associated with cusp forms. Comm. Pure Appl. Math. **9** (1976), no. 6, 783–804.

[Z] *D. Zagier*, Modular forms whose Fourier coefficients involve zeta–functions of quadratic fields, Modular Functions of One Variable VI, LNM 627, 105–170.

Department of Mathematics, Oklahoma State University, Stillwater, OK 74078

Department of Mathematics, Bucknell University, Lewisburg, PA 17837

A NEW MINOR-ARCS ESTIMATE FOR NUMBER FIELDS

Morley Davidson

Abstract. We re-examine Körner's number field version of a minor-arcs estimate introduced by Vinogradov into the study of Waring's function $G(k)$. Dependencies on both k and the degree n of the number field are reduced via improved mean value estimates for Weyl sums over 'smooth' algebraic integers.

1. Introduction

In this note we prove a generalized minor-arcs estimate for an exponential sum of type originally considered by Vinogradov (see [Vi]) and extended to number fields by Körner [K1]. When applied in conjunction with the mean value estimate proved in [D], it leads to a strengthening in bounds for $G_{\mathbb{K}}(k)$, Siegel's extension of the Waring function $G(k)$ to the number field $\mathbb{K}$.

Let $\mathcal{O}$ denote the integers of $\mathbb{K}$ and let J_k be the subring generated by the kth powers of the integers of $\mathbb{K}$. Then the Waring-Siegel function $G_{\mathbb{K}}(k)$ is defined as the least $s \geq 1$ such that, for all totally positive integers $\nu \in J_k$ of sufficiently large norm, the equation

$$\nu = \lambda_1^k + \cdots + \lambda_s^k$$

is soluble in totally nonnegative integers λ_i of $\mathbb{K}$ all of whose norms satisfy

$$N(\lambda_i) \leq cN(\nu)^{1/k} \quad (1 \leq i \leq s)$$

for some constant $c = c(k, \mathbb{K})$. In [D] we considered the case when $\mathbb{K}$ has class number one and extended the main result of [Wo]. By combining Eda's estimate with improved mean value estimates for Weyl sums over 'smooth' integers of $\mathbb{K}$ (defined below), it was proved that for large k, one has

$$G_{\mathbb{K}}(k) \leq k \log k + 3k \log\log k + c_n k,$$

where the o-constant depends only on k, and c_n is approximately $4n$. Using our new estimate (stated in Section 3) in place of Eda's, the coefficient of $\log\log k$ can be reduced from 3 to 1, and there is a slight improvement in the smaller-order terms of c_n. To conserve space in these proceedings, we omit the details of this application, as it is a straightforward deduction for experts in the Hardy-Littlewood method. We concentrate instead on the exponential sum estimate, whose proof is adapted from [K1] using ideas from [T] and Section 9 of [V2]. The sources of our improvements (in both k and n) are two primary changes in Körner's argument, which is otherwise mostly duplicated: we insert the improved mean value estimates from [D], and we use Hölder's inequality to better optimize our final estimates. We refer the reader to Wang's book [W] for background on the Waring-Siegel function and Siegel's generalization of the Hardy-Littlewood method.

S.D. Ahlgren et al. (eds.), Topics in Number Theory, 151–161.

2. Preliminaries and Statement of the Theorem

We write $n = r_1 + 2r_2$ for the degree of $\mathbb{K}$ over $\mathbb{Q}$, and denote the different (ramification) ideal by $\mathfrak{D}$ and its norm by D, so that $D = |\text{disc}\,\mathbb{K}|$. Fix an integral basis $\{\tau_1, \tau_2, \ldots, \tau_n\}$ and denote by $\{\rho_1, \rho_2, \ldots, \rho_n\}$ the dual basis of $\mathfrak{D}^{-1}$ satisfying

$$\text{Tr}(\rho_i \tau_j) = \begin{cases} 1, & \text{if } i = j \\ 0, & \text{otherwise,} \end{cases}$$

where Tr denotes the usual trace function $\text{Tr}_{\mathbf{K}/\mathbf{Q}}$. We define a real linear form

$$\xi = \sum_{1 \le i \le n} t_i \rho_i,$$

whose coefficients t_i play the role which for $\mathbb{Q}$ is commonly played by α (as in Vaughan's book [V1]).

For $\gamma \in \mathbb{K}$, let $\gamma^{(i)}$ $(1 \le i \le n)$ be the $\mathbb{K}$-conjugates of γ, and define a height

$$\|\gamma\| = \max_{1 \le i \le n} |\gamma^{(i)}|.$$

Also, we extend $\text{Tr}_{\mathbf{K}/\mathbf{Q}}$ to real linear forms in the natural way (see [W] for complete definitions). For μ either an element of $\mathbb{K}$ or a real linear form over $\mathbb{K}$, such as ξ defined above, let $E(\mu)$ denote $e^{2\pi i \text{Tr}(\mu)}$.

Given γ in $\mathbb{K}$, we can find unique integral ideals $\mathfrak{a}$ and $\mathfrak{b}$ with $(\mathfrak{a}, \mathfrak{b}) = 1$ such that $\gamma\mathfrak{D} = \mathfrak{b}/\mathfrak{a}$. This relation is denoted by $\gamma \to \mathfrak{a}_\gamma$, and we call $\mathfrak{a} = \mathfrak{a}_\gamma$ the denominator of γ.

Denote $[0,1]$ by $\mathbb{T}$, and for positive numbers T, a with $T > 1$ and $0 < a \le \frac{1}{4}$, define the Farey dissection parameters t and h by

$$t = T^{1-a}, \quad h = T^{k-1+a}. \tag{2.1}$$

The parameter T is to be thought of as being very large. Then $\Gamma(t)$ is defined as the set of γ which, when expressed via the dual basis as $\gamma = \sum_{i=1}^n u_i \rho_i$, satisfy

$$(u_1, \ldots, u_n) \in (\mathbb{T} \cap \mathbb{Q})^n \text{ and } \mathcal{N}(\mathfrak{a}_\gamma) \le t^n,$$

where $\mathcal{N}(\cdot)$ denotes the usual ideal norm in $\mathbb{K}$.

For $\gamma \in \Gamma(t)$, we define B_γ to be the set of all $\mathbf{t} \in \mathbb{T}^n$ which satisfy, for some $\gamma_0 \equiv \gamma \pmod{\mathfrak{D}^{-1}}$,

$$\prod_{i=1}^n \max\left(h|\xi^{(i)} - \gamma_0^{(i)}|, \frac{1}{t}\right) \le \frac{1}{\mathcal{N}(\mathfrak{a})}.$$

Siegel [S] defined the *basic domain* corresponding to h and t by

$$B = B(t, h) = \bigcup_{\gamma \in \Gamma(t)} B_\gamma,$$

generalizing the major-arcs for $\mathbb{Q}$. Analogous to the minor-arcs for $\mathbb{Q}$ is the complement

$$S = S(t,h) = \mathbb{T}^n \setminus B,$$

which he named the ***supplementary domain***. Taken together, the basic domain and the supplementary domain form a generalized Farey dissection of $\mathbb{T}^n$ into B and S with respect to (t,h).

We recall that a prime algebraic integer $\omega \in \mathcal{O}$ is one which generates a prime ideal $\langle\omega\rangle$. Given $T_0 > 1$ and $c > 1$, let $\mathcal{P}_c(T_0)$ be the set of all totally positive prime integers ω determined by the conditions

$$\frac{1}{2}cT_0 \le |\omega^{(l)}| \le cT_0 \qquad (1 \le l \le n),$$
$$|\arg \omega^{(m)}| \le \frac{\pi}{4k} \qquad (r_1 + 1 \le m \le r_1 + r_2).$$

By Mitsui's generalization of the prime number theorem (see p. 35 of [M1]) we have

$$1 \ll \operatorname{card} \mathcal{P}_c(T_0) T_0^{-n} \log T_0 \ll 1.$$

Here, constants implied by the symbols $\ll$ and $\gg$ may depend at most on $\mathbb{K}$. Later, implicit constants may depend also on k and/or ϵ as applicable.

Next, let $\mathcal{O}(T)$ and $\mathcal{O}^+(T)$ denote (respectively) the set of integers and totally positive integers $\gamma \in \mathcal{O}$ with $||\gamma|| \le T$. Fixing a positive constant u, define $\mathcal{A}(T,R)$ as a complete set of non-associated integers λ of $\mathbb{K}$ with

$$|\lambda^{(i)}| \asymp uT \quad (1 \le i \le n)$$

and such that λ is strictly a product of prime integers in $\mathcal{O}^+(R)$. See [D, Section 4.1] for a discussion of other viable definitions for $\mathcal{A}(T,R)$.

Given $0 < \eta < 1$ and a large positive number $T_0 = T^{1/2}$, define the generating function

$$h(\xi, T_0) = h(\xi; T_0, \eta) = \sum_{\omega \in \mathcal{P}(T_0,2)} \sum_{\gamma \in \mathcal{A}(T_0, T_0^{\eta})} E(\xi(\omega\gamma)^k).$$

Finally, let $b = \frac{n}{n^2+1}$. Then we have the following

Theorem. *Suppose that $\mathbb{K}$ has class number one. Then for each $r \in \mathbb{N}$ and any positive number ϵ, we have for T sufficiently large*

$$\sup_{t \in S(t,h)} |h(\xi, T_0)| \ll T^{n-\sigma+\epsilon},$$

where

$$\sigma = \sigma(r,k,n) = \frac{1}{8r}(b - n\Delta_{2r,k})$$

with $\Delta_{2r,k} = ke^{1-4r/k}$.

Remarks. The trivial upper estimate for $h(\xi, T_0)$ is T^n. Moreover, the value of σ is positive if and only if

$$r \ge \left\lceil \frac{k}{4} \log\left(ek(n^2+1)\right) \right\rceil.$$

Hence, for k large relative to n, one can achieve σ with

$$\sigma^{-1} = (2 + o_k(1))k(\log k)(n + 1/n).$$

If not for the single term n here, this savings would allow a practical resolution of Siegel's conjecture that bounds for $G_{\mathbb{K}}(k)$ should not depend on $\mathbb{K}$. Indeed, we would be able to show, roughly speaking, that current bounds for $G(k)$ (the rational case) hold for any $\mathbb{K}$. The only such uniform bound known is exponentially larger than these current bounds; it is due to independent work of Birch [B], Körner [K2], and Ramanujam [R], who gave the upper bound

$$G_{\mathbb{K}}(k) \leq \max\{8k^5, 2^k + 1\}.$$

Unfortunately, the term n in σ^{-1} seems stubbornly entrenched in our method, due to the form of Lemma 4 below.

The estimate of Eda [E, Theorem 5] applies to a modified exponential sum, and is stronger than Körner's estimate for large k. However, it has a savings with $(\log k)^3$ instead of the $\log k$ achievable in our σ. This accounts for the improvement in bounds for $G_{\mathbb{K}}(k)$ (when the class number is one) mentioned in the Introduction. Corresponding improvements when the class number exceeds one follow similarly, pending completion of our proof of [D, Lemma 4.13] in that case.

3. Proof of the Theorem

The following two lemmas are due to Siegel and will be needed later. The first is an extension of Dirichlet's theorem on diophantine approximation to number fields.

Lemma 1. *Fix a number field $\mathbb{K}$ and suppose that t, h are as given in (2.1). Then for any $\mathbf{t} \in \mathbb{T}^n$ there exist $\alpha \in \mathcal{O}$ and a number $\beta \in \mathfrak{D}^{-1}$ such that*

$$\|\alpha\xi - \beta\| < \frac{1}{h}, \quad t < \|\alpha\| \leq h,$$

$$\max(h|\alpha^{(i)}\xi^{(i)} - \beta^{(i)}|, |\alpha^{(i)}|) \geq \frac{1}{\sqrt{D}}, \quad (1 \leq i \leq n),$$

and

$$\mathcal{N}((\alpha, \beta\mathfrak{D})) \leq \sqrt{D}.$$

Proof. See [W, Theorem 3.1].

Lemma 2. *Suppose that γ_1 and γ_2 are distinct elements of $\Gamma(t)$. Then*

$$B_{\gamma_1} \cap B_{\gamma_2} = \emptyset.$$

Proof. See [W, Lemma 5.1].

For positive real parameters T_0, Y, and c, and complex numbers b_γ given for each $\gamma \in \mathcal{O}^+(Y)$, define the exponential sum

$$H(\xi) = H(\xi; T_0, Y, c) = \sum_{\omega \in \mathcal{P}_c(T_0)} \sum_{\gamma \in \mathcal{O}^+(Y)} b_\gamma E(\xi \omega^k \gamma).$$

Also, given $T \geq 1$ we define parameters

$$T_0 = T^{1/2},$$
$$b = \frac{n}{n^2+1},$$

and define a new basic and supplementary domain $\tilde{B}$ and $\tilde{S}$ corresponding to

$$(t_0, h_0) = (T_0^b, T_0^k t_0^{n-1} \log t_0).$$

It is easily seen that Lemmas 1 and 2 hold for this choice of the variables t and h.

Lemma 3. *Suppose that $Y \asymp T_0^k$. Then for any $\mathbf{t} \in S(t,h)$ and $\epsilon > 0$, for sufficiently large T one has*

$$|H(\xi)|^2 \ll T_0^{n(k+2)-b+\epsilon} \sum_{\gamma \in \mathcal{O}^+(Y)} |b_\gamma|^2.$$

The proof requires still more preliminaries. As in Section 2, we let $\{\tau_i\}_{i=1}^n$ denote a fixed integral basis of $\mathbb{K}$. Given $\mathbf{t} \in \mathbb{R}^n$, for each $1 \leq i \leq n$ we may write

$$t_i = \mathrm{Tr}(\xi \tau_i) = a_i + d_i, \quad -\frac{1}{2} < d_i \leq \frac{1}{2}$$

with rational integers a_i. Then we define

$$\vartheta = \vartheta(\xi) = \sum_{i=1}^n a_i \rho_i,$$
$$\zeta = \zeta(\xi) = \sum_{i=1}^n d_i \rho_i,$$

where $\{\rho_i\}_{i=1}^n$ is the dual basis to $\{\tau_i\}_{i=1}^n$ defined in Section 2. We clearly have

$$\xi = \vartheta + \zeta, \quad \mathfrak{D}^{-1} | \vartheta, \tag{3.1}$$

where $\mathfrak{D}$ is the different ideal.

Lemma 4. *Given $\mathbf{t} \in \mathbb{R}^n$, one has*

$$\sum_{\gamma \in \mathcal{O}^+(T)} E(\gamma \xi) \ll T^{n-1} \min(T, ||\zeta||^{-1}).$$

Proof. Hilfssatz 8 of Körner [K1].

Lemma 5. *Suppose $\mathfrak{a}$ is a nonzero fractional ideal of $\mathbb{K}$. Let T_i $(1 \leq i \leq n)$ be positive real numbers with $T_m = T_{m+r_2}$, and let a_m, b_m be real numbers with $0 < a_m - b_m \leq 2\pi$, where the index m runs over the range $r_1 + 1 \leq m \leq r_1 + r_2$. Let $\mathcal{O}(\mathfrak{a}, \mathbf{T}, \mathbf{a}, \mathbf{b})$ be the set of integers $\nu \in \mathfrak{a}$ satisfying*

$$|\nu^{(i)}| \leq T_i$$
$$b_m \leq \arg \nu^{(m)} \leq a_m,$$

where the indices i and m retain the same ranges as above. Then we have

$$\operatorname{card} \mathcal{O}(\mathfrak{a}, \mathbf{T}, \mathbf{a}, \mathbf{b}) = \frac{2^{r_1} \prod_m (a_i - b_i)}{\sqrt{D}\mathcal{N}(\mathfrak{a})} \prod_i T_i + E,$$

where

$$E \ll \left(\frac{\prod_i T_i}{\mathcal{N}(\mathfrak{a})}\right)^{1-1/n}.$$

Proof. This is Lemma 3.2 of [M2].

Lemma 6. *Suppose that Lemma 1 is applied with $t \in S(t_0, h_0)$ to obtain numbers α and β in $\mathbb{K}$ so that (in particular) $\mathfrak{D}^{-1}|\beta$. Let g_i $(1 \leq i \leq n)$ be rational integers, and choose positive real numbers Φ_i $(1 \leq i \leq n)$ with $\Phi_m = \Phi_{m+r_2}$ $(r_1 + 1 \leq m \leq r_1 + r_2)$. Given a prime element $\omega_0 \in \mathcal{P}_c(R)$ let $\mathcal{P}(\mathbf{g}, \Phi)$ be the set of prime elements $\omega \in \mathcal{P}_c(R)$ satisfying*

$$\begin{aligned} |\zeta^{(i)}| &\leq \Phi_i & &(1 \leq i \leq n), \\ g_j &\leq 4D^{1/n}|\alpha^{(j)}|\Re\zeta^{(j)} < g_j + 1 & &(1 \leq j \leq r_1 + r_2), \\ g_l &\leq 4D^{1/n}|\alpha^{(l)}|\Im\zeta^{(l)} < g_l + 1 & &(r_1 + r_2 + 1 \leq l \leq n), \end{aligned}$$

where $\zeta = \zeta(\xi(\omega^k - \omega_0^k))$, and $\Re$ and $\Im$ denote real and imaginary parts. Then for any $\epsilon > 0$ we have

$$\operatorname{card} \mathcal{P}(\mathbf{g}, \Phi) \ll |N(\alpha)|^\epsilon \left(1 + \prod_{j \in J_1} (h_0 \Phi_j R^{1-k}) \prod_{l \in J_2} (R|\alpha^{(l)}|^{-1})\right),$$

where J_1 is the set of indices $(1 \leq i \leq n)$ with $|\alpha^{(i)}| < D^{-1/2}$ and J_2 is the complementary set of indices.

Proof. This is Hilfssatz 11 of [K1].

Remark. The factor $|N(\alpha)|^\epsilon$ arises from the bound

$$M(\nu, \mathfrak{a}) \ll \mathcal{N}(\mathfrak{a})^\epsilon$$

for the number $M(\nu, \mathfrak{a})$ of solutions $\sigma \mod \mathfrak{a}$ of the congruence

$$\sigma^k \equiv \nu \mod \mathfrak{a}$$

for a given $\nu \in \mathcal{O}$ relatively prime to the integral ideal $\mathfrak{a}$; see Hilfssatz 5 of [K1]. Thus one can see similarity with the proofs of Lemma 5.4 of [Va] and the lemma in Section 3 of [T].

Proof of Lemma 3. We repeat the proof of [K1, Satz 3], changing it slightly to allow the mean value estimates from [D] to be incorporated later. Two cases are considered:

CASE 1. Suppose $\mathbf{t} \in S \cap \tilde{S}$. Abbreviate $\mathcal{P}_c(T_0)$ to $\mathcal{P}$. By Schwarz's inequality and Lemma 4 we have

$$|H(\xi)|^2 \leq \left(\sum_{\gamma \in \mathcal{O}+(Y)} |b_\gamma|^2 \right) \sum_{\omega, \omega_0 \in \mathcal{P}} \left| \sum_{\gamma \in \mathcal{O}+(Y)} E\left(\xi\gamma(\omega^k - \omega_0^k)\right) \right|$$

$$\ll \left(\sum_{\gamma \in \mathcal{O}+(Y)} |b_\gamma|^2 \right) Y^{n-1} \sum_{\omega_0 \in \mathcal{P}} H_1(\xi, \omega_0), \tag{3.2}$$

where

$$H_1(\xi, \omega_0) = \sum_{\omega \in \mathcal{P}} \min(Y, ||\zeta||^{-1}) \text{ with } \zeta = \zeta\left(\xi(\omega^k - \omega_0^k)\right).$$

We have

$$H_1(\xi, \omega_0) \leq \sum_{\substack{\omega \in \mathcal{P} \\ ||\zeta|| > Y^{-1} T_0^b}} Y T_0^{-b} + \sum_{\substack{\omega \in \mathcal{P} \\ ||\zeta|| \leq Y^{-1} T_0^b}} Y$$

$$= \sum\nolimits_1 + \sum\nolimits_2,$$

say. The trivial estimate implied by Lemma 5 gives

$$\sum\nolimits_1 \leq (\text{card } \mathcal{P}) Y T_0^{-b} \ll Y T_0^{n-b}.$$

For Σ_2 we apply Lemma 6 with $\Phi_i = Y^{-1} T_0^b$ $(1 \leq i \leq n)$. Let Z be the number of distinct $n-$tuplets $\mathbf{g} = (g_1, \dots, g_n)$ of rational integers which can arise as in Lemma 6 with $||\zeta|| \leq Y^{-1} T_0^b$. From the three conjugate-conditions in Lemma 6, we have

$$Z \ll \prod_{i \in J_2} (|\alpha^{(i)}| Y^{-1} T_0^b + 1),$$

and hence Lemma 6 implies that

$$\sum\nolimits_2 \leq Z(\zeta) \max_{\mathbf{g}} \mathcal{P}(\mathbf{g}, Y^{-1} T_0^b) Y$$

$$\ll \prod_{i \in J_2} (|\alpha^{(i)}| Y^{-1} T_0^b + 1) |N(\alpha)|^\epsilon \cdot X,$$

where

$$X = Y + Y \prod_{j\in J_1} (h_0\Phi_j T_0^{b+1}) \prod_{l\in J_2} (T_0|\alpha^{(l)}|^{-1}).$$

By Lemma 1, we have $t_0 < ||\alpha|| \le h_0$, and hence

$$\sum\nolimits_2 \ll YT_0^{\epsilon} \cdot X_1,$$

where X_1 is

$$\begin{aligned} &\ll (h_0Y^{-1}T_0^b)^n + \prod_{j\in J_1} (h_0Y^{-2}T_0^{b+1}) \prod_{l\in J_2} (Y^{-1}T_0^{b+1} + T_0|\alpha^{(l)}|^{-1}) \\ &\ll YT_0^{\epsilon}(T_0^{n^2b} + T_0^n||\alpha||^{-1}) \\ &\ll YT_0^{n-b+\epsilon}. \end{aligned}$$

Collecting our estimates for Σ_1 and Σ_2 and using the trivial estimate card $\mathcal{P} \ll T_0^n$ implied by Lemma 5 in (3.2), the desired conclusion follows in this case.

CASE 2. Suppose $\mathbf{t} \in S\cap\tilde{B}$. By Lemma 1 we have $\mathbf{t} \in S\cap\tilde{B}_{\tilde{\gamma}}$ for a unique $\tilde{\gamma} \in \Gamma(t_0)$. From the definition of $\tilde{B}_{\tilde{\gamma}}$ there exists $\tilde{\gamma}_0 \in \mathbb{K}$ such that $\tilde{\gamma}_0 \equiv \tilde{\gamma} \mod \mathfrak{D}^{-1}$ and $\tilde{\gamma}_0 \to \mathfrak{a}$ for some integral ideal $\mathfrak{a}$ with $\mathcal{N}(\mathfrak{a}) \le t_0^n$, and satisfying

$$||\xi - \tilde{\gamma}_0|| \le h_0^{-1} t_0^{n-1} \mathcal{N}(\mathfrak{a})^{-1}. \tag{3.3}$$

Since $t_0 < t$ one has $\mathbf{t} \notin B_{\tilde{\gamma}}$, and therefore

$$||\xi - \tilde{\gamma}_0|| > h^{-1} \mathcal{N}(\mathfrak{a})^{-1/n}. \tag{3.4}$$

If we let $\psi = \xi - \tilde{\gamma}_0$, then

$$H(\xi) = \sum_{\sigma \bmod \mathfrak{a}} H_\sigma(\xi),$$

where

$$H_\sigma(\xi) = \sum_{\gamma\in\mathcal{O}^+(Y)} E(\gamma\sigma^k\tilde{\gamma}_0) \sum_{\substack{\omega\in\mathcal{P} \\ \omega\equiv\sigma \bmod \mathfrak{a}}} E(\gamma\omega^k\psi).$$

As before, one has

$$|H_\sigma(\xi)|^2 \ll \left(\sum_{\gamma\in\mathcal{O}^+(Y)} |b_\gamma|^2 \right) Y^{n-1} \sum_{\substack{\omega_0\in\mathcal{P} \\ \omega_0\equiv\sigma \bmod \mathfrak{a}}} H_2(\xi, \omega_0),$$

where

$$H_2(\xi, \omega_0) = \sum_{\substack{\omega\in\mathcal{P} \\ \omega\equiv\omega_0 \bmod \mathfrak{a}}} \min(Y, ||\varsigma||^{-1})$$

with $\zeta = \zeta\left(\xi(\omega^k - \omega_0^k)\right)$. It is clear from $|\omega| \asymp T_0$ and (3.3) that

$$|\mathrm{Tr}\left((\omega^k - \omega_0^k)\psi\tau_j\right)| \ll h_0^{-1} t_0^{n-1} T_0^k,$$

and the right hand side approaches zero as T_0 tends to infinity, for each $j = 1, \ldots, n$. Hence, for sufficiently large T_0, it follows from (3.1) that

$$(\omega^k - \omega_0^k)\psi = \zeta. \tag{3.5}$$

We partition the set of primes $\omega \in \mathcal{P}$ with $\omega \equiv \omega_0 \mod \mathfrak{a}$ into two sets $\mathcal{P}_1$ and $\mathcal{P}_2$, where $\omega \in \mathcal{P}_1$ if

$$\min_{1 \le i \le n} |\omega^{(i)} - \omega_0^{(i)}| \ge \mathcal{N}(\mathfrak{a})^{1/n}.$$

Since the conditions defining $\mathcal{P}$ imply, for $i = 1, \ldots, n$, that

$$|(\omega^{(i)})^k - (\omega_0^{(i)})^k| \ge \left(\frac{c}{2} T_0\right)^{k-1} |\omega^{(i)} - \omega_0^{(i)}|,$$

it follows from (2.1), (3.4), and (3.5) that for $\omega \in \mathcal{P}_1$ one has

$$||\zeta||^{-1} \ll T_0^{1-k} \mathcal{N}(\mathfrak{a})^{-1/n} h \mathcal{N}(\mathfrak{a})^{1/n} = T_0^{k-1+2a}.$$

From Lemma 5 we see that

$$\begin{aligned} &\mathrm{card}\, \mathcal{P}_1 \ll T_0^n \mathcal{N}(\mathfrak{a})^{-1}, \\ &\mathrm{card}\, \mathcal{P}_2 \ll T_0^{n-1} \mathcal{N}(\mathfrak{a})^{-1+1/n} \ll T_0^{n-1+b} \mathcal{N}(\mathfrak{a})^{-1}. \end{aligned}$$

Collecting the last three estimates, one has

$$\begin{aligned} H_2(\xi, \omega_0) &\le \sum_{\omega \in \mathcal{P}_1} T_0^{k-1+2a} + \sum_{\omega \in \mathcal{P}_2} T_0^k \\ &\ll T_0^n \mathcal{N}(\mathfrak{a})^{-1} T_0^{k-1+2a} + T_0^{n-1+b} \mathcal{N}(\mathfrak{a})^{-1} T_0^k \\ &\ll Y T_0^{n-b} \mathcal{N}(\mathfrak{a})^{-1}. \end{aligned}$$

Applying Lemma 5 again we conclude that

$$|H(\xi)|^2 \le \left(\sum_{\gamma \in \mathcal{O}^+(Y)} |b_\gamma|^2\right) Y^{n-1} \cdot T_0^n \mathcal{N}(\mathfrak{a})^{-1} \left(Y T_0^{n-b} \mathcal{N}(\mathfrak{a})^{-1}\right),$$

from which the desired result follows immediately. This completes the proof of the lemma.

We are now in a position to prove the Theorem, restated below, as a consequence of Lemma 3 and the mean value estimate proved in [D].

Theorem. *Suppose that $\mathbb{K}$ has class number one. Then for T sufficiently large and any natural number r we have*

$$\sup_{t\in S(t,h)} |h(\xi,T_0)| \ll T^{n-\sigma+\epsilon},$$

where

$$\sigma = \sigma(r,k,n) = \frac{1}{8r}(b - n\Delta_{2r,k}),$$

with $\Delta_{2r,k} = ke^{1-4r/k}$.

Proof. Observe that by Hölder's inequality and Lemma 5, one has

$$\begin{aligned} |h(\xi,T_0)|^{2r} &\ll T_0^{n(2r-1)} \sum_{\omega\in\mathcal{P}_2(T_0)} \left| \sum_{\gamma\in\mathcal{A}(T_0,T_0^{\eta})} E(\xi(\omega\gamma)^k)\right|^{2r} \\ &\ll T_0^{n(2r-1)} \sum_{\omega\in\mathcal{P}_2(T_0)} \sum_{\gamma\in\mathcal{O}^+(rT_0^k)} c_\gamma E(\xi(\omega\gamma)^k), \end{aligned}$$

where c_γ is the number of solutions of

$$\gamma = (\lambda_1^k - \lambda_2^k) + (\lambda_3^k - \lambda_4^k) + \cdots + (\lambda_{2r-1}^k - \lambda_{2r}^k)$$

subject to $\lambda_j \in \mathcal{A}(T_0,T_0^{\eta})$, $(1\le j\le 2r)$. Note that this last step has made use of the convenient conditions

$$|\gamma^{(i)}| \asymp T_0 \qquad (1\le i\le n)$$

supplied in the definition of $\mathcal{A}(T_0,R)$. By Lemma 3, it follows that

$$\begin{aligned} \left|\sup_{t\in S(t,h)} h(\xi,T_0)\right|^{2r} &\ll T_0^{n(2r-1)} \left(T_0^{k(n-1)+\epsilon}\cdot T_0^{k+2n-b}\right)^{1/2} \\ &\quad \times \left(\sum_{\gamma\in\mathcal{O}^+(rT_0^k)} |c_\gamma|^2\right)^{1/2}. \end{aligned}$$

It is clear from the definition of c_γ that

$$\sum_{\gamma\in\mathcal{O}^+(rT_0^k)} |c_\gamma|^2 \ll S_{2r}(T_0,T_0^{\eta}),$$

where $S_s(T,R)$ denotes, as in [Wo] and [D], the number of solutions of the equation

$$x_1^k + \cdots + x_s^k = y_1^k + \cdots + y_s^k$$

where each variable x_i and y_j belongs to $\mathcal{A}(T,R)$. Hence, we conclude from [D, Lemma 4.13] that

$$\begin{aligned} &\sup_{t\in S(t,h)} \left|h(\xi,T^{1/2})\right| \\ &\ll T_0^{(2r-1)n/2r} T_0^{(k(n-1)+k+2n-b)/4r} T_0^{n(4r-k+\Delta_{2r,k}+\epsilon)/4r} \\ &= T^{n-\frac{1}{8r}(b-n\Delta_{2r,k})+\epsilon} \end{aligned}$$

as in the statement of the Theorem.

REFERENCES

[B] Birch, B. J., *Waring's problem in algebraic number fields*, Proc. Cambr. Phil. Soc. **57** (1961), 449–459.

[D] Davidson, M., *On Siegel's Conjecture in Waring's Problem*, To appear.

[E] Eda, Y., *On Waring's problem in an algebraic number field*, Rev. Colombiana Math. **9** (1975), 29–73.

[K1] Körner, O., *Über das Waringsche Problem in algebraischen Zahlkörpern*, Math. Ann. **144** (1961), 224–238.

[K2] ———, *Über Mittlewerte trigonometrischer Summen und ihre Anwendung in algebraischen Zahlkörpern* **147** (1962), 205–239.

[M1] Mitsui, T., *Generalized prime number theorem*, Jap. Journ. Math. **26** (1956), 1–42.

[M2] ———, *On the Goldbach problem in an algebraic number field I.*, J. Math. Soc. Japan **12** (1960), 290–324.

[R] Ramanujam, C. P., *Sums of m-th powers in p-adic rings*, Mathematika **10** (1963), 137–146.

[S] Siegel, C. L., *Generalization of Waring's problem to algebraic number fields*, American J. of Math. **66** (1944), 122–136.

[T] Thanigasalam, K., *Some new estimates for $G(k)$ in Waring's problem*, Acta. Arith. **42** (1982), 73–78.

[V1] Vaughan, R. C., *The Hardy-Littlewood method*, Cambridge University Press, Cambridge, 1981.

[V2] ———, *A new iterative method in Waring's problem*, Acta Math. **162** (1989), 1–71.

[Vi] Vinogradov, I. M., *The method of trigonometrical sums in the theory of numbers*, Interscience publishers, 1947.

[W] Wang, Y., *Diophantine Equations and Inequalities in Algebraic Number Fields*, Springer-Verlag, Berlin Heidelberg, 1991.

[Wo] Wooley, T. D., *Large improvements in Waring's problem*, Ann. Math. **135** (1992), 131–164.

DEPT. OF MATHEMATICS AND COMPUTER SCIENCE, KENT STATE UNIVERSITY, KENT, OH 44242

E-mail address: davidson@mcs.kent.edu

MODULAR FUNCTIONS, MAPLE AND ANDREWS' 10TH PROBLEM

FRANK GARVAN

ABSTRACT. We discuss how Andrews's 10th problem for generalized Frobenius partitions can be attacked using modular functions and a little help from Maple. Both elementary and function theoretic methods are employed. A multiplicative analogue of the Hecke operator U_q^* is constructed.

Dedicated to Basil Gordon on his sixty fifth birthday

1. INTRODUCTION

Given a partition π we may form the corresponding Frobenius symbol as follows: Firstly, we draw the Ferrers diagram of the partition and circle the main diagonal; then we read the number of nodes in each node to the right of the diagonal and the number of nodes in each column below the main diagonal. In this way we obtain two equal length sequences of decreasing nonnegative integers giving the desired Frobenius symbol:

$$\begin{pmatrix} a_1 & a_2 & a_3 & \cdots & a_r \\ b_1 & b_2 & b_3 & \cdots & b_r \end{pmatrix}.$$

This construction is illustrated below in FIGURE 1, using the partition $\pi = 7+7+5+4+4+2+2+1$.

The corresponding Frobenius symbol is

$$\begin{pmatrix} 6 & 5 & 2 & 0 \\ 7 & 5 & 2 & 1 \end{pmatrix}.$$

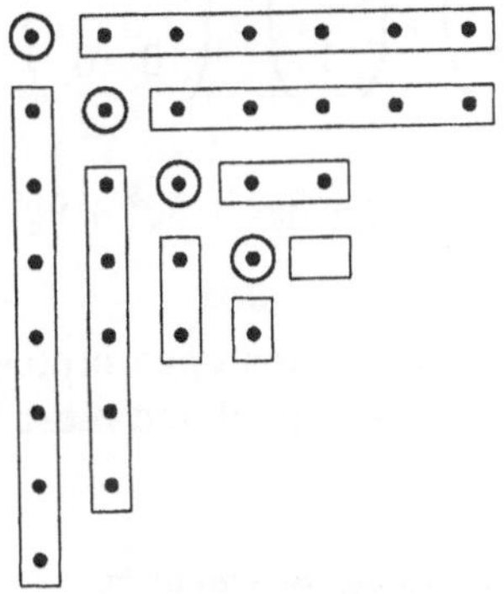

FIG. 1 Ferrers graph of π

S.D. Ahlgren et al. (eds.), Topics in Number Theory, 163–179.

Observe the number being partitioned can be obtained by adding the sum of the entries to the length of the symbol:

$$32 = 4 + (6 + 5 + 2 + 0) + (7 + 5 + 2 + 1).$$

One nice feature of the Frobenius symbol is that conjugation corresponds to swapping the rows of the symbol. In [2], Andrews considered generalizations of the Frobenius symbol, which originated in the solution of certain problems in statistical mechanics [3]. A ***generalized Frobenius partition*** (or more simply an F-partition) of n has the form

$$\begin{pmatrix} a_1 & a_2 & a_3 & \cdots & a_r \\ b_1 & b_2 & b_3 & \cdots & b_r \end{pmatrix},$$

where

$$a_1 \geq a_2 \geq \cdots \geq a_r \geq 0,$$
$$b_1 \geq b_2 \geq \cdots \geq b_r \geq 0,$$

and

$$n = r + \sum_{i=1}^{r} a_i + \sum_{i=1}^{r} b_i.$$

So a generalized Frobenius partition resembles an ordinary Frobenius symbol except repetition of parts is allowed.

Let $k > 0$. We define $\phi_k(n)$ to be the number of F-partitions of n in which entries on each row are repeated at most k times. According to Frobenius

$$\phi_1(n) = p(n),$$

the number of unrestricted partitions of n. We define

$$\Phi_k(q) := \sum_{n\geq 0} \phi_k(n) q^n.$$

EXAMPLE: There are five F-partitions of 3 with $k = 2$:

$$\begin{pmatrix} 2 \\ 0 \end{pmatrix}, \begin{pmatrix} 0 \\ 2 \end{pmatrix}, \begin{pmatrix} 1 \\ 1 \end{pmatrix}, \begin{pmatrix} 1 & 0 \\ 0 & 0 \end{pmatrix}, \begin{pmatrix} 0 & 0 \\ 1 & 0 \end{pmatrix},$$

and

$$\Phi_2(q) = 1 + q + 3q^2 + 5q^3 + 9q^4 + \cdots.$$

At the end of [2] Andrews proposed ten open problems. Many of these problems have been solved. See [9] and [4]. The final and most bewildering problem is given below.

PROBLEM 10, [2]. *Define integers a_n uniquely by*

$$\Phi_4(q) = \prod_{n=1}^{\infty} (1 - q^n)^{-a_n}.$$

Then

$$a_{10n-5} = 2. \tag{1.1}$$

for all n. All other a's seem to become large in absolute value for large subscript.

In this paper we prove (1.1) and give some related conjectures (see § 6). In § 2 we show how (1.1) may be reduced to the identity

$$\frac{X(q)}{X(-q)} = \prod_{n=1}^{\infty} \left[\frac{(1-q^{10n-5})(1+q^{2n-1})^5}{(1-q^{2n-1})^5(1+q^{10n-5})} \right]^2, \tag{1.2}$$

where

$$X(q) := \frac{\Phi(q^{1/5})\Phi(\zeta q^{1/5}) \cdots \Phi(\zeta^4 q^{1/5})}{\Phi(q)}, \tag{1.3}$$

$\Phi_4(q) = \Phi(q)$, and $\zeta = \exp(2\pi i/5)$.

We prove (1.2) by rewriting it as an equation between two modular functions on $\Gamma_0(40)$ (see equation (5.4)) and then checking that enough of their Fourier coefficients agree.

1.1. **Notation.** We use the standard notation

$$(a)_\infty = (a;q)_\infty = \prod_{n=1}^{\infty}(1 - aq^{n-1}),$$

where $|q| < 1$. Let $q := \exp(2\pi i \tau)$ (with $\Im\tau > 0$). As usual the Dedekind eta function is defined as

$$\eta(\tau) := \exp(\pi i \tau/12) \prod_{n=1}^{\infty}(1 - \exp(2\pi i n \tau)) \tag{1.4}$$

$$= q^{1/24} \prod_{n=1}^{\infty}(1 - q^n). \tag{1.5}$$

2. Reduction to a q-identity

In this section we show that the main result (1.1) is equivalent to a q-series identity (2.5). We let

$$\Phi(q) := \Phi_4(q) = \prod_{n=1}^{\infty} \frac{1}{(1-q^n)^{a_n}}$$

$$= \prod_{n \not\equiv 0 \pmod 5} \frac{1}{(1-q^n)^{a_n}} \prod_{n=1}^{\infty} \frac{1}{(1-q^{5n})^{a_{5n}}}$$

Now we let $\zeta = \exp(2\pi i/5)$ so that

$$\Phi(q)\Phi(\zeta q)\Phi(\zeta^2 q)\Phi(\zeta^3 q)\Phi(\zeta^4 q)$$

$$= \prod_{n \not\equiv 0 \pmod 5} \frac{1}{(1-q^{5n})^{a_n}} \prod_{n=1}^{\infty} \frac{1}{(1-q^{5n})^{5a_{5n}}}$$

We find that

$$X(q) := \frac{\Phi(q^{1/5})\Phi(\zeta q^{1/5})\cdots\Phi(\zeta^4 q^{1/5})}{\Phi(q)} = \prod_{n=1}^{\infty}\left[\frac{(1-q^{5n})}{(1-q^n)^5}\right]^{a_{5n}} \tag{2.1}$$

$$= \prod_{n=1}^{\infty}\left[\frac{(1-q^{10n})}{(1-q^{2n})^5}\right]^{a_{10n}} \prod_{n=1}^{\infty}\left[\frac{(1-q^{10n-5})}{(1-q^{2n-1})^5}\right]^{a_{10n-5}}. \tag{2.2}$$

Now we let

$$d_n = a_{10n-5},$$

for all $n \geq 1$. We have

$$\begin{aligned} \frac{X(q)}{X(-q)} &= \prod_{n=1}^{\infty}\left[\frac{(1-q^{10n-5})(1+q^{2n-1})^5}{(1-q^{2n-1})^5(1+q^{10n-5})}\right]^{d_n} \\ &= \prod_{n=1}^{\infty}\left[\frac{(1-q^{10n-5})^2}{(1-q^{2n-1})^{10}}\right]^{d_n} \prod_{n=1}^{\infty}\left[\frac{(1-q^{4n-2})^5}{(1-q^{20n-10})}\right]^{d_n}. \end{aligned} \tag{2.3}$$

LEMMA 2.1 *Let $X(q)$ be defined as above. Then*

$$a_{10n-5} = 2, \tag{2.4}$$

for all n, if and only if,

$$\frac{X(q)}{X(-q)} = \prod_{n=1}^{\infty}\left[\frac{(1-q^{10n-5})(1+q^{2n-1})^5}{(1-q^{2n-1})^5(1+q^{10n-5})}\right]^2 \tag{2.5}$$

Proof. We have already seen how (2.4) implies (2.5). Now we assume that (2.5) holds. We must show that

$$d_n = a_{10n-5} = 2,$$

for all n. By examining (2.3) we see that exponent of $(1 - q^{10n-5})$ in the product expansion of $X(q)/X(-q)$ is

$$2d_n - 10d_{5n-2}.$$

Similarly, when $k = 0, 1, 3$ or 4 we see that the exponent of $(1 - q^{2(5n-k)-1}) = (1 - q^{10n-2k-1})$ is

$$-10d_{5n-k}.$$

Hence we find that (2.5) implies that

$$d_n = \begin{cases} 2, & \text{if } n \not\equiv 3 \pmod 5, \\ \frac{1}{5}(8 + d_{(n+2)/5}), & \text{if } n \equiv 3 \pmod 5. \end{cases}$$

The result follows by induction on n. □

3. A palatable form for the generating function

In this section we show how the generating function $\Phi_4(q)$ may be written in terms of theta functions. We let $\zeta = \exp(2\pi i/5)$. From eq. (1) in [2] we have

$$\begin{aligned} f(q) &:= (q)_\infty^4 \Phi_4(q) \\ &= \sum_{m_1,m_2,m_3=-\infty}^{\infty} \zeta^{3m_1+2m_2+m_3} q^{m_1^2+m_2^2+m_3^2+m_1m_2+m_2m_3+m_1m_3}. \end{aligned}$$

We observe that the exponent of q above can be written neatly as

$$\begin{aligned} & m_1^2+m_2^2+m_3^2+m_1m_2+m_2m_3+m_1m_3 \\ &= \frac{1}{2}(m_1+m_2)^2 + \frac{1}{2}(m_2+m_3)^2 + \frac{1}{2}(m_1+m_3)^2. \end{aligned}$$

We change variables in the summation and find that

$$f(q) = \sum_{n_1+n_2+n_3 \equiv 0 \pmod 2} \zeta^{2n_1+n_3} q^{\frac{1}{2}(n_1^2+n_2^2+n_3^2)},$$

so that

$$\begin{aligned} 2f(q^2) = & \sum_{n_1,n_2,n_3} \zeta^{2n_1+n_3}(-1)^{n_1+n_2+n_3} q^{(n_1^2+n_2^2+n_3^2)} \\ & + \sum_{n_1,n_2,n_3} \zeta^{2n_1+n_3} q^{(n_1^2+n_2^2+n_3^2)}. \end{aligned} \tag{3.1}$$

We need Jacobi's triple product identity [1]

$$\sum_{n=-\infty}^{\infty} (-1)^n z^n q^{n^2} = (zq;q^2)_\infty (q/z;q^2)_\infty (q^2;q^2)_\infty. \tag{3.2}$$

Thus we may write the first sum on the right side of (3.1) as

$$\begin{aligned} & \sum_{n_1} (-1)^{n_1} \zeta^{2n_1} q^{n_1^2} \sum_{n_2} (-1)^{n_2} q^{n_2^2} \sum_{n_3} (-1)^{n_3} \zeta^{n_3} q^{n_3^2} \\ &= (\zeta^2 q;q^2)_\infty (q/\zeta^2;q^2)_\infty (q^2;q^2)_\infty (q;q^2)_\infty^2 (q^2;q^2)_\infty \\ &\quad (\zeta q;q^2)_\infty (q/\zeta;q^2)_\infty (q^2;q^2)_\infty \\ &= (q;q^2)_\infty^2 (q^5;q^{10})_\infty^2 (q^2;q^2)_\infty^3. \end{aligned}$$

Again via (3.2) we have

$$\begin{aligned} (q;q^2)_\infty (q^4;q^4)_\infty &= (q;q^4)_\infty (q^3;q^4)_\infty (q^4;q^4)_\infty \\ &= \sum_n (-1)^n q^{2n^2-n} \\ &= \sum_n q^{8n^2-2n} - q\sum_n q^{8n^2+6n}. \end{aligned}$$

Thus if we define

$$\alpha(q) := \sum_n q^{4n^2+n}, \qquad \beta(q) := \sum_n q^{4n^2+3n};$$

then we find that

$$(q;q^2)_\infty = \frac{1}{(q^4;q^4)_\infty}\left[\alpha(q^2) - q\beta(q^2)\right]$$

and after some simplification that

$$f(q) = (q)_\infty^4 \Phi_4(q) = \frac{(q)_\infty^3}{(q^2;q^2)_\infty(q^{10};q^{10})_\infty}\left\{\alpha(q)\alpha(q^5) + q^3\beta(q)\beta(q^5)\right\},$$

and

$$\Phi_4(q) = \frac{\alpha(q)\alpha(q^5) + q^3\beta(q)\beta(q^5)}{(q)_\infty(q^2;q^2)_\infty(q^{10};q^{10})}. \tag{3.3}$$

As a side remark it turns out that

$$q^{3/8}\left\{\alpha(-q)\alpha(-q^5) + q^3\beta(-q)\beta(-q^5)\right\} = \eta(4\tau)\,\eta(5\tau).$$

We omit the details.

4. Modular Functions

In this section we show how $\Phi_4(-q)$ can be written in terms of modular functions and set up the necessary results on modular functions to complete the proof.

4.1. Background theory. The necessary background theory of modular functions and modular forms may be found in [14], [16], [10] and [13]. Many of the results that we require are contained in [6], [7] and [8]. Let $\Gamma(1)$ denote the full modular group and as usual let

$$\Gamma_0(N) := \left\{ \begin{pmatrix} a & b \\ c & d \end{pmatrix} \in \Gamma(1) : c \equiv 0 \pmod N \right\}.$$

Let Γ be a subgroup of $\Gamma(1)$ with finite index. For a modular function f on Γ and a cusp ζ the order of f (mod Γ) at ζ is denoted by ORD $(f;\zeta;\Gamma)$ and the *invariant order* of f at ζ is denoted by ord $(f;\zeta)$. We have

$$\mathrm{ORD}\,(f;\zeta;\Gamma) = \kappa(\Gamma;\zeta)\,\mathrm{ord}\,(f;\zeta;\Gamma) \tag{4.1}$$

where $\kappa(\Gamma;\zeta)$ denotes the fan width of the cusp ζ (mod Γ), and

$$\mathrm{ord}\,(f \mid A;\zeta) = \mathrm{ord}\,(f;A\zeta), \tag{4.2}$$

for $A \in \Gamma(1)$. Here we use the usual stroke operator notation

$$(f \mid A)(\tau) := f(A\tau).$$

Any non-zero modular function f must satisfy the valence formula:

$$\sum_{s\in\mathcal{F}} \mathrm{ORD}\,(f;s;\Gamma) = 0, \tag{4.3}$$

where $\mathcal{F}$ is a fundamental set of Γ and the sum is taken over s with non-zero order so that this is a finite sum. For $s \in \mathcal{H}$ the order is interpreted in the usual sense. See [14] for more details. We shall use the valence formula to prove certain modular function identities. This is a standard technique.

4.2. Transformation Formulae. Following [15, p.6], [17] we define

$$\theta_{01}(v,\tau) := \prod_{n=1}^{\infty}(1-q^{n-1/2}e^{2\pi iv})(1-q^{n-1/2}e^{-2\pi iv})(1-q^n).$$

Here $q=\exp(2\pi i\tau)$. For odd k we define

$$\begin{aligned} U_k(\tau) &:= q^{k^2/16}\frac{\theta_{01}(k\tau,8\tau)}{\theta_{01}(0,8\tau)} \\ &= q^{k^2/16}\prod_{n=1}^{\infty}\frac{(1-q^{8n-4+k})(1-q^{8n-4-k})}{(1-q^{8n-4})^2}. \end{aligned} \tag{4.4}$$

We remark that

$$\alpha(-q) = \sum_n(-1)^n q^{4n^2+n} = \theta_{01}(\tau,8\tau), \tag{4.5}$$

$$\beta(-q) = \sum_n(-1)^n q^{4n^2+3n} = \theta_{01}(3\tau,8\tau). \tag{4.6}$$

Weber [17] has calculated transformation formulae for the theta-functions θ_{ij} under the action of each element of the full modular group Γ. Using these formulae it is a simple matter to show that

$$U_k(\tau)\,|\,A = \exp(\pi ik^2ab/8)\,U_{ka}(\tau), \tag{4.7}$$

where $A=\begin{pmatrix} a & b \\ c & d\end{pmatrix}\in\Gamma_0(8)$.

We need to write $\Phi_4(q)$ in terms of a modular function. The function we need is given in the following lemma.

LEMMA 4.1 *The function*

$$F(\tau) := U_1(\tau)U_3(\tau)\,[U_1(\tau)U_1(5\tau)-U_3(\tau)U_3(5\tau)] \tag{4.8}$$

is a modular function on the congruence subgroup $\Gamma_0(40)$.

Proof. Let $A=\begin{pmatrix} a & b \\ c & d\end{pmatrix}\in\Gamma_0(40)$. Then

$$A^* = \begin{pmatrix} a & 5b \\ c/5 & d\end{pmatrix}\in\Gamma_0(8).$$

Hence

$$U_1(5\tau)\,|\,A = (U_1(\tau)\,|\,A^*)\,(5\tau) = \exp(5ab\pi i/8)\,U_a(5\tau),$$

and

$$U_1(\tau)U_1(5\tau)\,|\,A = \exp(3ab\pi i/4)\,U_a(\tau)U_a(5\tau).$$

Similarly we find that

$$U_3(\tau)U_3(5\tau)\,|\,A = \exp(3ab\pi i/4)\,U_{3a}(\tau)U_{3a}(5\tau),$$

and

$$U_1(\tau)U_3(\tau)\,|\,A = \exp(5ab\pi i/4)\,U_a(\tau)U_{3a}(\tau).$$

Hence

$$F(\tau)\,|\,A = U_a(\tau)U_{3a}(\tau)\left[U_a(\tau)U_a(5\tau) - U_{3a}(\tau)U_{3a}(5\tau)\right].$$

From the definition (4.4) it is easy to show that

$$U_k(\tau) = -U_{k+8}(\tau), \qquad U_k(\tau) = U_{-k}(\tau),$$

for all integers k. The result

$$F(\tau)\,|\,A = F(\tau),$$

follows easily by considering the two cases $a \equiv \pm 1 \pmod 8$ and $a \equiv \pm 3 \pmod 8$. Thus F is a modular function on $\Gamma_0(40)$. □

By (3.3), and using (4.4)–(4.6), (4.8) we have

(4.9) $$\Phi_4(-q) = q^{1/6}\,\frac{\eta^2(20\tau)\eta^7(4\tau)}{\eta(40\tau)\eta(10\tau)\eta^5(8\tau)\eta^3(2\tau)}\,F(\tau).$$

4.3. Orders at cusps. We will need the following lemmas

LEMMA 4.2, [8]. *If $(r, s) = 1$, then the fan width of $\Gamma_0(N)$ at $\frac{r}{s}$ is*

$$\kappa\left(\Gamma_0(N); \frac{r}{s}\right) = \frac{N}{(N, s^2)}.$$

LEMMA 4.3, [8]. *If no square divides N other than 1 or 4, then a complete set of inequivalent cusps for $\Gamma_0(N)$ is*

$$\left\{\frac{1}{a} : a \mid n\right\}.$$

LEMMA 4.4, [8]. *If no square divides N other than 1 or 4, and if $(r, s) = (r', s') = 1$, then $\frac{r}{s}$ and $\frac{r'}{s'}$ are equivalent cusps modulo $\Gamma_0(N)$ if and only if $(s, N) = (s', N)$.*

THEOREM 4.5 *Suppose $4 \mid N$ and N has no other square divisors. If $f(\tau)$ is a modular function on $\Gamma_0(N)$ then the function $f(\tau + \frac{1}{2})$ is also a modular function on $\Gamma_0(N)$. Further*

1.

(4.10) $$\mathrm{ord}\,(f(\tau + \tfrac{1}{2}); 0) = \frac{1}{4}\mathrm{ord}\,(f; \tfrac{1}{2}).$$

2. *If $4 \mid c$ then*

(4.11) $$\mathrm{ord}\,(f(\tau + \tfrac{1}{2}); \tfrac{1}{c}) = \mathrm{ord}\,(f; \tfrac{1}{c}).$$

3. *If $c \equiv 2 \pmod 4$ then*

(4.12) $$\mathrm{ord}\,(f(\tau + \tfrac{1}{2}); \tfrac{1}{c}) = 4\,\mathrm{ord}\,(f; \tfrac{2}{c}).$$

4. *If c is odd then*

$$\operatorname{ord}(f(\tau+\tfrac{1}{2});\tfrac{1}{c}) = \frac{1}{4}\operatorname{ord}(f;\tfrac{1}{2c}). \tag{4.13}$$

Proof. Suppose $4 \mid N$ and N has no other square divisors, and suppose $f(\tau)$ is a modular function on $\Gamma_0(N)$. Let $A = \begin{pmatrix} a & b \\ c & d \end{pmatrix} \in \Gamma_0(N)$ so that $4 \mid c$. Since $ad-bc=1$, $a \equiv d \equiv 1 \pmod 2$. We let $B = \begin{pmatrix} 2 & 1 \\ 0 & 2 \end{pmatrix}$ and

$$C = BAB^{-1} = \begin{pmatrix} a+c/2 & \frac{1}{2}(d-a)+b-c/4 \\ c & d-c/2 \end{pmatrix},$$

so that $C \in \Gamma_0(N)$ and $BA = CB$. We have

$$\begin{aligned} f(\tau+\tfrac{1}{2}) \mid A &= (f \mid B) \mid A = f \mid (BA) \\ &= f \mid (CB) = (f \mid C) \mid B = f \mid B \quad &&\text{(since } f \text{ is a modular function on } \Gamma_0(N)\text{)} \\ &= f(\tau+\tfrac{1}{2}), \end{aligned}$$

and $f(\tau+\frac{1}{2})$ is a modular function on $\Gamma_0(N)$.

We will use (4.2) with $\zeta = \infty$ to prove (4.10)–(4.13). For a cusp $\frac{1}{c}$ we take $A = \begin{pmatrix} 1 & 1 \\ c & c+1 \end{pmatrix} \in \Gamma(1)$. For such a fixed cusp we show that there is a matrix $D = \begin{pmatrix} g & h \\ 0 & 4/g \end{pmatrix} \in \Gamma(1)$, where $g, h \in \mathbb{Z}$, $g \mid 4$ and such that $C = BAD^{-1} \in \Gamma(1)$. In this case

$$C = \begin{pmatrix} \frac{1}{g}(2+c) & \frac{g}{4}(3+c) - \frac{h}{4}(2+c) \\ \frac{2c}{g} & \frac{1}{2}((c+1)g - ch) \end{pmatrix},$$

$\det(C) = 1$ and $BA = CD$. Assuming $C \in \Gamma(1)$ we have

$$f(\tau+\tfrac{1}{2}) \mid A = (f \mid B) \mid A = (f \mid C) \mid D$$

so that

$$\operatorname{ord}(f(\tau+\tfrac{1}{2}); A\infty) = \frac{g^2}{4}\operatorname{ord}(f; C\infty). \tag{4.14}$$

There are four cases. In each case it is a simple matter to verify that $C \in \Gamma(1)$, and the corresponding desired result follows via Lemma 4.4 and equation (4.14).

Case 1. $4 \mid c$. We take $g = 2$, $h = 1$, and find that the cusp $C\infty$ is equivalent to $\frac{1}{c}$ and we have (4.11).

Case 2. $c \equiv 2 \pmod 4$. We take $g = 4$, $h = 2$, and find that the cusp $C\infty$ is equivalent to $\frac{1}{c/2}$ and we have (4.12).

CASE 3. $c \equiv 1 \pmod 4$. We take $g = 1$, $h = 4$, and find that the cusp $C\infty$ is equivalent to $\frac{1}{2c}$ and we have (4.13). We also note that (4.10) follows in this case since 0 and 1 are equivalent cusps.
CASE 4. $c \equiv 3 \pmod 4$. We take $g = 1$, $h = 2$, and find that the cusp $C\infty$ is equivalent to $\frac{1}{2c}$ and we have the remaining case of (4.13).
This completes the proof. □

Let p be a prime and suppose $p \mid N$. Then the Hecke operator U_p^* defined by

$$f \mid U_p^* := \sum_{j=0}^{p-1} f\left(\frac{\tau + j}{p}\right) \tag{4.15}$$

is well-known. The notation U_p^* is due to Atkin and Lehner [5, p.138]. If p is a prime, $p \mid N$ and $f(\tau)$ is a modular function on $\Gamma_0(N)$ then $f \mid U_p^*$ is also a modular function on $\Gamma_0(N)$. See [5, Lemma 6]. We need a multiplicative analog of this Hecke operator. For a modular function $f(\tau)$ and a prime p we define

$$f^{*(p)}(\tau) := \prod_{j=0}^{p-1} f\left(\frac{\tau + j}{p}\right). \tag{4.16}$$

We do not use the "stroke" notation here since this is clearly not a linear operator. We use the notation ${}^{*(p)}$ simply to indicate the dependence on p.
We have the following

THEOREM 4.6 *Suppose $4 \mid N$, N has no other square divisors and p is a fixed odd prime divisor of N. If $f(\tau)$ is a modular function on $\Gamma_0(N)$ then $f^{*(p)}(\tau)$ is also a modular function on $\Gamma_0(N)$. Further*

1. *if $p \nmid c$ then*

$$\mathrm{ord}\,(f^{*(p)}; \tfrac{1}{c}) = p\,\mathrm{ord}\,(f; \tfrac{1}{c}) + \tfrac{(p-1)}{p}\mathrm{ord}\,(f; \tfrac{1}{pc}); \quad \text{and} \tag{4.17}$$

2. *if $p \mid c$ then*

$$\mathrm{ord}\,(f^{*(p)}; \tfrac{1}{c}) = \mathrm{ord}\,(f; \tfrac{1}{c}). \tag{4.18}$$

Proof. Suppose N, p satisfy the conditions of the theorem and $f(\tau)$ is a modular function on $\Gamma_0(N)$. The proof that $f^{*(p)}(\tau)$ is also a modular function on $\Gamma_0(N)$ is completely analogous to the proof that the Hecke operator U_p^* has the same property. See [5, Lemma 6].

By Lemma 4.3, we need only consider cusps of the form $\frac{1}{c}$ where $c \mid N$. For $0 \le k \le p-1$ define

$$A_k = \begin{pmatrix} 1 & k \\ 0 & p \end{pmatrix}$$

and suppose $c \mid N$. Let

$$B = \begin{pmatrix} 1 & 1 \\ c & c+1 \end{pmatrix}.$$

For $a=1$ or $a=p$ we let

$$C=\begin{pmatrix} a & b \\ 0 & p/a \end{pmatrix}.$$

and

$$M=A_kBC^{-1}=\begin{pmatrix} \frac{1+kc}{a} & \frac{1}{p}((1+kc+k)a-b(1+kc)) \\ \frac{pc}{a} & a+ac-bc \end{pmatrix},$$

so that $\det(M)=1$. If $M\in\Gamma(1)$ we have

$$f\left(\tfrac{\tau+k}{p}\right)\mid B=f\mid A_kB=(f\mid M)\mid C.$$

Since C corresponds to a translation it follows that

$$\begin{aligned}(4.19)\qquad \operatorname{ord}(f(\tfrac{\tau+k}{p});B\infty)&=\operatorname{ord}(f(\tfrac{\tau+k}{p})\mid B;\infty)\\ &=\operatorname{ord}(f\mid MC;\infty)=\frac{a^2}{p}\operatorname{ord}(f;M\infty),\end{aligned}$$

again assuming that $M\in\Gamma(1)$.

CASE 1. $p\nmid c$. Suppose $0\le k\le p-1$. There exists a unique value of k, such that $kc\equiv -1\pmod{p}$. We denote this k by k_0.

- Suppose $k=k_0$. Then we take $a=p$ and $b=0$ so that

$$\operatorname{ord}(f(\tfrac{\tau+k_0}{p});\tfrac{1}{c})=p\operatorname{ord}(f;\tfrac{1}{c}),$$

since $M\infty=\frac{(1+k_0c)/p}{c}$ is equivalent to the cusp $\frac{1}{c}$ by Lemma 4.4.
- Suppose $k\ne k_0$. Therefore $p\nmid kc+1$ and we take $a=1$. This time we choose b so that $b(kc+1)\equiv 1+kc+k\pmod{p}$. Using (4.19) we find that

$$\operatorname{ord}(f(\tfrac{\tau+k}{p});\tfrac{1}{c})=\frac{1}{p}\operatorname{ord}(f;\tfrac{1}{pc}),$$

since $M\infty=\frac{1+k_0c}{pc}$ is equivalent to the cusp $\frac{1}{pc}$ by Lemma 4.4. Hence we have

$$\operatorname{ord}(f^{*(p)};\tfrac{1}{c})=\sum_{k=0}^{p-1}\operatorname{ord}(f(\tfrac{\tau+k}{p});\tfrac{1}{c})=p\operatorname{ord}(f;\tfrac{1}{c})+\tfrac{(p-1)}{p}\operatorname{ord}(f;\tfrac{1}{pc}),$$

which is (4.17).

CASE 2. $p\mid c$. Again we suppose $0\le k\le p-1$. We take $a=1$ and $b=1+k$ and find

$$M=\begin{pmatrix} 1+kc & -k^2c/p \\ pc & 1-kc \end{pmatrix}\in\Gamma(1).$$

Using (4.19) we find that

$$\operatorname{ord}(f(\tfrac{\tau+k}{p});\tfrac{1}{c})=\frac{1}{p}\operatorname{ord}(f;\tfrac{1+kc}{pc}).$$

Now $(1 + kc, pc) = (1, pc) = 1$ and $(pc, N) = (c, N) = c$ since $p \mid c$ and $p^2 \nmid N$. Therefore the two cusps $\frac{1+kc}{pc}$ and $\frac{1}{c}$ are equivalent by Lemma 4.4. For each k we have

$$\operatorname{ord}\left(f\left(\tfrac{\tau+k}{p}\right); \tfrac{1}{c}\right) = \frac{1}{p} \operatorname{ord}\left(f; \tfrac{1}{c}\right),$$

and (4.18) follows. □

We will need results on the orders at cusps of eta-products and theta-functions. Newman [12] has found necessary and sufficient conditions under which an eta-product is a modular function on $\Gamma_0(N)$. Let $N > 0$ be a fixed integer. Here an eta-product takes the form

$$f(\tau) = \prod_{d \mid N} \eta(d\tau)^{m_d}, \tag{4.20}$$

where each $d > 0$ and $m_d \in \mathbb{Z}$.

THEOREM 4.7, [12]. *The function $f(\tau)$ (given in (4.20)) is a modular function on $\Gamma_0(N)$ if and only if*

1. $\sum_{d \mid N} m_d = 0,$
2. $\sum_{d \mid N} d m_d \equiv 0 \pmod{24},$
3. $\sum_{d \mid N} \frac{N m_d}{d} \equiv 0 \pmod{24},$ *and*
4. $\prod_{d \mid N} d^{|m_d|}$ *is a square.*

Ligozat [11] has computed the order of the eta-product $f(\tau)$ at the cusps of $\Gamma_0(N)$.

THEOREM 4.8, [11]. *If the function $f(\tau)$ (given in (4.20)) is a modular function on $\Gamma_0(N)$, then its order at the cusp $s = \frac{b}{c}$ (assuming $(b, c) = 1$) is*

$$\operatorname{ord}(f(\tau); s) = \sum_{d \mid N} \frac{(d, c)^2 m_d}{24d}. \tag{4.21}$$

Biagoli [8] has computed the order of the theta function:

$$f_{n,\rho}(\tau) = q^{(n-2\rho)^2/(8n)} \prod_{m=1}^{\infty} (1 - q^{mn-\rho})(1 - q^{mn-(\rho-n)})(1 - q^{mn}), \tag{4.22}$$

at any cusp. Here $n, \rho \in \mathbb{Z}$ with $n \geq 1$ and $n \nmid \rho$.

THEOREM 4.9, [8, Lemma 3.2, p.285]. *The order at the cusp* $s = \frac{b}{c}$ *(assuming* $(b,c)=1$*) of the theta function* $f_{n,\rho}(\tau)$ *(defined above and assuming* $\rho \nmid n$*) is*

$$\operatorname{ord}(f_{n,\rho}(\tau); s) = \frac{e^2}{2n}\left(\frac{b\rho}{e} - \left[\frac{b\rho}{e}\right] - \frac{1}{2}\right)^2, \tag{4.23}$$

where $e = (n,c)$ *and* $[\ ]$ *is the greatest integer function.*

5. Proof of the main result

In section 2 we saw in Lemma 2.1 how the main result (2.4) is equivalent to the q-series identity (2.5). We now recast this identity in terms of modular functions and complete the proof.

Recall that $F(\tau)$ was defined in (4.8). We define

$$H(\tau) := \frac{\eta^6(10\tau)\eta^{10}(4\tau)\eta^{20}(\tau)}{\eta^2(20\tau)\eta^4(5\tau)\eta^{30}(2\tau)} \tag{5.1}$$

and

$$\Xi(\tau) := \frac{F^*(\tau)}{F(\tau)}. \tag{5.2}$$

Using (1.4), (1.5), it is straightforward to show that $H(\tau)$ is the reciprocal of the function on the right side of (2.5), and is a modular function on $\Gamma_0(40)$ by Theorem 4.7. From (4.9) we observe that the quotient of $\Phi(-q)$ and $F(\tau)$ is an even function of q. It follows that (2.5) is equivalent to the identity

$$\frac{\Xi(\tau)}{\Xi(\tau+\frac{1}{2})} = H(\tau); \tag{5.3}$$

i.e.

$$F^*(\tau)F(\tau+\tfrac{1}{2}) = H(\tau)\,F(\tau)F^*(\tau+\tfrac{1}{2}). \tag{5.4}$$

Thus we have been able to reduce our original problem to a modular function identity. Here $F^* = F^{*(p)}$ with $p=5$. Thus all the functions that appear in (5.4) are modular functions on $\Gamma_0(40)$ by Theorems 4.5 and 4.6.

Using Lemma 4.1 and Theorem 4.9 we may calculate lower bounds for the order of $F(\tau)$ at each cusp of $\Gamma_0(40)$. Using the techniques of [8] it is possible to calculate the exact order at each cusp with a little more work. We omit the details. These orders are given below in TABLE 1. A set of inequivalent cusps for $\Gamma_0(40)$ is

$$\left\{0, \tfrac{1}{40}, \tfrac{1}{20}, \tfrac{1}{10}, \tfrac{1}{8}, \tfrac{1}{5}, \tfrac{1}{4}, \tfrac{1}{2}\right\}.$$

Observe that from TABLE 1 we see that

$$\sum_{s\in S} \operatorname{ORD}(F; s; \Gamma_0(40)) = -2. \tag{5.5}$$

Here S is a set of inequivalent cusps for $\Gamma_0(40)$. (See first column of TABLE 1). This means, by the valence formula (4.3), that $F(\tau)$ has two inequivalent zeros (counted with multiplicity) in the upper-half plane $\mathcal{H}$. The author has found these zeros but this result and more general results will be left until a later paper.

Cusp s	$\operatorname{ord}(F;s)$	$\operatorname{ORD}(F;s;\Gamma_0(40))$
0	0	0
1/40	1	1
1/20	−1	−1
1/10	−1/2	−1
1/8	4/5	4
1/5	0	0
1/4	−3/5	−3
1/2	−1/5	−2

TABLE 1. Order of $F(\tau)$ at each inequivalent cusp of $\Gamma_0(40)$

Cusp s	$\operatorname{ord}(L;s) = \operatorname{ord}(R;s)$	ORD
0	−1/20	−2
1/40	2	2
1/20	−2	−2
1/10	−1/2	−1
1/8	28/5	28
1/5	−1/8	−1
1/4	−22/5	−22
1/2	−7/5	−14

TABLE 2. Orders of $L(\tau)$ and $R(\tau)$ at each inequivalent cusp of $\Gamma_0(40)$

We let $L(\tau)$ (resp. $R(\tau)$) denote the modular function on the left (resp. right) side of (5.4). We will use the valence formula (4.3) to prove that $L(\tau) - R(\tau)$ is identically zero. Since we know the exact order of $F(\tau)$ at each cusp we may compute the exact order of $L(\tau)$ and $R(\tau)$ using (4.1), and Theorems 4.5, 4.6 and 4.8. The results are given below in TABLE 2.

From the table we have

$$\sum_{s\in S} \operatorname{ORD}(L;s;\Gamma_0(40)) = \sum_{s\in S} \operatorname{ORD}(R;s;\Gamma_0(40)) = -12, \tag{5.6}$$

and

$$\sum_{s\in S^*} \mathrm{ORD}\,(L-R;s;\Gamma_0(40)) \geq -14. \tag{5.7}$$

Here $S^* = S\setminus\{1/40\}$.

We note that ∞ is equivalent to the cusp $1/40$. In view of the valence formula (4.3), our main result will follow if we can show

$$\mathrm{ORD}\,(L-R;\infty;\Gamma_0(40)) \geq 15.$$

We do this indirectly. Here it is easier to calculate a product expansion than a q-series expansion. Since $\kappa(\Gamma_0(40),\infty)=1$, and $\mathrm{ORD}\,(F;\infty;\Gamma_0(40)) = \mathrm{ord}\,(F;\infty) = \mathrm{ord}\,(F^*(\tau+\frac{1}{2})) = 1$ we need to show the q-expansion of both sides of (5.3) agree up through q^{13}. Now define a sequence of integers b_n by

$$\prod_{n=1}^{\infty}(1-q^n)^{b_n} = q^{-3/8}\left(U_1(\tau)U_1(5\tau) - U_3(\tau)U_3(5\tau)\right). \tag{5.8}$$

Using (4.8), the fact that

$$U_1(\tau)U_3(\tau) = q^{5/8}\prod_{n=1}^{\infty}\frac{(1-q^{2n-1})}{(1-q^{8n-4})^4}, \tag{5.9}$$

and an argument analogous to that used in the proof of Lemma 2.1, we need only show that

$$b_{10n-5} = 1, \tag{5.10}$$

for $n = 1$, 2, 3, 4, 5 and 6. We have been able to verify this using the symbolic algebra package MAPLE, thus proving our main result. In fact we have verified (5.10) for the first 20 values of n.

6. Conjectures

We ask whether the phenomenon of Andrews's 10th Problem occurs in any other naturally occurring q-series. For fixed odd p we define two sequences a_n and b_n by

$$\prod_{n=1}^{\infty}(1-q^n)^{a_n} = q^{-(p+1)/16}\left(U_1(\tau)U_1(p\tau) + U_3(\tau)U_3(p\tau)\right), \tag{6.1}$$

and

$$\prod_{n=1}^{\infty}(1-q^n)^{b_n} = q^{-(p+1)/16}\left(U_1(\tau)U_1(p\tau) - U_3(\tau)U_3(p\tau)\right). \tag{6.2}$$

Then we

CONJECTURE:

1. *If* $p \equiv 3 \pmod 8$ *then* $a_{2p(2n+1)} = 0$ *for all* n.
2. *If* $p \equiv 5 \pmod 8$ *then* $a_{p(2n+1)} = b_{p(2n+1)} = 1$ *for all* n.
3. *If* $p \equiv 7 \pmod 8$ *then* $a_{2(2n+1)} = 0$ *for all* n.

Andrews's 10th Problem is equivalent to the case for $b_{p(2n+1)}$ with $p = 5$. See (5.10). The conjectures for a_n and b_n have been checked for all odd $p \leq 91$ and $n \leq 994$, using Maple. For instance, $p = 91 \equiv 3 \pmod 8$ and we have checked that for this p, $a_{182} = a_{546} = a_{910} = 0$. On the face of it, this is quite remarkable when we observe that the other terms seem to grow rapidly in absolute value. For example,

$$\begin{aligned} a_{909} &= -92151123491267743803787143128 \\ &= -2^3 \cdot 11518890436408467975473392891. \end{aligned}$$

Each of the conjectures corresponds to a modular function identity analogous to (5.4). Since there are infinitely many such conjectures the methods of this paper are not sufficient. An alternative approach is to show that functions on each side of a particular identity have the same order for each cusp and have the same zeros in the upper-half plane. This author has made some progress in this direction. More details and hopefully a proof of the conjectures will appear in a later paper.

REFERENCES

[1] George E. Andrews. *The theory of partitions*. Addison-Wesley Publishing Co., Reading, Mass.-London-Amsterdam, 1976. Encyclopedia of Mathematics and its Applications, Vol. 2.

[2] George E. Andrews. Generalized Frobenius partitions. *Mem. Amer. Math. Soc.*, 49(301):iv+44, 1984.

[3] George E. Andrews. Use and extension of Frobenius' representation of partitions. In *Enumeration and design (Waterloo, Ont., 1982)*, pages 51–65. Academic Press, Toronto, Ont., 1984.

[4] George E. Andrews. *q-series: their development and application in analysis, number theory, combinatorics, physics, and computer algebra*, volume 66 of *CBMS Regional Conference Series in Mathematics*. Published for the Conference Board of the Mathematical Sciences, Washington, D.C., 1986.

[5] A. O. L. Atkin and J. Lehner. Hecke operators on $\Gamma_0(m)$. *Math. Ann.*, 185:134–160, 1970.

[6] Bruce C. Berndt, Anthony J. Biagioli, and James M. Purtilo. Ramanujan's "mixed" modular equations. *J. Ramanujan Math. Soc.*, 1(1-2):46–70, 1986.

[7] Bruce C. Berndt, Anthony J. Biagioli, and James M. Purtilo. Ramanujan's modular equations of "large" prime degree. *J. Indian Math. Soc. (N.S.)*, 51:75–110 (1988), 1987.

[8] Anthony J. F. Biagioli. A proof of some identities of Ramanujan using modular forms. *Glasgow Math. J.*, 31(3):271–295, 1989.

[9] F.G. Garvan. *Generalizations of Dyson's rank*. PhD thesis, The Pennsylvania State University, 1986.

[10] Marvin I. Knopp. *Modular functions in analytic number theory*. Markham Publishing Co., Chicago, Ill., 1970.

[11] Gérard Ligozat. *Courbes modulaires de genre* 1. Société Mathématique de France, Paris, 1975. Bull. Soc. Math. France, Mém. 43, Supplément au Bull. Soc. Math. France Tome 103, no. 3.

[12] Morris Newman. Construction and application of a class of modular functions. II. *Proc. London Math. Soc. (3)*, 9:373–387, 1959.

[13] Hans Rademacher. *Topics in analytic number theory*. Springer-Verlag, New York, 1973. Edited by E. Grosswald, J. Lehner and M. Newman, Die Grundlehren der mathematischen Wissenschaften, Band 169.

[14] Robert A. Rankin. *Modular forms and functions*. Cambridge University Press, Cambridge, 1977.

[15] Øystein Rødseth. Dissections of the generating functions of $q(n)$ and $q_0(n)$. *Årbok Univ. Bergen Mat.-Natur. Ser.*, 1969(12):12 pp. (1970).

[16] Bruno Schoeneberg. *Elliptic modular functions: an introduction*. Springer-Verlag, New York, 1974. Translated from the German by J. R. Smart and E. A. Schwandt, Die Grundlehren der mathematischen Wissenschaften, Band 203.

[17] Heinrich Weber. *Lehrbuch der Algebra*. Braunschweig, F. Vieweg und sohn, 1908. Vol. 3.

E-mail address: `frank@math.ufl.edu`

DEPARTMENT OF MATHEMATICS, UNIVERSITY OF FLORIDA, GAINESVILLE, FL 32611

HECKE CHARACTERS AND FORMAL GROUP CHARACTERS

LI GUO

ABSTRACT. Let E be an elliptic curve with complex multiplication by the ring of integers of an imaginary quadratic field K. The theory of complex multiplication associates E with a Hecke character ψ. The Hasse-Weil L-function of E equals the Hecke L-function of ψ, whose special value at $s = 1$ encodes important arithmetic information of E, as predicted by the Birch and Swinnerton-Dyer conjecture and verified by Rubin[**Ru**] when the special value is non-zero. For integers k, j, special values of the Hecke L-function associated to the Hecke character $\psi^k \bar{\psi}^j$ should encode arithmetic information of the Hecke character $\psi^k \bar{\psi}^j$, as predicted by the Bloch-Kato conjecture[**B-K**]. When p is a prime where E has good, ordinary reduction, the p-part of the conjecture has been verified when $j = 0$[**Ha**] and when $j \neq 0, p > k$[**Gu**]. To verify the conjecture in other cases, it is important to have an explicit description of the exponential map of Bloch and Kato. In this paper we provide such an explicit exponential map for the case $j \neq 0$, $p \nmid k$, by studying relation between the Hecke characters and formal group characters.

Let E be an elliptic curve over $\mathbb{Q}$ with complex multiplication by the ring of integers of an imaginary quadratic field. Let ψ be the Hecke character associated to E by the theory of complex multiplication. Let $\bar{\psi}$ be the complex conjugate character of ψ. For a pair of integers k and j, define the Hecke character $\varphi = \varphi(k, j) = \psi^k \bar{\psi}^j$. Let p be a prime where E has good, ordinary reduction. Let $\mathfrak{p}$ be a fixed prime of K dividing p. In this paper, we study properties of the Bloch-Kato exponential map for the p-adic Hecke character $\varphi_\mathfrak{p}$. In section 1, we establish a relation identifying $\varphi_\mathfrak{p}$ with the character from a Lubin-Tate formal group. This extends classical results on the Hecke character ψ associated to E and the p-adic cyclotomic character $\chi_p = \psi_\mathfrak{p} \bar{\psi}_\mathfrak{p}$. Then, in section 2, we use this relation, together with results of Fontaine [**Fo2**] and Harrison [**Ha**], to give an explicit description of the exponential map of the p-adic Galois representation given by $\varphi_\mathfrak{p}$, extending theorems of Bloch-Kato [**B-K**] and Harrison [**Ha**]. We then describe the image of the exponential map in section 3. Such an description of the exponential map is important in studying the arithmetic properties of the Hecke character φ, for example in verifying the Tamagawa number conjecture of Bloch and Kato for the motive associated to φ. For another approach to explicit exponential maps, see the papers of Kato [**Ka**] and Han [**Han**]. The current paper and Han's paper were

1991 *Mathematics Subject Classification.* MSC Number: Primary 11S31; Secondary 11G45, 11R23, 14G10.

Key words and phrases. Formal groups, Hecke characters, exponential maps.

This research is supported in part by NSF Grant # DMS-9525833.

The author would like to thank K. Rubin for helpful discussions.

S.D. Ahlgren et al. (eds.), Topics in Number Theory, 181–192.

written independently, and the contexts, results, as well as methods, of the two papers are different.

1. Hecke characters and formal group characters

Let E be an elliptic curve over $\mathbb{Q}$ with complex multiplication by the ring of integers $\mathcal{O}_K$ of an imaginary field K. Let ψ be the Hecke character associated to E by the theory of complex multiplication. Fix an embedding $i_p : \bar{\mathbb{Q}} \hookrightarrow \bar{\mathbb{Q}}_p$ and let v_p be the associated place of $\bar{\mathbb{Q}}$ above p. Let $\mathfrak{p}$ be the prime of K with $v_p \mid \mathfrak{p}$. Let $\varphi_\mathfrak{p}$ be the $\mathfrak{p}$-adic character associated to φ, i.e., the Weil realization of ψ at $\mathfrak{p}$. Let $\psi(\mathfrak{p}) = \pi_E$. Then π_E is a generator of the principle ideal $\mathfrak{p}$. $\psi_\mathfrak{p}$ could be obtained from the action of G_K on $E[\pi_E^\infty] \stackrel{\text{def}}{=} \cup_{n=1}^\infty E[\pi_E^n]$, the group of π_E-power division points on the elliptic curve E.

Let $K_\mathfrak{p}$ be the completion of K at $\mathfrak{p}$ and let $\mathcal{O}_\mathfrak{p}$ be the ring of integers of K. Let $\bar{K}_\mathfrak{p}$ be the algebraic closure of $K_\mathfrak{p}$, and let $K_\mathfrak{p}^{\text{un}}$ be the maximal unramified extension of $K_\mathfrak{p}$ in $\bar{K}_\mathfrak{p}$. Denote $\mathbb{C}_\mathfrak{p}$ (resp. $\mathbb{D}_\mathfrak{p}$) for the completion of $\bar{K}_\mathfrak{p}$ (resp. $K_\mathfrak{p}^{\text{un}}$) and denote $\mathcal{O}_{K_\mathfrak{p}}$ and $\mathcal{O}_{\mathbb{D}_\mathfrak{p}}$ for the corresponding rings of integers. Let $G_\mathfrak{p}$ be the Galois group of $\bar{K}_\mathfrak{p}$ over $K_\mathfrak{p}$. Let π be a uniformizer of $K_\mathfrak{p}$ and let $\mathbf{F}$ be a one-dimensional Lubin-Tate module over $\mathcal{O}_\mathfrak{p}$ of height one for π. Let $\mathfrak{m}_p$ be the maximal ideal of $\mathcal{O}_{\mathbb{C}_p}$ and let $\mathbf{F}_{\pi^n}$ be the group of π^n-torsion points of $\mathbf{F}(\mathfrak{m}_p)$. The action of $G_{K_\mathfrak{p}}$ on $\mathbf{F}_{\pi^\infty} \stackrel{\text{def}}{=} \cup_{n=1}^\infty \mathbf{F}_{\pi^n}$ defines the formal group character $\kappa_\pi : G_{K_\mathfrak{p}} \to \mathcal{O}_\mathfrak{p}^\times$.

It is well-known that the restriction of $\psi_\mathfrak{p}$ to $G_\mathfrak{p}$ is the same as the formal group character κ_{π_E} [de]. This relation is important in the study of elliptic curves with complex multiplication. It is analogous to the fact that the restriction to $G_p = \text{Gal}(\bar{\mathbb{Q}}_p/\mathbb{Q}_p)$ of the p-adic cyclotomic character $\chi_p : G_\mathbb{Q} \to \mathbb{Z}_p^\times$ induced from the action of $G_\mathbb{Q}$ on $\mu_{p^\infty} \stackrel{\text{def}}{=} \cup_{n=1}^\infty \mu_{p^n}$, is the same as the formal group character $\kappa_p : G_p \to \mathbb{Z}_p^\times$ induced from the action of G_p on $(\mathbb{G}_m)_{p^\infty} \stackrel{\text{def}}{=} \cup_{n=1}^\infty (\mathbb{G}_m)_{p^n}$, the group of p-power division points of the multiplicative formal group $\mathbb{G}_m$, which is the Lubin-Tate formal group for p.

Let k and j be a pair of integers, we have the Hecke character $\varphi \stackrel{\text{def}}{=} \psi^k \bar{\psi}^{-j}$. We will only consider the case when E has good, ordinary reduction at $p \geq 2$. Then p splits into two distinct primes $\mathfrak{p}$ and $\mathfrak{p}^*$ in $\mathcal{O}_K$ and $K_\mathfrak{p} = \mathbb{Q}_p$. For $u \in \mathbb{Z}_p^\times$, let $< u >$ be the projection of u onto the direct factor $1 + \mathbb{Z}_p$ of $\mathbb{Z}_p^\times$.

Theorem 1. 1. *Let π be a uniformizer of $\mathbb{Z}_p$. The formal group character κ_π equals to $\psi_\mathfrak{p} \psi_{\mathfrak{p}^*}^r |_{G_\mathfrak{p}}$, where $r = \frac{\log_p(\pi/p)}{\log_p(p/\pi_E)} + 1$.*

2. *Let (k, j) be a pair of integers and let p be a prime number with $p \nmid k$. The restriction to $G_\mathfrak{p}$ of the p-adic Hecke character $\psi_\mathfrak{p}^k \psi_{\mathfrak{p}^*}^j$ is equal to the formal group character κ_π^k, where $\pi = \pi_E < \bar{\psi}^j(\mathfrak{p}) >^{1/k}$.*

Proof: Since $\mathbf{F}_\pi$ and $\hat{E} = \mathbf{F}_{\pi_E}$ are isomorphic over the maximal unramified extension of $\mathbb{Q}_p$, $\kappa_\pi \psi_\mathfrak{p}^{-1}$ is an unramified character $G_{\mathbb{Q}_p} \to \mathbb{Z}_p^\times$, hence induces a character $\text{Gal}(\mathbb{Q}_p^{un}/\mathbb{Q}_p) \cong \hat{\mathbb{Z}} \to \mathbb{Z}_p^\times$. Since $\psi_{\mathfrak{p}^*} : G_{\mathbb{Q}_p} \to \mathbb{Z}_p^\times$ is unramified and induces surjective map $\psi_{\mathfrak{p}^*} : \text{Gal}(\mathbb{Q}_p^{nr}/\mathbb{Q}_p) \cong \hat{\mathbb{Z}} \to \mathbb{Z}_p^\times$, there is $f : \mathbb{Z}_p^\times \to \mathbb{Z}_p^\times$ such that

$f \circ \psi_{\mathfrak{p}^*} = \kappa_\pi \psi_{\mathfrak{p}}^{-1}$. Since $\mathbb{Z}_p^\times$ is pro-cyclic, f is of the form $f : x \mapsto x^r$ for some $r \in \mathbb{Z}_p^\times$. Thus $\kappa_\pi = \psi_{\mathfrak{p}} \psi_{\mathfrak{p}^*}^r$. By local class field theory,

$$\kappa_\pi(\pi) = \psi_{\mathfrak{p}}(\pi_E) = \psi_{\mathfrak{p}^*}(u) = 1.$$

By the theory of complex multiplication,

$$\psi_{\mathfrak{p}^*}(\pi_E) = p/\pi_E.$$

Thus

$$\psi_{\mathfrak{p}^*}(\pi) = \psi_{\mathfrak{p}^*}(\pi_E) = p/\pi_E$$

and

$$\psi_{\mathfrak{p}}(\pi) = \psi_{\mathfrak{p}^*}(\pi)^{-r} = (p/\pi_E)^{-r}.$$

On the other hand, Since $\psi_{\mathfrak{p}}(\pi_E) = 1$, we have $\psi_{\mathfrak{p}}(\pi) = \psi_{\mathfrak{p}}(\pi/\pi_E)$. Since $\hat{E}$ and $\mathbb{G}_m$ are isomorphic over the maximal unramified extension of $\mathbb{Q}_p$, $\psi_{\mathfrak{p}} \chi_p^{-1}$ is unramified. Thus we have $\psi_{\mathfrak{p}}(\pi/\pi_E) = \chi_p(\pi/\pi_E)$. By class field theory[Ne, p.63],

$$\zeta_{p^n}^{\chi_p(\pi/\pi_E)} = (\pi/\pi_E, \mathbb{Q}_p)\zeta_{p^n} = [(\pi/\pi_E)^{-1}]_{\mathbb{G}_m} \zeta_{p^n} = \zeta_{p^n}^{(\pi/\pi_E)^{-1}}.$$

So $\chi_p(\pi/\pi_E) = (\pi/\pi_E)^{-1}$. Thus $\frac{up}{\pi_E} = \frac{\pi}{\pi_E} = \psi_{\mathfrak{p}}(\pi) = (\frac{p}{\pi_E})^r = (\frac{p}{\pi_E})^r$ and $u = (p/\pi_E)^{r-1}$. Taking the base p logarithm proves the first part of the theorem.

Now let (k, j) be a pair of integers with $p \nmid k$. Then raising to the k-th power gives an isomorphism of $1 + p\mathbb{Z}_p$. So taking the k-root is a well-defined map. Let $\pi = \psi(\mathfrak{p}) < \bar{\psi}^{-j}(\mathfrak{p}) >^{1/k}$. Let κ_π be the corresponding formal group character. Since $p = \psi(\mathfrak{p})\bar{\psi}(\mathfrak{p})$ and $\psi(\mathfrak{p}) = \pi_E$, we have $\bar{\psi}(\mathfrak{p}) = p/\pi_E$. By the first part of the theorem, if $\pi = p(p/\pi_E)^r$ with $r \in \mathbb{Z}_p^\times$, then $\kappa_\pi = \psi_{\mathfrak{p}} \psi_{\mathfrak{p}^*}^r$. Let $r' \in \mathbb{Z}_p^\times$ be chosen with $(\frac{p}{\pi_E})^{r'} =< \frac{p}{\pi_E} >$, then we get

$$\pi = \psi(\mathfrak{p}) < \bar{\psi}^{-j}(\mathfrak{p}) >^{1/k} = p(\frac{\pi_E}{p})(\frac{\pi_E}{p})^{r'(-j/k)}.$$

Thus $\kappa_\pi = \psi_{\mathfrak{p}} \psi_{\mathfrak{p}^*}^{r'(-j/k)}$ and $\kappa_\pi^k = \psi_{\mathfrak{p}}^k (\psi_{\mathfrak{p}^*}^{r'})^{(-j)}$. Note that multiplication by r' is the projection of $\mathbb{Z}_p^\times$ onto the factor $1 + \mathbb{Z}_p$. Thus $\psi_{\mathfrak{p}^*}^{r'}$ is trivial on the prime to p part of the Galois group $\mathrm{Gal}(\mathbb{Q}_p(E_{(\mathfrak{p}^*)^\infty})/\mathbb{Q}_p)$ and is the identity on the pro-p-part of this group. Therefore $\phi_{\mathfrak{p}^*}^{r'}$ is the same as $\psi_{\mathfrak{p}^*}$ regarded as characters on $\mathbb{Q}_p(E_{\mathfrak{p}^*})$. Thus $\kappa_\pi^k = \psi_{\mathfrak{p}}^k \psi_{\mathfrak{p}^*}^{-j}$. ∎

2. Exponential maps

Now we apply Theorem 1 to to get an explicit exponential map for the Hecke character $\varphi = \psi^k \bar{\psi}^{-j}$ at p where the associated elliptic curve $E/\mathbb{Q}$ has good, ordinary reduction.

First recall some notations and basic properties. Assume $p \nmid k$ and let

$$\pi = \pi_\varphi = \psi(\mathfrak{p}) < \bar{\psi}^{-j}(\mathfrak{p}) >^{1/k}.$$

Let L be the finite unramified extension of $\mathbb{Q}_p$ of degree d. Let F be the Frobenius automorphism of L. $L(\mathbf{F}_{\pi^\infty})/L$ is a totally ramified $\mathbb{Z}_p^\times$-extension. The formal group character κ_π restricts to a character $\kappa_\pi : G_L \to \mathbb{Z}_p^\times$ with kernel $G_{L(\mathbf{F}_{\pi^\infty})}$. Let $\mathbb{Q}_p(\kappa^n)$ be the 1-dimensional G_L-representation over $\mathbb{Q}_p$ induced from κ^n, $n \in \mathbb{Z}$.

The multiplicative formal group $\mathbb{G}_m$ is isomorphic to $\mathbf{F}$ over $\mathcal{O}_{\mathbf{D}_p}$. Let $\theta(T) \in \mathcal{O}_{\mathbf{D}_p}[[T]]$ be such an isomorphism. Fix a coherent set of primitive p-power roots of unity $(\zeta_n)_{n\geq 0}$, that is, $\zeta_0 = 1, \zeta_1 \neq 1, \zeta_{n+1}^p = \zeta_n$ for $n \geq 0$. Define $\pi_n = \theta^{F^{-n}}(\zeta_n - 1), n \geq 0$. Then we have $\pi_n \in \mathbf{F}_{\pi^n}/\mathbf{F}_{\pi^{n-1}}$ and $[\pi]_{\mathbf{F}}(\pi_n) = \pi_{n-1}$. Let $\lambda_{\mathbf{F}}(T) \in \mathrm{Hom}(\mathbf{F}, \mathbb{G}_a)$ be the formal logarithm of $\mathbf{F}$ and let $\log(1+T) = \sum_{i=0}^{\infty} \frac{(-1)^{i-1}}{i} T^i$ be the formal logarithm of $\mathbb{G}_m$. Then $\lambda_{\mathbf{F}}(\theta(T)) = \Omega_p \log(1+T)$. Here $\Omega_p \in \mathcal{O}_{\mathbf{D}_p}$ satisfies $\Omega_p^\sigma = (\kappa(\sigma)/\chi_p(\sigma))\Omega_p$, $\sigma \in G_{\mathbf{Q}_p}$, and χ_p is the p-cyclotomic character associated to the formal group $\mathbb{G}_m$.

Denote $U^1_{L(\mathbf{F}_{\pi^n})}$ for the units of $\mathcal{O}_{L(\mathbf{F}_{\pi^n})}$ which are congruent to 1 modulo the maximal ideal. Define $U^1_\infty = \varprojlim U^1_{L(\mathbf{F}_{\pi^n})}$ where the inverse limit is with respect to the norm maps. For a norm coherent sequence $u = (u_n) \in U^1_\infty$ of 1-units, let $g_u(X) \in \mathcal{O}_L[[X]]^\times$ be the Coleman power series such that $g_u^{F^{-i}}(\pi_i) = u_i$, $i \geq 1$ and $g_u(0)^{1-F^{-1}} = u_0$.

For $k \geq 1$, the k-th Coates-Wiles homomorphism

$$\phi^k_{CW} = \phi^k_{CW,\pi} : U^1_\infty \to L$$

is defined by

$$\phi^k_{CW}(u) = \left[\frac{1}{\lambda'_{\mathbf{F}}(Y)}\frac{d}{dY}\right]^k \log g_u(Y)|_{Y=0}. \tag{1}$$

ϕ^k_{CW} is a $\mathbf{Z}_p$-homomorphism and $\phi^k_{CW}(u^\sigma) = \kappa^k(\sigma)\phi^k_{CW}(u)$ for $\sigma \in G_L$.

Let $\mathbb{Q}_p(\varphi_\mathfrak{p})$ be the one-dimensional representation of G_L over $\mathbb{Q}_p$ defined by $\varphi_\mathfrak{p}$. Let T_φ be a basis of $\mathbb{Q}_p(\varphi_\mathfrak{p})$. Consider the fundamental exact sequence of Bloch and Kato [**B-K**, 1.17]

$$0 \to \mathbb{Q}_p \xrightarrow{\alpha} B^{F=1}_{crys} \oplus B^+_{DR} \xrightarrow{\beta} B_{DR} \to 0 \tag{2}$$

Here

- $B_{\mathrm{DR}} = B_{\mathrm{DR},p}$ is a ring of Fontaine with an action by $G_{\mathbf{Q}_p}$ and a decreasing filtration $Fil^i B_{\mathrm{DR}}$, $i \in \mathbf{Z}$;
- $B^+_{\mathrm{DR}} \overset{\mathrm{def}}{=} F^0 B_{\mathrm{DR}}$;
- $B_{\mathrm{crys}} = B_{\mathrm{crys},p}$ is a subring of B_{DR} with the induced $G_{\mathbf{Q}_p}$-action, the induced filtration and a Frobenius endomorphism f;
- $B^{f=1}_{\mathrm{crys}} \overset{\mathrm{def}}{=} \{x \in B_{\mathrm{crys}} \mid x^f = x\}$;
- α is given by $\alpha(x) = (x, x)$;
- β is given by $\beta(x, y) = (x - y)$.

Tensoring the exact sequence (2) with $\mathbb{Q}_p(\varphi_\mathfrak{p})$, we get

$$0 \to \mathbb{Q}_p(\varphi_\mathfrak{p}) \xrightarrow{\alpha} (B^{f=1}_{crys} \oplus B^+_{DR}) \otimes \mathbb{Q}_p(\varphi_\mathfrak{p}) \xrightarrow{\beta} B_{DR} \otimes \mathbb{Q}_p(\varphi_\mathfrak{p}) \to 0. \tag{3}$$

The exponential map $\mathrm{Exp}_L = \mathrm{Exp}_L(\mathbb{Q}_p(\varphi_\mathfrak{p}))$ of Bloch and Kato is defined to be the connecting homomorphism when we take G_L-invariants of the exact sequence (3). By definition,

$$DR_L(\mathbb{Q}_p(\varphi_\mathfrak{p})) = (B_{DR} \otimes \mathbb{Q}_p(\varphi_\mathfrak{p}))^{G_L}.$$

Thus

$$\mathrm{Exp}_L : DR_L(\mathbb{Q}_p(\varphi_\mathfrak{p})) \to H^1(L, \mathbb{Q}_p(\varphi_\mathfrak{p})).$$

Theorem 2. *Let p be a prime where E has good, ordinary reduction with $p \nmid k$. Let $\pi = \psi(\mathfrak{p}) < \bar{\psi}^{-j}(\mathfrak{p}) >^{1/k}$. Let $L \supseteq \mathbb{Q}_p(E^{\otimes(-j)}_{\bar{\psi}(\mathfrak{p})})$ be an unramified extension. Identify L with $DR_L(\mathbb{Q}_p(\varphi_\mathfrak{p}))$ via the map $a \mapsto at_\pi^{-k} \otimes T_\varphi$. Then the exponential map*

$$\mathrm{Exp}_L : L = DR_L(\mathbb{Q}_p(\varphi_\mathfrak{p})) \to H^1(L, \mathbb{Q}_p(\varphi_\mathfrak{p}))$$

of Bloch and Kato has the following explicit description. There is a canonical isomorphism

$$\eta_\varphi : H^1(L, \mathbb{Q}_p(\varphi_\mathfrak{p})) \to \mathrm{Hom}_G(U^1_\infty, \mathbb{Q}_p(\varphi_\mathfrak{p}))$$

such that, for $a \in L = DR_L(\mathbb{Q}_p(\varphi_\mathfrak{p}))$, the element

$$f_a = \eta_\varphi \circ \mathrm{Exp}_L(a) \in \mathrm{Hom}_G(U^1_\infty, \mathbb{Q}_p(\varphi_\mathfrak{p}))$$

is given by

$$f_a(u) = \frac{1}{(k-1)!} \mathrm{Tr}^L_{\mathbb{Q}_p}[(a - \frac{a^f}{\pi^k})\phi^k_{CW}(u)]T_\varphi.$$

Proof: Theorem 1 enable us to reduce the proof of Theorem 2 to an explicit reciprocity law for $\mathbb{Q}_p(\kappa^n)$ constructed by Fontaine [**Fo2**] and Harrison [**Ha**]. We will need more notation to formulate this explicit reciprocity law. For more details see [**de**], [**Fo2**] and [**Ha**].

We first define the reciprocity map for $\mathbb{Q}_p(\kappa_\pi)$. Define $t_\pi = \Omega_p t \in B^+_{\mathrm{crys}}$ where t is the usual element of B_{crys} on which $G_{\mathbb{Q}_p}$ acts via the p-adic cyclotomic character. Then we have $t_\pi^F = \pi t_\pi$ and $t_\pi^\sigma = \kappa(\sigma) t_\pi$, $\sigma \in G_L$. Multiplying each term of the exact sequence

$$0 \to \mathbb{Q}_p(k) \to J_{\mathbb{Q}}^{[k]} \overset{1-p^{-k}f}{\longrightarrow} B^+_{\mathrm{crys}} \to 0$$

from [**B-K**, (1.13)] by Ω_p^k we get the exact sequence

$$0 \to \mathbb{Q}_p t_\pi^k \to \mathrm{Fil}^k B^+_{\mathrm{crys}} \overset{\pi^{-k}f-1}{\longrightarrow} B^+_{\mathrm{crys}} \to 0$$

Note that $\mathbb{Q}_p t_\pi^k = \mathbb{Q}_p(\kappa^k)$ as G_L-representations. Thus the cohomology coboundary map from this exact sequence gives the reciprocity law

$$L = (B^+_{\mathrm{crys}})^{G_L} \longrightarrow H^1(L, \mathbb{Q}_p(\kappa^k)) \tag{4}$$

Let $G = \mathrm{Gal}(L_\infty/L)$. By inflation-restriction map and local class field theory we also have the isomorphism

$$\begin{aligned} \eta_k : \quad & H^1(L, \mathbb{Q}_p(\kappa^k)) \cong H^1(L_\infty, \mathbb{Q}_p(\kappa^k))^G \\ & \cong \mathrm{Hom}_G(G^{\mathrm{ab}}_{L_\infty}, \mathbb{Q}_p(\kappa^k)) \cong \mathrm{Hom}_G(U^1_\infty, \mathbb{Q}_p(\kappa^k)) \end{aligned}$$

where $G^{\mathrm{ab}}_{L_\infty}$ is the largest abelian quotient of G_{L_∞}.

Harrison [**Ha**] proved the following explicit description of the reciprocity law from the equation (4), generalizing Fontaine's work [**Fo2**] on the multiplicative formal group.

Theorem 3. *(Harrison) Let L be a finite unramified extension of $\mathbb{Q}_p$ and $r \geq 1$. Then, identifying $\mathbb{Q}_p(\kappa^k)$ with $\mathbb{Q}_p t_\pi^k \subseteq B^+_{\text{crys}}$, the reciprocity law map from (4)*

$$\eta_k \circ \partial^k : L = (B^+_{\text{crys}})^{G_L} \to H^1(L, \mathbb{Q}_p(\kappa^k)) \cong \text{Hom}(U^1_\infty, \mathbb{Q}_p(\kappa^k))$$

is given by

$$(\eta_k \circ \partial^k)(a) = \{u \mapsto \frac{1}{(k-1)!} \text{Tr}^L_{\mathbb{Q}_p}(a\phi^k_{CW}(u))t^k_\pi, u \in U^1_\infty\}, a \in L$$

where $\text{Tr}^L_{\mathbb{Q}_p}$ is the trace map from L to $\mathbb{Q}_p$.

We are now ready to give the explicit exponential map for $\mathbb{Q}_p(\varphi_\mathfrak{p})$. From Theorem 1, for $\pi = \pi_\varphi = \psi(\mathfrak{p}) < \bar{\psi}^{-j}(\mathfrak{p}) >^{1/k}$, we have $\kappa^k_\pi = \psi^k_\mathfrak{p}\psi^{-j}_{\mathfrak{p}^*}$ on L. It follows that $t^{-k}_\pi \otimes T_\varphi$ is an element in $DR_L(\mathbb{Q}_p(\varphi_\mathfrak{p}))$ and we could identify L with $DR_L(\mathbb{Q}_p(\varphi_\mathfrak{p}))$, as claimed in the theorem.

Given $a \in L$, identified with $DR_L(\mathbb{Q}_p(\varphi_\mathfrak{p}))$ as above. From the exact sequence (3)

$$0 \to \mathbb{Q}_p(\varphi_\mathfrak{p}) \xrightarrow{\alpha} (B^{f=1}_{crys} \otimes \mathbb{Q}_p(\varphi_\mathfrak{p})) \oplus (B^+_{DR} \otimes \mathbb{Q}_p(\varphi_\mathfrak{p})) \xrightarrow{\beta} B_{DR} \otimes \mathbb{Q}_p(\varphi_\mathfrak{p}) \to 0.$$

we could choose $x \in B^{f=1}_{crys}(\varphi_\mathfrak{p})$ and $y \in B^+_{DR}$ such that $x - y = at^{-k}_\pi$. Then $y = x - at^{-k}_\pi \in B_{crys}$.

Let $x_1 = t^k_\pi x$, $y_1 = t^k_\pi y$. Since $y \in Fil^0 B_{crys} \subseteq B_{crys}$ and $t^k_\pi \in B^+_{crys}$, we have

$$y_1 \in B_{crys} \cap Fil^k B_{DR} = Fil^k B_{crys}.$$

By definition, $x^f_1 = \pi^k x_1$. From $x = y + at^{-k}_\pi$ we get

$$x_1 = y_1 + a \in Fil^k B_{crys} + B^+_{crys} \subseteq Fil^0 B_{crys}$$

Also,

$$\begin{aligned} y_1^{f^m} &= x_1^{f^m} - a^{f^m} = \pi^{km} x_1 - a^{f^m} \\ &\in Fil^{km} B_{crys} + B^+_{crys} \subseteq Fil^0 B_{crys}, \ \forall m \geq 0. \end{aligned}$$

Now we show that $y_1 \in B^+_{crys}$. This would give $y_1 \in Fil^k B^+_{crys}$. As $B_{crys} = \cup_{s \geq 0} t^{-s} B^+_{crys}$, there is $s \geq 0$ with $y_1 \in t^{-s} B^+_{crys}$. If $s = 0$ we are done. If $s > 0$, then $t^s y_1 \in B^+_{crys}$ and

$$(t^s y_1)^{f^m} = p^{ms} t^s y_1^{f^m} \in \text{Fil}^s B_{crys} \subseteq Fil^1 B_{crys}.$$

By [**Fo1**, Prop. 4.14], this shows $t^s y_1 \in tB^+_{crys}$. Hence $t^{s-1} y_1 \in B^+_{crys}$. Repeat this process s times, we get $y_1 \in B^+_{crys}$, as needed.

By the definition of Exp_L, for $a \in L$ and $g \in G_L$,

$$\begin{aligned} \text{Exp}_L(at^{-k}_\pi \otimes T_\varphi)(g) &= (x \otimes T_\varphi, y \otimes T_\varphi)^g - (x \otimes T_\varphi, y \otimes T_\varphi) \\ &= ((x \otimes T_\varphi)^g - x \otimes T_\varphi, (y \otimes T_\varphi)^g - y \otimes T_\varphi). \end{aligned}$$

which is mapped to zero via β. Hence it is in the image of α. Thus $(x \otimes T_\varphi)^g - x \otimes T_\varphi = (y \otimes T_\varphi)^g - y \otimes T_\varphi \in \mathbb{Q}_p(\varphi_p)$ and

$$\begin{aligned}\mathrm{Exp}_L(at_\pi^{-k} \otimes T_\varphi)(g) &= (y \otimes T_\varphi)^g - y \otimes T_\varphi \\ &= \varphi_p^{-1}(g) t_\pi^{-k} y_1^g \varphi_p \otimes T_\varphi - t_\pi^{-k} y_1 \otimes T_\varphi \\ &= t_\pi^{-k}(y_1^g - y_1) \otimes T_\varphi\end{aligned}$$

It is shown above that $y_1 \in Fil^k B^+_{crys}$, hence $(\pi^{-k} f - 1) y_1 \in B^+_{crys}$ in the exact sequence

$$0 \to \mathbb{Q}_p t_\pi^k \to \mathrm{Fil}^k B^+_{\mathrm{crys}} \xrightarrow{\pi^{-k} f - 1} B^+_{\mathrm{crys}} \to 0.$$

So $y_1^g - y_1$ is the image of $(\pi^{-k} f - 1) y_1$ under the boundary map ∂_π^k. But

$$\pi^{-k} y_1^f = \pi^{-k}(x_1 - a)^f = \pi^{-k} x_1^f - \pi^{-k} a^f = x_1 - \pi^{-k} a^f.$$

So

$$\begin{aligned}(\pi^{-k} f - 1) y_1 &= \pi^{-k}(x_1 - a)^f - (x_1 - a) \\ &= x_1 - \pi^{-k} a^f - (x_1 - a) \\ &= a - \pi^{-k} a^f.\end{aligned}$$

Thus

$$\begin{aligned}\mathrm{Exp}_L(at_\pi^{-k} \otimes T_\varphi)(g) &= t_\pi^{-k}(y_1^g - y_1) \otimes T_\varphi \\ &= t_\pi^{-k} \partial_\pi^k((\pi^{-k} f - 1) y_1)(g) \\ &= t_\pi^{-k} \partial_\pi^k(a - \pi^{-k} a^f)(g).\end{aligned}$$

Composing this with the isomorphism $\eta_\varphi \stackrel{\text{def}}{=} \eta_{\kappa^k}$ and applying Theorem 3, we get, for $u \in U^1_\infty$,

$$\mathrm{Exp}_L(at_\pi^{-k} \otimes T_\varphi)(u) = \frac{1}{(k-1)!} \mathrm{Tr}^L_{\mathbb{Q}_p}[(a - \frac{a^f}{\pi^k}) \phi^k_{CW}] T_\varphi. \quad \blacksquare$$

3. Image of the exponential map

We now apply Theorem 1 and Theorem 2 to describe the image of the exponential map. Let $L/\mathbb{Q}_p$ be an unramified extension of degree $d|p-1$. For application to the Bloch-Kato conjecture, L will be $\mathbb{Q}_p(E^{\otimes(-j)}_{(\pi_E)})$. The degree is $(p-1)/gcd(p-1, j)$. By local class field theory and Kummer theory, $L = \mathbb{Q}_p(\alpha)$ where $\alpha = u^{1/d}$ with u in $\mathbb{Q}_p^\times$ but not in $(\mathbb{Q}_p^\times)^r$ for any proper divisor r of d. In other words, u is of order d in $\mathbb{Q}_p^\times/(\mathbb{Q}_p^\times)^d$. For $1 \le n \le \infty$, define $L_n = L(\mathbf{F}_{\pi^n})$. Define $\mathcal{G} = \mathrm{Gal}(L_\infty/\mathbb{Q})$, $G = \mathrm{Gal}(L_\infty/L)$, $\Delta = \mathrm{Gal}(L(\mathbf{F}_\pi)/L)$, $\Delta_1 = \mathrm{Gal}(L(\mathbf{F}_\pi)/\mathbb{Q}_p)$ and $\Delta' = \mathrm{Gal}(L/\mathbb{Q}_p)$. Then we have a canonical isomorphisms $\mathcal{G} \cong G \times \Delta' \cong \Gamma \times \Delta_1$, $G \cong \Gamma \times \Delta$, $\Delta_1 \cong \Delta \times \Delta'$. The related fields are shown in the following diagram.

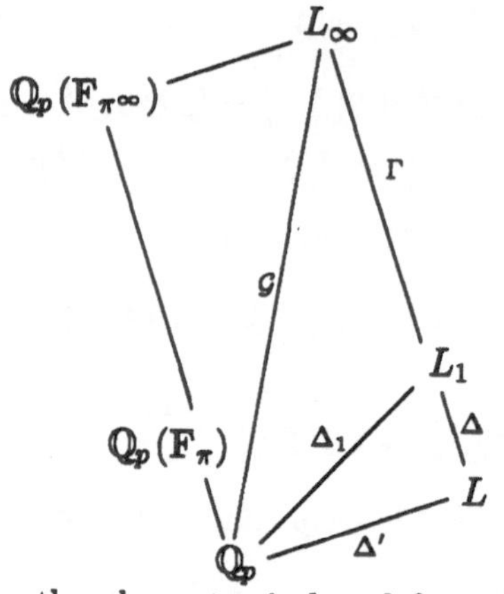

Let $\omega = \kappa_\pi|_\Delta$. Let ϵ be the character induced from the action of Δ' on α. It is in fact the cyclotomic character of order d. These characters can be regarded as characters of $\mathcal{G}$ acting trivially on Γ.

Proposition 3.1. *Let L be the unramified extension of $\mathbb{Q}_p$ of degree $d \mid p-1$. Let F be the Frobenius element of $\mathrm{Gal}(L/\mathbb{Q}_p)$. Let ϵ be the cyclotomic character of order d. For any integer $N > 0$, the following statements are equivalent.*

1. $\mu_{p^N} \subseteq L(\mathbf{F}_{\pi^\infty})$.
2. $(\frac{\pi}{p})^{[L:\mathbb{Q}_p]} \equiv 1 \bmod p^N$.
3. $\frac{\pi}{p} \equiv \epsilon(F)^S \bmod p^N$, *for some* $S \geq 0$.

If one (hence all) of the above conditions holds then we also have

4. $\chi_N(\sigma) \equiv \kappa(\sigma) \bmod p^N$, *for* $\sigma \in \mathrm{Gal}(L(\mathbf{F}_{\pi^\infty})/L)$, *where* $\chi_N : \mathcal{G} \to (\mathbb{Z}/p^N\mathbb{Z})^\times$ *denotes the character giving the action of* $\mathrm{Gal}(L(\mathbf{F}_{\pi^\infty})/L)$ *on* μ_{p^N}.
5. $\chi_N(\sigma) \equiv \epsilon^S(\sigma) \bmod p^N$, *for* $\sigma \in \mathrm{Gal}(L(\mathbf{F}_{\pi^\infty})/L)$ *and for the integer S in statement 3.*

We define the index of *anomality* of $\mathbf{F}_\pi$ over L to be the largest integer N in the proposition. By local class field theory, this N exists for $\mathbf{F}_\pi \neq \mathbb{G}_m$.

Proof of the Proposition: (1) $\Longleftrightarrow$ (2) is a consequence of local class field theory[**de**, I.3.7].

(2) $\Longleftrightarrow$ (3) follows from the fact that μ_d is a subset of $\mu_{p-1} \subseteq \mathbb{Z}_p^\times$ and hence are distinct modulo p.

(1) $\Longrightarrow$ (4) is obtained from lifting σ to $\tilde{\sigma} \in \mathrm{Gal}(\bar{\mathbb{Q}}_p/\mathbb{Q}_p^{\mathrm{un}})$ and applying it to the equation $\pi_N = \theta^{F^{-N}}(\zeta_N - 1)$. Similarly for (5). For more details, see [**de, Ha**]. ■

From our choice of d, μ_d is contained in $\mu_{p-1} \subseteq \mathbb{Z}_p^\times$ and so are distinct modulo p. It follows from Proposition 3.1 that the index of anomality N of L is the largest N such that $\pi/p \equiv \epsilon(F)^S \bmod p^N$ for some S. If $N > 0$ we choose S such that $0 \leq S \leq d-1$.

We have a canonical decomposition $U_\infty^1 \cong \oplus_{(s,t) \bmod (d,p-1)} (U_\infty^1)^{(s,t)}$ of U_∞^1 on which Δ_1 acts via $\epsilon^s\omega^t$, and we can decompose $L_1 \cong \oplus_{0 \leq t \leq p-2} L_1^{(t)}$ into the 1-dimensional L-eigenspaces of L_1 with eigen-character ω^t. Choose a L-basis e_t of

$L_1^{(t)}$ such that $e_t = 1$ if $t = 0$ and $0 \le v_{\mathfrak{P}}(e_t) \le p-2$ for any t, where $\mathfrak{P}$ is the maximal ideal of $\mathcal{O}_{L_1}$. This could be done by changing e_t by a power of p if necessary. Now $\mathbf{D}_p(\mathbf{F}_\pi) = \mathbf{D}_p(\mu_p)$ and if we identify $\mathrm{Gal}(\mathbf{D}_p(\mathbf{F}_\pi)/\mathbf{D}_p)$ with $\mathrm{Gal}(L(\mathbf{F}_\pi)/L) = \Delta$ then ω is the cyclotomic character from the action on μ_p. Thus we can define the Gauss sum

$$G(t) = \begin{cases} \frac{1}{p-1}\sum_{\sigma\in\Delta}\omega^{-t}(\sigma)\zeta_1^{\sigma}, & t \not\equiv 0 \\ 1, & t = 0 \end{cases}$$

It is well-known that $v_{\mathfrak{P}_1}(G(t)) = t$, $0 \le t \le p-2$, where $\mathfrak{P}_1$ is the maximal ideal of $\mathbf{D}_p(\mathbf{F}_\pi)$. Since $G(t)$ and e_t are both in $\mathbf{D}_p(\mathbf{F}_\pi)^{(t)}$ and have $\mathfrak{P}_1$-adic valuation between 0 and $p-2$, it follows that they must differ by a unit: $G(t) = u_t e_t$, $u_t \in \mathcal{O}_{\mathbf{D}_p}^{\times}$, $0 \le t \le p-2$. As $L \cong \oplus_{0\le s\le d-1}\mathbb{Q}_p\alpha^s$, we get the decomposition $L_1 \cong \oplus_{(s,t)\bmod(d,p-1)}L_1^{(s,t)}$ of L_1 into eigenspaces $L_1^{(s,t)} = \mathbb{Q}_p\alpha^s e_t$ of L_1 with eigen-character $\epsilon^s\omega^t$.

Define

$$h_u^k(X) = \left(\frac{1}{\lambda'_{\mathbf{F}}(X)}\frac{d}{dX}\right)^k \log g_u(X) \in \mathcal{O}_L[[X]].$$

Then the following equations hold [**Ha**]. For $k \ge 1$,

$$\phi_{CW}^k(u) - \frac{\pi^k}{p}[\phi_{CW}^k(u)]^F = \Omega_p^{-k}\int_G \kappa(\sigma)^k d\mu_u(\sigma) \tag{5}$$

$$\pi^{-k}(h_u^k)^{F^{-1}}(\pi_1) = (p\Omega_p)^{-k}\int_G \kappa(\sigma)^k\zeta_1^{\kappa(\sigma)}d\mu_u^{F^{-1}}(\sigma) + p^{-1}\phi_{CW}^k(u) \tag{6}$$

Here ζ_1 is a primitive p-th root of unity. For $k \ge 1$, define

$$\phi_k : U_\infty^1 \to L_1,\ u \mapsto \pi^{-k}(h_u^k)^{F^{-1}}(\pi_1).$$

Here $\pi_1 = \theta^{F^{-1}}(\zeta_1 - 1)$, and $\theta(X) \in \mathcal{O}_{\mathbf{D}_{\checkmark}}[[T]]$ is an $\mathbb{Z}_p$-isomorphism from $\mathbb{G}_m$ onto $\mathbf{F}_\pi$. ϕ_k is a $\mathbb{Z}_p$-homomorphism. We will explicitly determine its image.

Theorem 4. *Let $k \ge 2$. Define $N_k = \min\{N, 1 + v_p(k-1)\}$. Then*

1. $\phi_k(U_\infty^1) = p^{-k}\left[\oplus_{s\not\equiv -S \text{ or } t\not\equiv 1-k}\mathbb{Z}_p\alpha^s e_t \oplus p^{N_k}\mathbb{Z}_p\alpha^{-S}e_{1-k}\right]$.
2. *If $\mathcal{L}_k$ is the lattice of L_1 which is precisely the exact annihilator to $\frac{1}{(k-1)!}\phi_k(U_\infty^1)$ under the pairing*

$$Tr_{\mathbb{Q}_p}^{L_1} : L_1 \times L_1 \to \mathbb{Q}_p/\mathbb{Z}_p$$

then

$$\mathcal{L}_k = \begin{cases} p^k(k-1)!\left[\bigoplus_{t\equiv 0(p-1)}\mathbb{Z}_p\alpha^s e_t \oplus \bigoplus_{t\not\equiv 0(p-1)}p^{-1}\mathbb{Z}_p\alpha^s e_t\right], & \text{if } N = 0 \\ p^k(k-1)!\Big[\bigoplus_{t\equiv 0(p-1)}\mathbb{Z}_p\alpha^s e_t & \\ \quad \oplus \bigoplus_{t\not\equiv 0,k-1(p-1),\text{or } t\equiv k-1, s\not\equiv S}p^{-1}\mathbb{Z}_p\alpha^s e_t & \\ \quad \bigoplus p^{-N_k-1}\mathbb{Z}_p\alpha^S e_{k-1}\Big], & \text{if } N \ne 0 \end{cases}$$

Proof: As in [de, I.3.5.3], the properties of g_u imply that, for $\sigma \in \mathcal{G}$, $\phi_k(u^\sigma) = \kappa(\sigma)^k[\phi_k(u)]^\sigma$, where we consider κ as a character on $\mathcal{G}$ that is trivial on Δ'. Therefore if $\sigma \in \Delta$, we have $\phi_k(u^\sigma) = \omega^k(\sigma)[\phi_k(u)]^\sigma$ and so $\phi_k((U^1_\infty)^{(s,t)})$ is contained in $L_1^{(s,t-k)} = \mathbb{Q}_p\alpha^s e_{t-k}$.

As ϕ_k is $\mathbf{Z}_p$-linear we have $\phi_k((U^1_\infty)^{(s,t)}) = p^{r_{s,t}}\mathbf{Z}_p\alpha^s e_{t-k}$ for some $r_{s,t} \in \mathbf{Z}$ and so by $G(t) = u_t e_t$, $u_t \in \mathcal{O}^\times_{\mathbf{D}_p}$

$$\phi_k((U^1_\infty)^{(s,t)})\mathcal{O}_{\mathbf{D}_p} = p^{r_{s,t}}G(t-k)\mathcal{O}_{\mathbf{D}_p}$$

as a subset of $\mathbf{D}_p(\mathbf{F}_\pi)$. We only need to compute the integers $r_{s,t}$.

First decompose $\mathcal{O}_{\mathbf{D}_p}[[G]]$ into eigenspaces for the eigen-characters ω^t of Δ

$$\mathcal{O}_{\mathbf{D}_p}[[G]] = \oplus_{0\le t\le p-2}\mathcal{O}_{\mathbf{D}_p}[[\Gamma]]E_t,$$

with $E_t = \frac{1}{p-1}\sum_{\sigma\in\Delta}\omega^{-t}(\sigma)\sigma \in \mathbf{Z}_p[\Delta]$. This enables us to write μ_u as

$$\mu_u = \sum E_t\mu_{u,t},\ \mu_{u,t} \in \mathcal{O}_{\mathbf{D}_p}[[\Gamma]],$$

considered as elements in $\mathcal{O}_{\mathbf{D}_p}[[G]]$. For each $0 \le t \le p-2$, define

$$\psi_t : U^1_\infty \to \mathcal{O}_{\mathbf{D}_p},\ u \mapsto \int_\Gamma \kappa(\sigma)^k d\mu_{u,t}(\sigma).$$

This map is $\mathbf{Z}_p$-linear and hence can be extended to a $\mathcal{O}_{\mathbf{D}_p}$-linear map from $U^1_\infty\hat{\otimes}_{\mathbf{Z}_p}\mathcal{O}_{\mathbf{D}_p}$ to $\mathcal{O}_{\mathbf{D}_p}$. The decomposition of U^1_∞ provides us with a decomposition of $U^1_\infty\hat{\otimes}_{\mathbf{Z}_p}\mathcal{O}_{\mathbf{D}_p}$ into a sum of eigenspaces $(U^1_\infty)^{(s,t)}\hat{\otimes}_{\mathbf{Z}_p}\mathcal{O}_{\mathbf{D}_p}$. From $\mu_{u^\sigma} = \sigma\mu_u$, $\sigma \in G$, we see that for $\sigma \in \Delta$, $\mu_{u^\sigma,t} = \omega^t(\sigma)\mu_{u,t}$ and $\psi_t(u^\sigma) = \omega^t(\sigma)\psi_t(u)$. Thus for those $u \in (U^1_\infty)^{(s,t)}\hat{\otimes}_{\mathbf{Z}_p}\mathcal{O}_{\mathbf{D}_p}$, we have $\psi_{t_1}(u) = 0$ unless $t_1 = t$. Similarly from $\phi^k_{CW}(u^\sigma) = \kappa^k(\sigma)\phi^k_{CW}(u) = \omega^k(\sigma)\psi^k_{CW}(u)$, $\sigma \in \Delta$, we obtain that $\phi^k_{CW}(u) = 0$ for $u \in (U^1_\infty)^{(s,t)}\hat{\otimes}_{\mathbf{Z}_p}\mathcal{O}_{\mathbf{D}_p}$ unless $k \equiv t \bmod p-1$.

By equation (6),

$$\begin{aligned}\phi_k(u) &= (p\Omega_p)^{-k}\int_G \kappa(\sigma)^k\zeta_1^{\kappa(\sigma)}d\mu_u^{F^{-1}}(\sigma) + p^{-1}\phi^k_{CW}(u)\\ &= (p\Omega_p)^{-k}\frac{1}{p-1}\sum_t\sum_{\sigma\in\Delta}\omega^{-t}(\sigma)\int_\Gamma \kappa(\sigma\gamma)^k\zeta_1^{\kappa(\sigma\gamma)}d\mu_{u,t}^{F^{-1}}(\gamma) + p^{-1}\phi^k_{CW}(u)\end{aligned}$$

Since $\kappa(\sigma\gamma) = \kappa(\sigma)\kappa(\gamma) = \omega(\sigma)\kappa(\gamma)$, and for $\gamma \in \Gamma$, we have $\kappa(\gamma) \equiv 1 \bmod p$ and $\zeta_1^{\kappa(\sigma\gamma)} = \zeta_1^{\omega(\sigma)}$, we obtain

$$\phi_k(u) = (p\Omega_p)^{-k}\sum_t G(t-k)[\psi_t(u)]^{F^{-1}} + p^{-1}\phi^r_{CW}(u) \tag{7}$$

Now if $u \in (U^1_\infty)^{(s,t)}\hat{\otimes}_{\mathbf{Z}_p}\mathcal{O}_{\mathbf{D}_p}$ with $t \not\equiv k \bmod p-1$, then $\phi^k_{CW}(u) = 0$ and $\psi_{t_1}(u) = 0$ for $t_1 \ne t$. So the previous equation implies that $\phi_k(u) = (p\Omega_p)^{-k}G(t-k)[\psi_t(u)]^{F^{-1}}$. Hence

$$\phi_k((U^1_\infty)^{(s,t)})\mathcal{O}_{\mathbf{D}_p} = p^{-k}G(t-k)[\psi_t((U^1_\infty)^{(s,t)}\hat{\otimes}_{\mathbf{Z}_p}\mathcal{O}_{\mathbf{D}_p})]^{F^{-1}}.$$

On the other hand, if $u \in (U_\infty^1)^{(s,t)}\hat{\otimes}_{\mathbf{Z}_p}\mathcal{O}_{\mathbf{D}_p}$ with $t \equiv k \bmod p-1$, then it is easy to see that $\phi_{CW}^k(u^\tau) = [\phi_{CW}^k(u)]^\tau$ for $\tau \in \Delta' = G(L/\mathbf{Q}_p) \cong G(L(\mathbf{F}_{\pi^\infty})/\mathbf{Q}_p(\mathbf{F}_{\pi^\infty})))$, so $\phi_{CW}^k(u) \in L^s$, the eigenspace for the character χ^s, and

$$[\phi_{CW}^k(u)]^F = \chi^s(F)\phi_{CW}^k(u) = \rho^s\phi_{CW}^k(u).$$

Then equation (5) shows that

$$(1-\frac{\pi^k}{p}\rho^s)\phi_{CW}^k(u) = \Omega_p^{-k}\int_G \kappa(\sigma)^k d\mu_u(\sigma) = \Omega_p^{-k}\psi_\tau(u) = \Omega_p^{-k}\psi_t(u).$$

Then equation (7) gives

$$\phi_k(u) = (p\Omega_p)^{-k}[\psi_t(u)]^{F^{-1}} + p^{-1}\Omega_p^{-k}(1-\frac{\pi^k}{p})^{-1}\psi_t(u).$$

We now compute $\psi_t((U_\infty^1)^{(s,t)}\hat{\otimes}_{\mathbf{Z}_p}\mathcal{O}_{\mathbf{D}_p})$. Since $\chi_N(\sigma) \equiv \epsilon^{-S}(\sigma) \bmod p^N$ by Proposition 3.1, the fundamental exact sequence [de, I.3.7]

$$0 \to U_\infty^1\hat{\otimes}_{\mathbf{Z}_p}\mathcal{O}_{\mathbf{D}_p} \xrightarrow{i} \mathcal{O}_{\mathbf{D}_p}[[\mathcal{G}]] \xrightarrow{j} (\mathcal{O}_{\mathbf{D}_p}/p^N)(\chi_N) \to 0$$

induces exact sequences of $\mathcal{O}_{\mathbf{D}_p}[[\Gamma]]$-modules

$$0 \to (U_\infty^1)^{(s,t)}\hat{\otimes}_{\mathbf{Z}_p}\mathcal{O}_{\mathbf{D}_p} \to \mathcal{O}_{\mathbf{D}_p}[[\Gamma]] \to 0$$

if $(s,t) \not\equiv (-S,1) \bmod (d,p-1)$ and

$$0 \to (U_\infty^1)^{(s,t)}\hat{\otimes}_{\mathbf{Z}_p}\mathcal{O}_{\mathbf{D}_p} \to \mathcal{O}_{\mathbf{D}_p}[[\Gamma]] \to (\mathcal{O}_{\mathbf{D}_p}/p^N)(\kappa) \to 0$$

if $(s,t) \equiv (-S,1) \bmod (d,p-1)$. If $(s,t) \not\equiv (-S,1) \bmod (d,p-1)$, then

$$\psi_t((U_\infty^1)^{(s,t)}\hat{\otimes}_{\mathbf{Z}_p}\mathcal{O}_{\mathbf{D}_p}) = \mathcal{O}_{\mathbf{D}_p}.$$

So $\phi_k((U_\infty^1)^{(s,t)})\mathcal{O}_{\mathbf{D}_p} = p^{-k}G(t-k)\mathcal{O}_{\mathbf{D}_p}$ which means that

$$\phi_k((U_\infty^1)^{(s,t)}) = p^{-k}\mathbf{Z}_p\alpha^s e_{t-k}.$$

If $(s,t) \equiv (-S,1) \bmod (d,p-1)$, then the above exact sequence shows that

$$\begin{aligned}&\psi_1((U_\infty^1)^{(-S,1)}\hat{\otimes}_{\mathbf{Z}_p}\mathcal{O}_{\mathbf{D}_p})\mathcal{O}_{\mathbf{D}_p}\\ &= \{P(\kappa(\gamma_0)^k-1) \mid P(X) \in \mathcal{O}_{\mathbf{D}_p}[[X]], P(\kappa(\gamma_0)-1) \equiv 0 \bmod p^N\}\end{aligned}$$

when we identify $\mathcal{O}_{\mathbf{D}_p}[[\Gamma]]$ with $\mathcal{O}_{\mathbf{D}_p}[[X]]$ by choosing a generator γ_0 of Γ with $\kappa(\gamma_0) = 1+p$ and identifying γ_0 with $1+X$. Let S_n be the set of polynomials of degree n in $\mathcal{O}_{\mathbf{D}_p}[[X]]$ such that $P(p) \equiv 0 \bmod p^N$, then it is easily shown that $P((1+p)^k-1) \equiv 0 \bmod p^{N_k}$ for any $P \in S_n$ and that there is a $P \in S_N$ such that $P((1+p)^k-1) \not\equiv 0 \bmod p^{N_k+1}$. Since every element of $\mathcal{O}_{\mathbf{D}_p}[[X]]$ can be written as a product of a polynomial and a unit power series, the same is true for the set of power series whose polynomial part has degree n. Therefore $\psi_1((U_\infty^1)^{(-S,1)}\hat{\otimes}_{\mathbf{Z}_p}\mathcal{O}_{\mathbf{D}_p}) = p^{N_k}$. Thus $\phi_k((U_\infty^1)^{(-S,1)}) = p^{N_k-k}\mathbf{Z}_p\alpha^{-S}e_{1-k}$. This proves the first statement.

To prove the second statement, first note that $Tr_{\mathbf{Q}_p}^{L_1}(\alpha^s e_t) = 0$ unless $s = t \equiv 0$. It then follows that $Tr_{\mathbf{Q}_p}^{L_1}(\alpha^s e_t \cdot \alpha^{s_1}e_{t_1}) = 0$ unless $s \equiv -s_1$ and $t \equiv -t_1$ since $\alpha^s e_t \cdot \alpha^{s_1}e_{t_1} \in \alpha^{s+s_1}e_{t+t_1}$ by definition. Finally, $e_1 = 1$, and, for $t \neq 0$, since $v_\mathfrak{p}(e_t) = v_\mathfrak{p}(G(t)) = t$, where $\mathfrak{p}$ is the prime of L_1, we get $v_\mathfrak{p}(e_t \cdot e_{-t}) = p-1$ and hence $v_p(e_t \cdot e_{-t}) = 1$. The second statement follows from these facts and the first statement. ■

References

[B-K] S. Bloch and K. Kato, *L-functions and Tamagawa numbers of motives,* The Grothendieck Festschrift, Vol 1. Birkhäuser(1990), 333–400.

[de] E. de Shalit, *The Iwasawa theory of elliptic curves with complex multiplication,* Perspect. Math. Vol.3 (1987).

[Fo1] J.-M. Fontaine, *Sur certaine types de représentations p-adiques du groupe de Galois d'un corps local; Construction d'un anneau de Barsotti-Tate,* Ann. Math. **115** (1982), 529–577.

[Fo2] J.-M. Fontaine, *Appendice: Sur un théorème de Bloch et Kato (lettre à B. Perrin-Riou),* Invent. Math. **115** (1994), 151-161.

[Gu] L. Guo, *On the Bloch-Kato conjecture for Hecke L-functions,* J. of Number Theory **57** (1996), 340-365.

[Han] B. Han, *On Bloch-Kato conjecture of Tamagawa numbers for Hecke characters of imaginary quadratic number fields,* Ph.D. thesis, University of Chicago, 1997.

[Ha] M. Harrison, *On the conjecture of Bloch-Kato for Grössencharacters over* $\mathbb{Q}(i)$, Ph.D. thesis, Cambridge University, 1993.

[Ka] K. Kato, *Lectures on the approach to Iwasawa theory for Hasse-Weil L-function via* B_{dR}, In: Lecture Notes in Math, vol.1553, Springer-Verlag, 50-163.

[Ne] J. Neukirch, *Class field theory,* Grund. Der Math. Wissen. **280**, Springer-Verlag, 1985.

[Ru] K. Rubin, *The "main conjecture" of Iwasawa theory for imaginary quadratic fields,* Invent. Math. **103** (1991), 25–68.

[Wi] A. Wiles, *Higher explicit reciprocity laws,* Ann. Math. **107**(1978), 235-254.

Department of Mathematics and Computer Science, Rutgers University at Newark Newark, NJ 07102

E-mail address: `liguo@newark.rutgers.edu`

On certain Gauss periods with common decomposition field

Dedicated to Basil Gordon on his 65th *birthday*

S. Gurak

Abstract

Fix $f > 1$ and a positive integer r prime to f, and consider a prime p congruent to $r \pmod{f}$. The decomposition field K of the prime p in the cyclotomic field of f-roots of unity has conductor $f_0|f$. Suppose that $f_0|f'|f$ for some positive integer f' having the same prime divisors as f. Let $\mathbf{F}_q$ be the finite field of q elements, where $q = ef + 1$ is a power of p, and $\mathbf{F}_{q'}$ a subfield with $q' = e'f' + 1$ elements. In this work I investigate the extent to which the Gauss periods of order e for $\mathbf{F}_q$ and those of order e' for $\mathbf{F}_{q'}$ are related. The results generalize those I previously noticed in the case $K = \mathbf{Q}$.

1. Introduction

Let $q = p^a$ be a power of a prime, and e and f positive integers such that $ef + 1 = q$. Let $\mathbf{F}_q$ denote the field of q elements, $\mathbf{F}_q^*$ its multiplicative group and g a fixed generator of $\mathbf{F}_q^*$. Let $\mathrm{Tr} : \mathbf{F}_q \to \mathbf{F}_p$ be the usual trace map and set $\theta = \exp(2\pi i/p)$, a primitive p-th root of unity. Put $\delta = \gcd(\frac{q-1}{p-1}, e)$ and $R = \frac{q-1}{\delta(p-1)} = \frac{f}{\gcd(p-1,f)}$, and let C_e denote the group of e-th powers in $\mathbf{F}_q^*$. The Gauss periods of order e for $\mathbf{F}_q$ are given by

$$\eta_j = \sum_{i=1}^{f} \theta^{\mathrm{Tr}\, g^{e(i-1)+j}} \quad (1 \le j \le e) \tag{1}$$

(if $j > e$ then the subscripts in η_j are taken modulo e), and satisfy the period polynomial

$$\Phi(x) = \prod_{j=1}^{e} (x - \eta_j). \tag{2}$$

[0]1991 Mathematics Subject Classification. Primary 11T22,11T24

S.D. Ahlgren et al. (eds.), Topics in Number Theory, 193–205.

G. Myerson [8] showed that $\Phi(x)$ splits over $\mathbf{Q}$ into δ factors, each of degree e/δ. To be precise

$$\Phi(x) = \prod_{w=1}^{\delta} \Phi^{(w)}(x), \tag{3}$$

where

$$\Phi^{(w)}(x) = \prod_{k=0}^{e/\delta-1} (x - \eta_{w+k\delta}) \ \ (1 \le w \le \delta). \tag{4}$$

The coefficients $a_r = a_r(w)$ of the factor

$$\Phi^{(w)}(x) = x^{e/\delta} + a_1 x^{e/\delta-1} + \cdots + a_{e/\delta}, \tag{5}$$

or equivalently of

$$F^{(w)}(X) = X^{e/\delta}\Phi^{(w)}(X^{-1}) = 1 + a_1 X + \cdots + a_{e/\delta}X^{e/\delta}, \tag{6}$$

are expressed in terms of the symmetric power sums

$$S_n = S_n(w) = \sum_{k=0}^{e/\delta-1} (\eta_{w+k\delta})^n \ \ (n \ge 0) \tag{7}$$

through Newton's identities

$$S_r + a_1 S_{r-1} + \cdots + a_{r-1}S_1 + ra_r = 0 \ \ (1 \le r \le e/\delta). \tag{8}$$

Set $t_w(0) = 1$ and for $n > 0$, let $t_w(n)$ count the number of tuples $(x_1, \ldots, x_n)$ with $x_i \in C_e$ $(1 \le i \le n)$ for which $\mathrm{Tr}(g^w(x_1 + \cdots + x_n)) = 0$. The $S_n(w)$ can be computed [5] using

$$S_n(w) = (pt_w(n) - f^n)/\gcd(p-1, f). \tag{9}$$

Now fix f and select a positive integer r prime to f, say with $ord_f\ r = b$, and consider primes $p \equiv r(\bmod f)$. One finds $\frac{e}{\delta} = \frac{p-1}{\gcd(p-1,f)}$ and $R = \frac{f}{\gcd(p-1,f)}$. Further, all such primes p have common decomposition field K in $\mathbf{Q}(\zeta_f)$, where $\zeta_f = \exp(2\pi i/f)$ and $[\mathbf{Q}(\zeta_f) : K] = b$. (The field K is that subfield of $\mathbf{Q}(\zeta_f)$ fixed by the action $\zeta_f \to \zeta_f^r$.) For some selection of factor $w = w(p)$, $1 \le w \le \delta$, suppose $b_w(n)$ is a counting function (independent of p) which for $n \ge 0$, coincides with $t_w(n)$ for all sufficiently large primes $p \equiv r(\bmod f)$. I shall say such a counting function $b_w(n)$ **agrees with $t_w(n)$ almost everywhere**. (Necessarily $b_w(0) = 1$.) Such counting functions $b_w(n)$ were given for the factor $w = \delta$ and $w = \delta/2$ (when δ is even) by the author in previous work. In [5], $b_\delta(n)$ is selected to count the number of times a sum $x_1 + \cdots + x_n = 0$ for choice of x_i $(1 \le i \le n)$ taken from the traces

$$\mathrm{Tr}_{\mathbf{Q}(\zeta_f)/K}\ \zeta_f^{\nu} \ \ (1 \le \nu \le f). \tag{10}$$

For δ even, one may set $b_{\delta/2}(n) = f^n$ when b is odd. For even b, $b_{\delta/2}(n)$ is selected in [7] to count the number of times a sum $x_1 + \cdots + x_n = 0$ for choice of x_i $(1 \le i \le n)$ taken from the values

$$\omega_\nu = \zeta_f^\nu + \zeta_{2^u}\zeta_f^{r\nu} + \cdots + \zeta_{2^u}^{1+r+\cdots+r^{b-2}}\zeta_f^{r^{b-1}\nu} \quad (1 \le \nu \le f). \tag{11}$$

(Here the integer $u > 0$ is determined by the condition $2^{u-1}||R$; and $\zeta_{2^u} = \exp(2\pi i/2^u)$.) In general, finding a suitable $b_w(n)$ seems quite difficult except if $w = j\delta/m$, where $m|\delta$ and $1 \le j \le m$. (Details concerning the case $w = j\delta/m$ will appear elsewhere in a subsequent paper.) On the other hand, if $b_w(n)$ agrees with $t_w(n)$ almost everywhere, then for any $n \ge 0$, it follows from (9) that the expression $\frac{1}{p}(\frac{f}{R}S_n(w(p)) + f^n)$ becomes constant equal to $b_w(n)$ for all sufficiently large $p \equiv r(\text{mod } f)$. It is also reasonable to expect that for any $n > 0$, that the finite set $\xi_n = \xi_n(w)$ of exceptions be computable. This was the case for the specific $b_w(n)$ above when $w = \delta$ and $w = \delta/2$.

Assuming $b_w(n)$ agrees with $t_w(n)$ almost everywhere, except for primes in $\xi_n = \xi_n(w)$ $(n > 0)$, the proofs of Theorems 1 and 2 in [5] extend to yield the following results.

Theorem 1 *Suppose $b_w(n)$ agrees with $t_w(n)$ almost everywhere. Let h be the smallest positive integer for which $b_w(h) \neq 0$. For all primes $p \not| a$, $p \equiv r(\text{mod } f)$, but not in $\xi_n(n \le s)$, the coefficient a_s for $\Phi^{(w)}(x)$ in (5) (or $F^{(w)}(X)$ in (6)) satisfies $a_s = \vartheta_s(p)$, where ϑ_s is a polynomial of degree $[s/h]$.*

The above has an alternative formulation in terms of the rational power series

$$C_w(X) = \exp(-\frac{R}{f}\sum_{n=1}^{\infty} b_w(n)X^n/n), \tag{12}$$

defined in terms of the counting function $b_w(n)$.

Theorem 2 *For any $s > 0$ and prime $p \not| a$, $p \equiv r(\text{mod } f)$ but $p \notin \xi_n$ $(n \le s)$,*

$$F^{(w)}(X) \equiv \frac{C_w(X)^p}{(1 - fX)^{R/f}} \quad \text{mod } X^{s+1} \tag{13}$$

The congruence actually holds in $\mathbf{Z}[[X]]$, since the right side of (13) is seen to have an integral power series expansion.

In a previous paper [6], I studied Gauss periods over finite fields $\mathbf{F}_q$, where p has decomposition field $K = \mathbf{Q}$. I discovered that certain such Gauss periods were integer translates of others, and found algebraic relations among their corresponding power series $C(X)$ defined in (12). In special cases I explicitly found

$C(X)$. The goal of the present paper is to generalize these results for certain Gauss periods with common decomposition field $K \neq \mathbf{Q}$. This is described in the next section. Some explicit formulas appear in the third and concluding section.

2. Gauss periods with common decomposition field

Here I retain the notation of the introductory section, considering the Gauss periods η in (1) for the finite field $\mathbf{F}_q$ of $q = p^a$ elements, where $p \equiv r (\mathrm{mod}\ f)$ has decomposition field K in $\mathbf{Q}(\zeta_f)$. Let f_0 be the smallest positive integer dividing f for which $K \subseteq \mathbf{Q}(\zeta_{f_0})$, say with index $[\mathbf{Q}(\zeta_{f_0}) : K] = b_0$. (In the language of classfield theory f_0 is known as the conductor of the extension $K/\mathbf{Q}$.) Select a positive integer f' with $f_0 | f' | f$, where f and f' have the same prime divisors, say with $[\mathbf{Q}(\zeta_{f'}) : K] = b'$. Then $b = ord_f\ r = \frac{\phi(f)}{\phi(f')} ord_{f'}\ r = \frac{f}{f'} b'$. Write $q' = p^{b'} = e'f' + 1$ and put $\delta' = \gcd(e', \frac{q'-1}{p-1})$ and $R' = \frac{q'-1}{\delta'(p-1)}$. Consider the Gauss periods η' in (1) of order e' for the finite field $\mathbf{F}_{q'}$. Their corresponding period polynomial $\Phi'(x)$ factors as a product of δ' factors, $\Phi'^{(w)}(x)$ $(1 \leq w \leq \delta')$, each of degree e'/δ'. The coefficients $a'_r = a'_r(w)$ of the factor

$$\Phi'^{(w)}(x) = x^{e'/\delta'} + a'_1 x^{e'/\delta' - 1} + \cdots + a'_{e'/\delta'}, \tag{14}$$

or equivalently of

$$F'^{(w)}(X) = X^{e'/\delta'} \Phi'^{(w)}(X^{-1}) = 1 + a'_1 X + \cdots + a'_{e'/\delta'} X^{e'/\delta'}, \tag{15}$$

are similarly expressed in terms of symmetric power sums $S'_n = S'_n(w)$ $(n \geq 0)$ through Newton's identities. In addition, the S'_n can be computed using (9)

$$S'_n(w) = (pt'_w(n) - f'^n)/\gcd(p-1, f') \tag{16}$$

in terms of the corresponding counting function $t'_w(n)$.

In the simplest case $K = \mathbf{Q}$ (so $f_0 = 1$), r must be a primitive root of f and f'. Then the only possibilities with $f' \neq f$ are (i) $f = 4, f' = 2$ (ii) $f = l^v, f' = l^\alpha$ where $v > \alpha > 0$ and (iii) $f = 2l^v, f' = 2l^\alpha$ where $v > \alpha > 0$. In [6] I essentially showed for each of the cases (i) - (iii), that if $p \not| \frac{a}{b}$ then $\eta_\delta = \bar{\eta}' + (f - f')$ for some conjugate $\bar{\eta}'$ of $\eta'_{\delta'}$ so that $\Phi^{(\delta)}(x) = \Phi'^{(\delta')}(x - (f - f'))$. Similarly, if $p \not| \frac{a}{b}$ one finds that $\eta_{\delta/2} = \bar{\eta}' + (f - f')$ for some conjugate $\bar{\eta}'$ of $\eta'_{\delta'/2}$ in each of the cases (i) - (iii) from Theorem 3, its Corollary and Proposition 6 in [7], so $\Phi^{(\delta/2)}(x) = \Phi'^{(\delta'/2)}(x - (f - f'))$.

It is natural to ask if the above instances suggest a more general relationship between certain Gauss periods of order e for $\mathbf{F}_q$ and those of order e' for $\mathbf{F}_{q'}$. Indeed, this is the case.

Theorem 3 *Suppose $f' | f$ as in the above with $p \equiv r (\mathrm{mod}\ f)$ having decomposition field $K \subseteq \mathbf{Q}(\zeta_{f'})$ and $p \not| \frac{a}{b}$. Then for $1 \leq w \leq \delta'$, $\eta_{(q-1)w/(q'-1)} = \bar{\eta}' +$*

$(f - f')$ for some conjugate $\bar{\eta}'$ of η'_w. In particular, the factor $\Phi^{((q-1)w/(q'-1))}(x) = \Phi'^{(w)}(x - (f - f'))$.

Before demonstrating Theorem 3 it is necessary to mention the following lemmas.

Lemma 1 *Suppose $f_0|f'|f$ as described in the beginning of this section with $p \equiv r (mod\ f)$ having decomposition field K in $\mathbf{Q}(\zeta_f)$ of conductor f_0. Then (i)* $\gcd(p-1, f') = \gcd(p-1, f)$ *and (ii)* $\gcd(e', f/f') = 1$.

Proof: Clearly $\gcd(p-1, f') | \gcd(p-1, f)$. Suppose a prime l divides the quotient $\frac{\gcd(p-1,f)}{\gcd(p-1,f')}$. The prime l must divide both f' and f so $p \equiv 1 (mod\ l^\alpha)$ for some $\alpha > 1$ with $l^\alpha | f$ but $l^\alpha \nmid f'$. But then K is not contained in $\mathbf{Q}(\zeta_{f'})$ contrary to the hypothesis that $f_0 | f'$ with $K \subseteq \mathbf{Q}(\zeta_{f_0})$. Hence $\gcd(p-1, f') = \gcd(p-1, f)$.

To establish (ii), write $\gcd(p^{b'} - 1, f) = f'u$, where $u | \frac{f}{f'}$. Then $ord_{f'u}\ p = b'$. Since p also has decomposition field K in $\mathbf{Q}(\zeta_{f'u})$, one has $[\mathbf{Q}(\zeta_{f'u}) : K] = b'$. Then $\phi(f') = \phi(f'u) = \phi(f')u$, so $u = 1$ and $\gcd(p^{b'} - 1, f) = f'$. Thus $\gcd(e', \frac{f}{f'}) = \gcd(\frac{p^{b'}-1}{f'}, \frac{f}{f'}) = 1$.

Lemma 2 *Under the same hypotheses as in the lemma above, let G be any generator for $\mathbf{F}^*_{p^{b'}}$. Then $\mathbf{F}_{p^b} = \mathbf{F}_{p^{b'}}(u)$ where $u^{f/f'} = G$. In particular,* $\mathrm{Tr}_{\mathbf{F}_{p^b}/\mathbf{F}_{p^{b'}}} u^m = 0$ *for $m \not\equiv 0 (mod\ f/f')$.*

Proof: The lemma follows from the fact that $[\mathbf{F}_{p^b} : \mathbf{F}_{p^{b'}}] = b/b' = f/f'$ here.

Proof of Theorem 3: Set $G = g^{\frac{q-1}{q'-1}}$ to generate $\mathbf{F}^*_{p^{b'}}$. From (1), $\eta_{\frac{q-1}{q'-1}w} = \sum_{i=1}^{f} \theta^{\mathrm{Tr}_{\mathbf{F}_q/\mathbf{F}_p} g^{ei+(q-1)w/(q'-1)}} = \sum_{i=1}^{f} \theta^{\mathrm{Tr}_{\mathbf{F}_{p^b}/\mathbf{F}_p} g^{ei} \mathrm{Tr}_{\mathbf{F}_q/\mathbf{F}_{p^b}} G^w}$ $= \sum_{i=1}^{f} \theta^{\frac{a}{b}\mathrm{Tr}_{\mathbf{F}_{p^{b'}}/\mathbf{F}_p}(G^w \mathrm{Tr}_{\mathbf{F}_{p^b}/\mathbf{F}_{p^{b'}}} g^{ei})}$ since g^e lies in $\mathbf{F}_{p^b}$. Further, as $e = e'\frac{(q-1)f'}{(q'-1)f}$, one may choose $u = g^{e/e'}$ in Lemma 2 above. Then the last expression is just $\sum_{i=1}^{f} \theta^{\frac{a}{b}\mathrm{Tr}_{\mathbf{F}_{p^{b'}}/\mathbf{F}_p}(G^w \mathrm{Tr}_{\mathbf{F}p^b/\mathbf{F}_{p^{b'}}} u^{e'i})} = (f - f')$ $+ \sum_{i=1}^{f'} \theta^{\frac{a}{b}\mathrm{Tr}_{\mathbf{F}_{p^{b'}}/\mathbf{F}_p} G^w \frac{b}{b'} G^{e'i}} = (f - f') + \sum_{i=1}^{f'} \theta^{\frac{a}{b'}\mathrm{Tr}_{\mathbf{F}_{p^{b'}}/\mathbf{F}_p} G^{w+e'i}}$ since $\gcd(e', f/f') = 1$ from Lemma 1. In particular, $\eta_{(q-1)w/(q'-1)} = (f - f') + \bar{\eta}'$, where $\bar{\eta}'$ is a conjugate of η'_w, so $\Phi^{((q-1)w/(q'-1))}(x) = \Phi'^{(w)}(x - (f - f'))$.

Now suppose $b'_w(n)$ is a suitably chosen counting function in that $b'_w(n)$ agrees with $t'_w(n)$ almost everywhere (but for an exceptional set $\xi'_n(w)$.) Then by Theorem 2, if $p \nmid a$ and $p \notin \xi'_n$ $(n \leq s)$

$$F'^{(w)}(X) \equiv \frac{C'_w(X)^p}{(1 - f'X)^{R'/f'}} \bmod X^{s+1} \tag{17}$$

holds in $\mathbf{Z}[[X]]$, where

$$C'_w(X) = \exp(-\frac{R'}{f'}\sum_{n=1}^{\infty} b'_w(n)X^n/n) \tag{18}$$

The following result shows for the situation at hand that it is always possible to find a function $b_{(q-1)w/(q'-1)}(n)$ which agrees with $t_{(q-1)w/(q'-1)}(n)$ almost everywhere, and how the corresponding power series $C_{(q-1)w/(q'-1)}(X)$ in (12) is algebraically related to $C'_w(X)$.

Proposition 1 *The function*

$$b_{\frac{q-1}{q'-1}w}(n) = \sum_{i=0}^{n}(f-f')^{n-i}\binom{n}{i}b'_w(i) \tag{19}$$

equals $t_{\frac{q-1}{q'-1}w}(n)$ *except if* $p \in \xi'_i(w)$ *for some* i, $1 \leq i \leq n$, *or if* $p|\frac{a}{b}$. *Moreover, the corresponding power series* $C_{\frac{q-1}{q'-1}w}(X)$ *in (12) satisfies*

$$C_{\frac{q-1}{q'-1}w}(X) = (1-(f-f')X)^{R/f}C'_w(X/(1-(f-f')X)). \tag{20}$$

Proof: Computing from (9) one easily obtains $t_{(q-1)w/(q'-1)}(n) = \frac{1}{p}(\frac{f}{R}S_n(\frac{q-1}{q'-1}w(p)) + f^n) = \sum_{i=0}^{n}\binom{n}{i}(f-f')^{n-i}\frac{1}{p}(\frac{f}{R}S'_i(w(p)) + f'^i)$, since $S_n(\frac{q-1}{q'-1}w) = \sum_{k=0}^{e/\delta-1}(\eta'_{w+k\delta'} + (f-f'))^n = \sum_{i=0}^{n}\binom{n}{i}(f-f')^{n-i}S'_i(w)$ in view of Theorem 3. But $f/R = f'/R'$ here, so $t_{\frac{q-1}{q'-1}w}(n) = \sum_{i=0}^{n}\binom{n}{i}(f-f')^{n-i}t'_w(i)$ if $p \not| \frac{a}{b}$. Since $b'_w(i) = t'_w(i)$ except for p in $\xi'_i(w)$, it follows that $b_{\frac{q-1}{q'-1}w}(n)$ in (19) equals $t_{\frac{q-1}{q'-1}w}(n)$ whenever $p \not| \frac{a}{b}$ and $p \notin \xi'_n(w)(i \leq n)$. To establish (20) one may argue as in the proof of Proposition 3 in [7].

It is worth noting that through straightforward combinatorial manipulation (19) is alternatively expressed in terms of generating functions by

$$\sum_{n=0}^{\infty} b_{\frac{q-1}{q'-1}w}(n)\frac{X^n}{n!} = \exp((f-f')X)\sum_{n=0}^{\infty} b'_w(n)\frac{X^n}{n!}. \tag{21}$$

Example: Consider the case $f = 2^v$ with $v > 2$ and $p \equiv 3(\text{mod } f)$. Here $K = \mathbf{Q}(\sqrt{-2})$ with $b = 2^{v-2}$, and $f_0 = 8$. Choosing $f' = 8$ one finds $b' = 2$ with $e'/\delta' = e/\delta = \frac{p-1}{2}$, $R' = 4$ and $R = 2^{v-1}$. Since $q' = p^2$ one also has $e' = \frac{p^2-1}{8}$ and $\delta' = \frac{p+1}{4}$ both odd. The first few terms for $b'_{\delta'}(n)$ are computed to be $b'_{\delta'}(1) = 2$, $b'_{\delta'}(2) = 14$, $b'_{\delta'}(3) = 68$, $b'_{\delta'}(4) = 454$, with exceptional sets $\xi_2 = \xi_3 = \{3\}$ and $\xi_4 = \{3, 11, 19\}$ (chiefly, Example 2 in [5].) Thus $C'_{\delta'}(X) = \exp(-\frac{1}{2}\sum_{n=1}^{\infty} b'_{\delta'}(n)\frac{X^n}{n}) = 1 - X - 3X^2 - 8X^3 - 41X^4 - \cdots$ From Theorem 2, the first few coefficients of the factor $\Phi'^{(\delta')}(x)$ are $a'_1 = -(p-4)$ for $p > 3$, $a'_2 = \frac{1}{2}(p^2 - 15p + 48)$ for $p > 3$, $a'_3 = -\frac{1}{6}(p^3 - 33p^2 + 296p - 960)$ for $p > 3$, and $a'_4 = \frac{1}{24}(p^4 - 58p^3 + 1043p^2 - 8306p + 26880)$ for $p > 19$. From Proposition 1, $b_\delta(n) = \sum_{i=0}^{n}(2^v - 8)^{n-i}\binom{n}{i} b'_{\delta'}(n)$ agrees with $t_\delta(n)$ almost everywhere, and $C_\delta(X) = \sqrt{1 - (2^v - 8)X}\, C'_{\delta'}(X/(1 - (2^v - 8)X))$. In particular with $f = 16$ and $q = p^4$ one finds $C_\delta(X) = 1 - 5X - 15X^2 - 100X^3 - 881X^4 - \cdots$, and $\Phi^{((p^2+1)w)}(x) = \Phi'^{(w)}(x - 8)$ for $1 \le w \le \frac{p+1}{4}$. The first few coefficients of the factor $\Phi^{(\delta)}(x)$ are found to be $a_1 = -(5p - 8)$ for $p > 3$, $a_2 = \frac{1}{2}(25p^2 - 231p + 192)$ for $p > 3$, $a_3 = -\frac{1}{6}(125p^3 - 1425p^2 + 5500p - 7680)$ for $p > 3$, and $a_4 = \frac{1}{24}(625p^4 - 12250p^3 + 90275p^2 - 307154p + 430080)$ for $p > 19$.

3. Some explicit formulas when K is a quadratic field

In this concluding section I give a systematic treatment for the factors $\Phi^{(w)}(x)$, where $w = \delta$ (and $w = \delta/2$ when δ is even), for those cases with decompostion field K quadratic. Throughout the section, l, l_1 and l_2 will denote odd primes. Since $\mathbf{Q}(\zeta_{f_0})/K$ must be cyclic, at most two distinct primes may divide the conductor f_0. For prime power conductor f_0, the only possibilities are $K = \mathbf{Q}(\sqrt{-1})$, $\mathbf{Q}(\sqrt{2})$, $\mathbf{Q}(\sqrt{-2})$ and $\mathbf{Q}(\sqrt{l^*})$, where $l^* = (-1)^{(l-1)/2}l$. The remaining possibilities are readily seen to be $K = \mathbf{Q}(\sqrt{-l^*})$ for $f_0 = 4l$ and $K = \mathbf{Q}(\sqrt{l_1^* l_2^*})$ for $f_0 = l_1 l_2$ with $\gcd(l_1 - 1, l_2 - 1) = 2$. I now derive generating functions for the counting functions $b_\delta(n)$ and $b_{\delta/2}(n)$ in all cases but the last. Unfortunately, there seems to be no nice closed form expressions for the corresponding power series $C(X)$ in (12) (except for the trivial case when $b_w(n) = f^n$ and $C_w(X) = (1 - fX)^{R/f}$.)

(1) The case $K = \mathbf{Q}(\sqrt{-1})$ arises when (a) $f = 2^v, v \ge 2$ with $p \equiv 1 (\mathrm{mod}\ 4)$ and $b = 2^{v-2}$, or (b) $f = 4l^v, v > 0$ with $p \equiv 1 (\mathrm{mod}\ 4)$ and a primitive root for l^v. With $f' = 4$ in case (a), one has

$$b'_{\delta'}(n) = \begin{cases} \binom{n}{n/2}^2 & \text{if } 2|n \\ 0 & \text{otherwise,} \end{cases}$$

and $b'_{\delta'/2}(n) = 4^n$ when δ' is even. Thus, from Proposition 1, it follows that

$$\sum_{n=0}^{\infty} b_\delta(n)\frac{X^n}{n!} = \exp((2^v - 4)X)\left(\sum_{n=0}^{\infty}\frac{X^{2n}}{(n!)^2}\right)^2$$

and that $b_{\delta/2}(n) = 2^{nv}$ when δ is even, with $\Phi^{(\delta)}(x) = \Phi'^{(\delta')}(x - (2^v - 4))$ and $\Phi^{(\delta/2)}(x) = (x - 2^v)^{\frac{p-1}{4}}$, respectively.

For case (b) with $f' = 4l$ one computes the traces $\mathrm{Tr}_{\mathbf{Q}(\zeta_{4l})/\mathbf{Q}(i)}\, \zeta_{4l}^{\nu}$ $(1 \le \nu \le 4l)$ in (10) to find $l-1$ occurrences each of $1, -1, i$ and $-i$ and one occurrence each of $l-1$, $-(l-1)$, $(l-1)i$ and $-(l-1)i$. Since the set $\{1, i\}$ is a basis for $K/\mathbf{Q}$, it readily follows that

$$\sum_{n=0}^{\infty} b'_{\delta'}(n)\frac{X^n}{n!} = (\sum_{n=0}^{\infty} \beta_{2l}(n)\frac{X^n}{n!})^2,$$

where $\beta_{2l}(n)$ is given by formula (16) in [6]. Thus

$$\sum_{n=0}^{\infty} b_{\delta}(n)\frac{X^n}{n!} = \exp((4l^v - 4l)X)(\sum_{n=0}^{\infty} \beta_{2l}(n)\frac{X^n}{n!})^2$$

from Proposition 1. Similarly, one computes the values ω_ν $(1 \le \nu \le 4l)$ in (11) when $f' = 4l$ to find $l-1$ occurrences each of $1, -1, i$ and $-i$ and four zero values. Thus

$$\sum_{n=0}^{\infty} b'_{\delta'/2}(n)\frac{X^n}{n!} = \exp(4X)(\sum_{n=0}^{\infty} \frac{((l-1)X)^{2n}}{(n!)^2})^2,$$

so

$$\sum_{n=0}^{\infty} b_{\delta/2}(n)\frac{X^n}{n!} = \exp(4(l^v - l + 1)X)(\sum_{n=0}^{\infty} \frac{((l-1)X)^{2n}}{(n!)^2})^2$$

when δ' and δ are even.

(2) The case $K = \mathbf{Q}(\sqrt{-2})$ arises when $f = 2^v$, $v \ge 3$ with $p \equiv 3 \pmod 8$ and $b = 2^{v-2}$. With $f' = 8$ one computes the traces $\mathrm{Tr}_{\mathbf{Q}(\zeta_8)/\mathbf{Q}(\sqrt{-2})}\, \zeta_8^{\nu}$ $(1 \le \nu \le 8)$ in (10) to find two occurrences each of 0, $\sqrt{-2}$ and $-\sqrt{-2}$ and one occurrence each of 2 and -2. Thus

$$\sum_{n=0}^{\infty} b'_{\delta'}(n)\frac{X^n}{n!} = \exp(2X)(\sum_{n=0}^{\infty} \frac{X^{2n}}{(n!)^2})(\sum_{n=0}^{\infty} \frac{(2X)^{2n}}{(n!)^2}),$$

so

$$\sum_{n=0}^{\infty} b_{\delta}(n)\frac{X^n}{n!} = \exp((2^v - 6)X)(\sum_{n=0}^{\infty} \frac{X^{2n}}{(n!)^2})(\sum_{n=0}^{\infty} \frac{(2X)^{2n}}{(n!)^2})$$

from Proposition 1.

Similarly, one computes the values ω_ν $(1 \le \nu \le 8)$ in (11) to find $\pm\omega_+$, $\pm i\omega_+$, $\pm\omega_-$ and $\pm i\omega_-$, where $\omega_\pm = \zeta_8 \pm 1$. As $\{\omega_+, \omega_-\}$ is a basis for $\mathbf{Q}(\zeta_8)/K$ one finds easily that

$$\sum_{n=0}^{\infty} b'_{\delta'/2}(n)\frac{X^n}{n!} = (\sum_{n=0}^{\infty} \frac{X^{2n}}{(n!)^2})^4.$$

Again

$$\sum_{n=0}^{\infty} b_{\delta/2}(n)\frac{X^n}{n!} = \exp((2^v - 8)X)(\sum_{n=0}^{\infty}\frac{X^{2n}}{(n!)^2})^4$$

from Proposition 1.

(3) The case $K = \mathbf{Q}(\sqrt{2})$ arises only when $f = 8$ and $p \equiv 7 (\text{mod } 8)$. Computing the traces $\text{Tr}_{\mathbf{Q}(\zeta_8)/\mathbf{Q}(\sqrt{2})}\, \zeta_8^{\nu}$ $(1 \le \nu \le 8)$ in (10), one finds two occurrences each of 0, $\sqrt{2}$ and $-\sqrt{2}$ and one occurrence each of 2 and -2. The generating functions for $b_\delta(n)$ is thus the same as that for $b'_{\delta'}(n)$ as in case (2) above.

Computing the values ω_ν $(1 \le \nu \le 8)$ in (11) yields two occurrences each of $\pm\omega_1$ and $\pm\omega_2$, where $\omega_1 = 1 + \zeta_8$ and $\omega_2 = i(1 - \zeta_8)$. Since $\{\omega_1, \omega_2\}$ is a basis for $\mathbf{Q}(\zeta_8)/K$ one obtains the generating function

$$\sum_{n=0}^{\infty} b_{\delta/2}(n)\frac{X^n}{n!} = (\sum_{n=0}^{\infty}\frac{(2X)^{2n}}{(n!)^2})^2$$

for $b_{\delta/2}(n)$.

(4) The case $K = \mathbf{Q}(\sqrt{l^*})$ arises when $f = l^v$ or $2l^v$, $v > 0$, with odd p generating the squares modulo l^v. With $f' = l$ one has that

$$b_{\delta'}(n) = \begin{cases} (\frac{l-1}{2})^{n(l-1)/l}\frac{n!}{(n/l)!(\frac{(l-1)n}{2l}!)^2} & \text{if } l|n \\ 0 & \text{otherwise} \end{cases}$$

from Proposition 4 in [5]. Thus for $f = l^v$,

$$\sum_{n=0}^{\infty} b_\delta(n)\frac{X^n}{n!} = \exp((l^v - l)X)\sum_{n=0}^{\infty}\frac{(\frac{l-1}{2})^{n(l-1)}X^{ln}}{n!((l-1)n/2)!((l-1)n/2)!}.$$

With $f' = 2l$, one finds $\frac{l-1}{2}$ occurrences each of $\frac{\pm1\pm\sqrt{l^*}}{2}$ and one occurrence each of $\pm\frac{l-1}{2}$ among the traces $\text{Tr}_{\mathbf{Q}(\zeta_l)/K}\, \zeta_{2l}^{\nu}$ $(1 \le \nu \le 2l)$. The contribution to the count $b_{\delta'}(n)$ arising from a sum of n such traces consisting of exactly j groups of $\frac{l-1}{2}$ occurrences of $\frac{1+\sqrt{l^*}}{2}$, $\frac{l-1}{2}$ occurrences of $\frac{1-\sqrt{l^*}}{2}$ and one $-\frac{l-1}{2}$, of i additional occurrences of $-\frac{l-1}{2}$ each paired with a $+\frac{l-1}{2}$, and of k additional occurrences of $\frac{1+\sqrt{l^*}}{2}$ each paired with a $-\frac{1+\sqrt{l^*}}{2}$ with the remaining $n - jl - 2i - 2k$ traces containing an equal number of $\pm\frac{1-\sqrt{l^*}}{2}$, is

$$\frac{n!((l-1)/2)^{n-j-2i}}{i!(i+j)!k!\frac{n-jl-2i-2k}{2}!\frac{2k+j(l-1)}{2}!\frac{n-jl-2i-2k+j(l-1)}{2}!}$$

or

$$\binom{n}{\frac{n-jl}{2}}\binom{\frac{n-jl}{2}}{i}\binom{\frac{n+jl}{2}}{i+j}\binom{\frac{n-jl-2i}{2}}{k}\binom{\frac{n+jl-2i-2j}{2}}{k+j\frac{l-1}{2}}\left(\frac{l-1}{2}\right)^{n-j-2i}.$$

Necessarily, $j \equiv n \pmod 2$ here. The number of all such sums with exactly j groups of $\frac{l-1}{2}$ traces $\frac{1+\sqrt{l^*}}{2}$, $\frac{l-1}{2}$ traces $\frac{1-\sqrt{l^*}}{2}$ and one $-\frac{l-1}{2}$ with remaining $n-jl$ values paired so that each appears with its negative is thus $\binom{n}{\frac{n-jl}{2}}$ times the sum

$$\sum_{i=0}^{\frac{n-jl}{2}} \binom{\frac{n-jl}{2}}{i}\binom{\frac{n+jl}{2}}{i+j}\left(\frac{l-1}{2}\right)^{n-j-2i} \sum_{k=0}^{\frac{n-jl}{2}-i} \binom{\frac{n-jl-2i}{2}}{k}\binom{\frac{n+jl-2i-2j}{2}}{k+j\frac{l-1}{2}}.$$

Now when $j > 0$ there is an equal contribution from such sums of n values consisting of exactly j groups of $\frac{l-1}{2}$ traces $-\frac{1+\sqrt{l^*}}{2}$, $\frac{l-1}{2}$ traces $-\frac{1-\sqrt{l^*}}{2}$ and one $\frac{l-1}{2}$. Since any contribution to $b'_\delta(n)$ must arise from a sum of the form described above with $0 \le j \le [n/l]$ and $j \equiv n \pmod 2$, one finds that

$$b'_{\delta'}(n) = \sum A_j \binom{n}{\frac{n-jl}{2}} \sum_{i=0}^{\frac{n-jl}{2}} \binom{\frac{n-jl}{2}}{i}\binom{\frac{n+jl}{2}}{i+j}\left(\frac{l-1}{2}\right)^{n-j-2i} \cdot \qquad (22)$$

$$\sum_{k=0}^{\frac{n-jl}{2}-i} \binom{\frac{n-jl-2i}{2}}{k}\binom{\frac{n+jl-2i-2j}{2}}{k+j\frac{l-1}{2}},$$

with $A_0 = 1$ and $A_j = 2$ for $j > 0$, where the outer sum is taken over $j \equiv n \pmod 2$ satisfying $0 \le j \le [n/l]$. In general

$$\sum_{n=0}^{\infty} b_\delta(n)\frac{X^n}{n!} = \exp(2(l^v - l)X)\sum_{n=0}^{\infty} b'_{\delta'}(n)\frac{X^n}{n!}$$

here, with $b'_{\delta'}(n)$ as in (22).

The determination of $b_{\delta/2}(n)$ is similar for $f = l^v$ or $2l^v$, unlike the situation for $w = \delta$, but quite different whether $l \equiv 1$ or $3 \pmod 4$. The result for $l \equiv 1 \pmod 4$ was treated in [7] where

$$b_{\delta/2}(n) = \left(\frac{f}{R}\right)^n \sum_{i=0}^{[n/2]} \binom{n}{2i}\binom{2i}{i}^2\left(\frac{l-1}{4}\right)^{2i}(l^v - l + 1)^{n-2i},$$

so I will not repeat it here. For $l \equiv 3 \pmod 4$, b is odd so that by Proposition 1 in [7], $b_{\delta/2}(n) = f^n$ and $\Phi^{(\delta/2)}(x) = (x-f)^{e/\delta}$, whenever δ is even.

(5) The case $K = \mathbf{Q}(\sqrt{-l^*})$ arises when $f = 4l^v$, $v > 0$, and $p \equiv 3 \pmod 4$ is a primitive root of l^v. With $f' = 4l$, one computes the traces $\mathrm{Tr}_{\mathbf{Q}(\zeta_{4l})/K}\, \zeta_{4l}^{\nu}$ $(1 \le \nu \le 4l)$ in (10) to find $l-1$ occurrences each of $\pm\sqrt{-l^*}$ and ± 1, one

occurrences each of $\pm(l-1)$ and two zero values. Thus

$$\sum_{n=0}^{\infty} b'_{\delta'}(n)\frac{X^n}{n!} = \exp(2X)\left(\sum_{n=0}^{\infty}\beta_{2l}(n)\frac{X^n}{n!}\right)\left(\sum_{n=0}^{\infty}\frac{((l-1)X)^{2n}}{(n!)^2}\right)$$

so

$$\sum_{n=0}^{\infty} b_{\delta}(n)\frac{X^n}{n!} = \exp((4l^v-4l+2)X)\sum_{n=0}^{\infty}\beta_{2l}(n)\frac{X^n}{n!}\cdot\sum_{n=0}^{\infty}\frac{((l-1)X)^{2n}}{(n!)^2}.$$

For $w=\delta/2$, one computes the values ω_ν $(1\le\nu\le 4l)$ in (11) to find $l-1$ occurrences each of $\pm(\frac{1+\sqrt{l^*}}{2}+i\frac{1-\sqrt{l^*}}{2})$ and $\pm(\frac{1-\sqrt{l^*}}{2}+i\frac{1+\sqrt{l^*}}{2})$, and two occurrences each of $\pm\frac{l-1}{2}(1+i)$. The contribution to the count $b'_{\delta'/2}(n)$ arising from a sum of n such values consisting of exactly j groups of $\frac{l-1}{2}$ values $\frac{1+\sqrt{l^*}}{2}+i\frac{1-\sqrt{l^*}}{2}$, $\frac{l-1}{2}$ values $\frac{1-\sqrt{l^*}}{2}+i\frac{1+\sqrt{l^*}}{2}$ and one $-\frac{l-1}{2}(1+i)$, of i additional occurrences of $\frac{1+\sqrt{l^*}}{2}+i\frac{1-\sqrt{l^*}}{2}$ paired with as many $-(\frac{1+\sqrt{l^*}}{2}+i\frac{1-\sqrt{l^*}}{2})$ and k additional occurrences of $\frac{1-\sqrt{l^*}}{2}+i\frac{1+\sqrt{l^*}}{2}$ paired with as many $-(\frac{1-\sqrt{l^*}}{2}+i\frac{1+\sqrt{l^*}}{2})$ with the remaining $n-jl-2i-2k$ values containing an equal number of $\pm\frac{l-1}{2}(1+i)$ is

$$\frac{n!((l-1)/2)^{2i+2k+j(l-1)}2^{n-2i-2k-j(l-1)}}{\frac{2i+j(l-1)}{2}!\,i!\,\frac{2k+j(l-1))}{2}!\,k!\,\frac{n-jl-2i-2k+2j}{2}!\,\frac{n-jl-2i-2k}{2}!}$$

or

$$2^n\binom{n}{\frac{n-jl}{2}}\binom{\frac{n-jl}{2}}{i}\binom{\frac{n+jl}{2}}{i+j\frac{l-1}{2}}\binom{\frac{n-jl}{2}-i}{k}\binom{\frac{n+l}{2}-l}{k+j\frac{l-1}{2}}\left(\frac{l-1}{4}\right)^{\gamma},$$

where $\gamma=2i+2k+j(l-1)$. Arguing in a manner similar for the case (4) above when $w=\delta'$, one obtains the following expression for $b'_{\delta'/2}(n)$. Namely,

$$b'_{\delta'/2}(n) = 2^n\sum A_j\binom{n}{\frac{n-jl}{2}}\sum_{i=0}^{\frac{n-jl}{2}}\binom{\frac{n-jl}{2}}{i}\binom{\frac{n+jl}{2}}{i+j\frac{l-1}{2}}\cdot \tag{23}$$

$$\sum_{k=0}^{\frac{n-jl}{2}-i}\binom{\frac{n-jl}{2}-i}{k}\binom{\frac{n+l}{2}-i}{k+j\frac{l-1}{2}}\left(\frac{l-1}{4}\right)^{\gamma},$$

with $A_0=1$ and $A_j=2$ for $j>0$, where the outer sum is taken over $j\equiv n\pmod 2$ satisfying $0\le j\le[n/l]$. In general,

$$\sum_{n=0}^{\infty} b_{\delta/2}(n)\frac{X^n}{n!} = \exp(4(l^v-l)X)\sum_{n=0}^{\infty} b'_{\delta'/2}(n)\frac{X^n}{n!}$$

here, with $b'_{\delta'/2}(n)$ as in (23).

(6) The case $K = \mathbf{Q}(\sqrt{l_1^* l_2^*})$ arises when $f = l_1^{v_1} l_2^{v_2}$ or $2l_1^{v_1} l_2^{v_2}$, $v_1, v_2 > 0$, $\gcd(\phi(l_1^{v_1}), \phi(l_2^{v_2})) = 2$ and p is an odd primitive root of both $l_1^{v_1}$ and $l_2^{v_2}$. With $f' = l_1 l_2$ one can show that there are $\frac{(l_1-1)(l_2-1)}{2}$ occurrences each of $\frac{1+\sqrt{l_1^* l_2^*}}{2}$ and $\frac{1-\sqrt{l_1^* l_2^*}}{2}$, l_1-1 occurrences of $-\frac{l_2-1}{2}$, l_2-1 occurrences of $-\frac{l_1-1}{2}$ and one occurrence of $\frac{(l_1-1)(l_2-1)}{2}$ among the f' traces in (10), whereas there are an equal number of these and their negatives for $f' = 2l_1 l_2$. The details are left to the reader. Unfortunately, there seems to be no nice expression for a generating function of $b_\delta(n)$ here. With $f' = l_1 l_2$ one can also show that there are $\frac{(l_1-1)(l_2-1)}{4}$ occurrences each of $\frac{1}{2}(\pm\sqrt{l_1^*} \pm \sqrt{l_2^*})$, $\frac{l_1-1}{2}$ occurrences each of $\pm\frac{l_2-1}{2}\sqrt{l_1^*}$, $\frac{l_2-1}{2}$ occurrences each of $\pm\frac{l_1-1}{2}\sqrt{l_2^*}$, and one zero among the f' values ω_ν in (11), whereas twice such numbers for $f' = 2l_1 l_2$. Again the details are left to the reader. There seems to be no nice expression for a generating function of $b_{\delta/2}(n)$ here either.

Before concluding, I wish to give a natural generalization of the first case $K = \mathbf{Q}(\sqrt{-1})$ to that when $K = \mathbf{Q}(\zeta_{2^\alpha})$, $\alpha > 2$. In particular, the decomposition field $K = \mathbf{Q}(\zeta_{2^\alpha})$ arises when (a) $f = 2^v$, $v \geq \alpha$, with $p \equiv 1 (\mathrm{mod}\ 2^\alpha)$ and $b = 2^{v-\alpha}$, or (b) $f = 2^\alpha l^v$, $v > 0$ with $p \equiv 1 (\mathrm{mod}\ 2^\alpha)$ a primitive root of l^v.

With $f' = 2^\alpha$ in case (a) one finds that

$$\sum_{n=0}^{\infty} b'_{\delta'}(n) \frac{X^n}{n!} = \left(\sum_{n=0}^{\infty} \frac{X^{2n}}{(n!)^2} \right)^{2^{\alpha-1}}$$

from equation (4) in [2], and $b'_{\delta'/2}(n) = 2^{\alpha n}$ when δ' is even as $b' = 1$(chiefly Proposition 1 in [7].) Thus from Proposition 1 it follows that

$$\sum_{n=0}^{\infty} b_\delta(n) \frac{X^n}{n!} = \exp((2^v - 2^\alpha)X) \left(\sum_{n=0}^{\infty} \frac{X^{2n}}{(n!)^2} \right)^{2^{\alpha-1}}$$

and $b_{\delta/2}(n) = 2^{vn}$ when δ is even, with $\Phi^{(\delta)}(x) = \Phi'^{(\delta')}(x - (2^v - 2^\alpha))$ and $\Phi^{(\delta/2)}(x) = (x - 2^v)^{\frac{p-1}{2^\alpha}}$, respectively.

With $f' = 2^\alpha l$ in case (b), one finds $l - 1$ occurrences each of the values $\pm\zeta_{2^\alpha}^j$ ($0 \leq j < 2^{\alpha-1}$) and one occurrence each of $\pm(l-1)\zeta_{2^\alpha}^j$ ($0 \leq j < 2^{\alpha-1}$) among the traces $\mathrm{Tr}_{\mathbf{Q}(\zeta_{2^\alpha l})/\mathbf{Q}(\zeta_{2^\alpha})} \zeta_{2^\alpha l}^\nu$ $(1 \leq \nu \leq 2^\alpha l)$ in (10). Since the set $\{\zeta_{2^\alpha}^j \mid 0 \leq j < 2^{\alpha-1}\}$ forms a basis for $\mathbf{Q}(\zeta_{2^\alpha})/\mathbf{Q}$,

$$\sum_{n=0}^{\infty} b'_{\delta'}(n) \frac{X^n}{n!} = \left(\sum_{n=0}^{\infty} (\beta_{2l}(n) \frac{X^n}{n!} \right)^{2^{\alpha-1}}$$

so

$$\sum_{n=0}^{\infty} b_\delta(n)\frac{X^n}{n!} = \exp(2^\alpha(l^v - l)X)\left(\sum_{n=0}^{\infty}\beta_{2l}(n)\frac{X^n}{n!}\right)^{2^{\alpha-1}}.$$

For $w = \delta'/2$, one can show that there are $l-1$ occurrences each of the values $\pm\zeta_{2^\alpha}^j\sqrt{l}$ for $0 \le j < 2^{\alpha-1}$ and 2^α zeros among the $2^\alpha l$ values ω_ν in (11). Thus

$$\sum_{n=0}^{\infty} b'_{\delta'/2}(n)\frac{X^n}{n!} = \exp(2^\alpha X)\left(\sum_{n=0}^{\infty}\frac{((l-1)X)^{2n}}{(n!)^2}\right)^{2^{\alpha-1}}$$

so

$$\sum_{n=0}^{\infty} b_{\delta/2}(n)\frac{X^n}{n!} = \exp(2^\alpha(l^v - l + 1)X)\left(\sum_{n=0}^{\infty}\frac{((l-1)X)^{2n}}{(n!)^2}\right)^{2^{\alpha-1}}.$$

REFERENCES

[1] Z. Borevich and I. Shafarevich, *Number Theory*, Academic Press, New York, 1966

[2] S. Gupta and D. Zagier, On the coefficients of the minimal polynomial of Gaussian periods, Math. Comp. 60 (1993), pp. 385-398.

[3] S. Gurak, Minimal polynomials for Gauss circulants and cyclotomic units, Pac. J. Math. 102, no. 3 (1982), pp. 347-353.

[4] S. Gurak, Factors of period polynomials for finite fields, II, *Comtemporary Math.*, Amer. Math. Soc., 168 (1994), pp. 127-138.

[5] S. Gurak, On the last factor of the period polynomial for finite fields, Acta Arith. 71 (1995), pp. 391-400.

[6] S. Gurak, On the minimal polynomials for certain Gauss periods over finite fields, *Finite Fields and their Applications*, edited by S. Cohen & H. Niederreiter, pp. 85-96, Cambridge University Press, 1996.

[7] S. Gurak, On the middle factor of the period polynomial for finite fields (to appear).

[8] G. Myerson, Period polynomials and Gauss sums for finite fields, Acta Arith. 39 (1981), pp. 251-264.

DEPARTMENT OF MATHEMATICS AND COMPUTER SCIENCE, UNIVERSITY OF SAN DIEGO, SAN DIEGO, CA 92110-2492

E-mail address: gurak@pwa.acusd.edu

THEOREMS AND CONJECTURES INVOLVING ROOK POLYNOMIALS WITH ONLY REAL ZEROS

JAMES HAGLUND, KEN ONO, AND DAVID G. WAGNER

Dedicated to Basil Gordon on his sixty fifth birthday.

ABSTRACT. Let $A = (a_{ij})$ be a real $n \times n$ matrix with non-negative entries which are weakly increasing down columns. If $B = (b_{ij})$ is the $n \times n$ matrix where $b_{ij} := a_{ij} + z$, then we conjecture that all of the roots of the permanent of B, as a polynomial in z, are real. Here we establish several special cases of the conjecture.

1. INTRODUCTION AND STATEMENT OF RESULTS

Throughout $A = (a_{ij})$ will denote an $n \times n$ matrix. A placement of rooks on the squares of A is non-attacking if no two rooks are in the same column, and no two are in the same row. We define the *weight* of such a placement to be the product of the entries in A which are under the rooks, and we define the kth rook number $r_k(A)$ to be the sum of these weights over all non-attacking placements of k rooks on A. Furthermore by convention $r_0(A) := 1$. When $n = 2$ these rook numbers are

$$r_2(A) = a_{11}a_{22} + a_{12}a_{21}, \quad r_1(A) = a_{11} + a_{12} + a_{21} + a_{22} \quad \text{and} \quad r_0(A) = 1.$$

Note that $r_n(A)$ equals per(A), the permanent of A, defined as

$$\text{per}(A) := \sum_{\sigma \in S_n} \prod_{i=1}^{n} a_{i,\sigma_i}.$$

If each a_{ij} is zero or one, then A is called a *board*, and $r_k(A)$ is the number of placements of k rooks on the non-zero entries of A.

We call $R(z; A) := \sum_{k=0}^{n} r_k(A)z^k$ the *rook polynomial* of A. Nijenhuis [Nij] proved that if the a_{ij} are non-negative real numbers, then all the roots of $R(z; A)$ are real. One application is that the rook numbers $r_k(A)$ are log-concave (i.e.

1991 *Mathematics Subject Classification.* Primary 05C70, 05A20.

Key words and phrases. rook polynomial, matching polynomial, permanent, real roots.

The first author is supported by NSF grant DMS-9627432. The second author is supported by NSF grants DMS-9508976 and DMS-9304580, and NSA grant MSPR-97Y012. The third author is supported by the Natural Sciences and Engineering Research Council of Canada under operating grant OGP0105392.

S.D. Ahlgren et al. (eds.), Topics in Number Theory, 207–221.

$r_i(A)^2 \geq r_{i-1}(A)r_{i+1}(A)$). This follows from the fact that if $f(z) = \sum_k b_k z^k$ has only real roots , then the b_k are log-concave. In fact more is true, it turns out that [p. 52, HLP]

$$b_k^2 \geq b_{k-1}b_{k+1}\left(1+\frac{1}{k}\right)\left(1+\frac{1}{n-k}\right).$$

E. Bender noted that Nijenhuis' result follows from the Heilmann-Lieb Theorem [HeLi] which asserts that all of the roots of matching polynomials of simple graphs with non-negative weights are real (the Heilmann-Lieb Theorem has been extensively generalized in Theorem 3.3 of [Wag3]). To see this consider the complete bipartite graph G from n vertices on top to n vertices below, where the edge between vertex i above and vertex j below is assigned weight a_{ij}. Given a *matching* M with k edges (i.e. any selection of k edges no two of which share a common vertex) one obtains a rook placement by placing a rook on a_{ij} in A if and only if M contains the edge connecting i above to j below. So the weighted matching polynomial of G is just the rook polynomial of A.

Now we briefly describe various combinatorial investigations from which it is known that all of the roots of certain *natural* polynomials are real. A multiset is a set in which elements can occur more than once. A permutation σ of a multiset L of positive integers is a linear list $\sigma_1\sigma_2\cdots$ of the elements of L, and a *descent* of such a σ is a positive integer i such that $\sigma_i > \sigma_{i+1}$. Let $N_k(\mathbf{v})$ denote the number of permutations of the multiset

$$\mathbf{v} := \{1^{v_1}2^{v_2}\cdots m^{v_m}\}$$

(i^{v_i} denotes v_i copies of i) with exactly $k-1$ descents. Simion proved that all the roots of

$$\sum_{k\geq 1} N_k(\mathbf{v})z^k \tag{1}$$

are real [Sim], and consequently that the $N_k(\mathbf{v})$ are log-concave. For example, if $\mathbf{v} = \{1^3 2^1\}$, the 4 permutations are

1112 1121 1211 2111,

and so (1) is $3z^2 + z$. Simion's result proves a special case of the famous Neggers-Stanley "Poset Conjecture" [Bre], [Wag2], which says that a certain polynomial associated to a labelled poset has only real zeros.

The *hit polynomial* $T(z; A)$ of A is

$$T(z;A) = \sum_{k=0}^{n} k!\, r_{n-k}(A)(z-1)^{n-k}.$$

If A is a board, then its coefficient of z^k is called the kth *hit number*, the number of ways of placing n non-attacking rooks on A where exactly k rooks lie on non-zero entries. In their seminal paper on rook theory [KaRi] (see also [Rio]), Riordan and Kaplansky showed that

$$\prod_{i=1}^{m} v_i! \sum_{k\geq 1} N_k(\mathbf{v}) z^{k-1}$$

is the hit polynomial for the "Simon Newcomb board" (see Figure 1). This is the matrix where the entries in the first v_1 columns are all zero, and the next v_2 columns have ones in the bottom v_1 rows and zeros above, etc. Therefore Simion's result implies that all the roots of these polynomials are real, and of course that the hit numbers of these boards are log-concave.

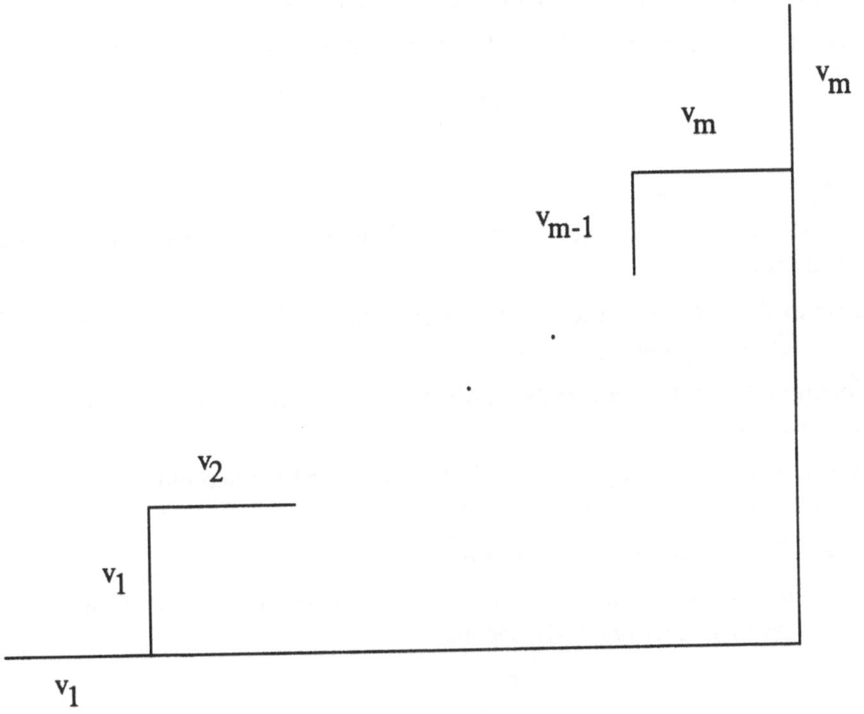

FIGURE 1. The Simon Newcomb for the multiset $\{1^{v_1}\cdots m^{v_m}\}$.

By Laguerre's Theorem, that if $\sum_k b_k z^k$ has only real roots then $\sum_k \frac{b_k}{k!} z^k$ also has only real roots ([Lag]; see also [Th. 11, GaWa]), one can deduce that if $T(z;A)$ has only real roots, then so too does $R(z;A)$. The converse of Laguerre's Theorem is false, and in particular the hit polynomials of boards do not generally have only real roots. However we will show that this is true for Ferrers boards.

Definition 1. *Let A be a board with the property that if $a_{ij} = 1$, then $a_{st} = 1$ whenever $i \le s \le n$ and $j \le t \le n$. Then A is called a Ferrers board.*

Theorem 1. *All the roots of the hit polynomial of a Ferrers board are real.*

An r-descent of a multiset permutation σ is a value of i such that $\sigma_i - \sigma_{i+1} \ge r$. Let $N_k(\mathbf{v}, r)$ be the number of permutations of the multiset $\mathbf{v} = \{1^{v_1} \cdots m^{v_m}\}$ containing exactly $k-1$ r-descents. For example, if $\mathbf{v} = \{1^1 2^1 3^1\}$, then the permutations are

$$123 \qquad 132 \qquad 213 \qquad 231 \qquad 312 \qquad 321.$$

Of these, 231 and 312 have one 2-descent, so $N_2(\{1^1 2^1 3^1\}, 2) = 2$.

The numbers $N_k(\mathbf{v}, r)$ are the hit numbers for special Ferrers boards, ones generalizing the Simon Newcomb boards [Hag1,Hag2]. Thus Theorem 1 implies the following, which reduces to Simion's result when $r = 1$.

Corollary 1. *Let r be a positive integer, and let $\mathbf{v}$ be a multiset of positive integers. Then all the roots of*

$$\sum_k N_k(\mathbf{v}, r) z^k$$

are real.

It is natural to seek a weighted version of Theorem 1 analogous to the Heilmann-Lieb theorem.

Definition 2. *Let A be a real $n \times n$ matrix. We call A a Ferrers matrix if $a_{ij} \le a_{st}$ whenever $i \le s \le n$ and $j \le t \le n$.*

Conjecture 1. *If A is a Ferrers matrix, then all the roots of the hit polynomial $T(z; A)$ are real.*

Note that if $a_{ij} \in \{0, 1\}$, then Conjecture 1 holds by Theorem 1. In section 2 we prove Theorem 1 and also the next result.

Theorem 2. *Conjecture 1 is true if $n \le 3$.*

Let J_n denote the $n \times n$ matrix, all of whose entries equal one. When expanded out in powers of z, $\mathrm{per}(zJ_n + A)$ becomes

$$\sum_{k=0}^{n} k! r_{n-k}(A) z^k.$$

Conjecture 1 is thus equivalent to the assertion that $\mathrm{per}(zJ_n + A)$ has all real zeros whenever A is a Ferrers matrix.

Expressing Conjecture 1 (and also Conjecture 2 below) in terms of permanents has several advantages. One implication is that if $z > -\min_{ij} a_{ij}$ (resp. $z < -\max_{ij} a_{ij}$) all the entries of $zJ_n + A$ are positive (resp. negative) in which case $\mathrm{per}(zJ_n + A)$ cannot be zero. Hence if A is Ferrers all real roots of $\mathrm{per}(zJ_n + A)$ are in the interval $[-a_{nn}, -a_{11}]$. It is also easy to see that in Conjecture 1 we can assume without loss of generality that

<u>C1</u> $\quad 0 \le a_{ij} \le 1$ for all $1 \le i, j \le n$.

Less trivially, we can develop recursions by expanding the permanent in terms of permanental minors, a method which plays a crucial role in the broadest special case of Conjecture 1 we can prove. We require some additional hypotheses. Define $a_{0j} := 0$ for all $0 \leq j \leq n+1$, $a_{i0} := 0$ for all $0 \leq i \leq n+1$, $a_{n+1,j} := 1$ for all $1 \leq j \leq n+1$, and $a_{i,n+1} := 1$ for all $1 \leq i \leq n+1$. Also define conditions

C2 For all $1 \leq r < s \leq n$ and $0 \leq t < u < v \leq n+1$:

$$(a_{ru} - a_{rt})(a_{sv} - a_{su}) \leq (a_{rv} - a_{ru})(a_{su} - a_{st}),$$

and

C3 For all $0 \leq t < u < v \leq n+1$ and $1 \leq r < s \leq n$:

$$(a_{ur} - a_{tr})(a_{vs} - a_{us}) \leq (a_{vr} - a_{ur})(a_{us} - a_{ts}).$$

Theorem 3. *If A is a Ferrers matrix satisfying C1, C2, and C3, then $\mathrm{per}(zJ_n+A)$ has all its zeros in the interval $[-1,0]$.*

In section 3 we prove Theorem 3 using mathematical induction and the method of interlacing roots. By the same method, we also prove that Conjecture 1 holds for any $n \times n$ Ferrers matrix A provided $a_{i,n} \leq a_{i+1,1}$ for all $1 \leq i \leq n-1$.

By running a program written in Maple which constructs matrices using a random number generator, the authors and E. R. Canfield have verified the following stronger form of Conjecture 1 holds for over 100,000 matrices of sizes ranging from 3×3 to 13×13.

Conjecture 2. *If A is a matrix with real entries where $a_{ij} \leq a_{sj}$ when $s \geq i$, then all the roots of the hit polynomial $T(z; A)$ are real.*

Since permuting the columns of a matrix doesn't change the rook numbers, the $a_{ij} \in \{0,1\}$ case of Conjecture 2 also follows from Theorem 1.

Example 1. If $a_{ij} := q_j$, then $\mathrm{per}(zJ_n + A) = n!(z+q_1)(z+q_2)\cdots(z+q_n)$. Thus the set of polynomials with only real roots and leading coefficient $n!$ are all examples of the phenomenon asserted by Conjectures 1 and 2. Furthermore Conjecture 1 is equivalent to the claim that for any Ferrers matrix A there exists another matrix A' which is constant down columns and *rook equivalent* to A (i.e. $r_k(A) = r_k(A')$ for $0 \leq k \leq n$).

2. The Tableau Method

We now turn to the proofs of Theorems 1 and 2.

Proof of Theorem 1: Goldman, Joichi, and White [GJW] proved that

$$\sum_{k=0}^{n} x(x-1)\cdots(x-k+1)r_{n-k}(A) = \prod_{i=1}^{n}(x+c_i-i+1), \tag{2}$$

where c_i is the number of ones in (the height of) the ith column of the Ferrers board A. This is equivalent to

$$\sum_{k=0}^{n} \binom{x+k}{n} t_k(A) = \prod_{i=1}^{n}(x+c_i-i+1), \tag{3}$$

where $t_k(A)$ is the coefficient of z^k in $T(z;A)$. The result now follows easily by [Bre, p. 43].

Theorem. *(Brenti) Let $f(x) = \sum_{k=0}^{n} \binom{x+k}{n} b_k$ be a polynomial with all real zeros, with smallest root $\lambda(f)$ and largest root $\Lambda(f)$. If all the integers in the intervals $[\lambda, -1]$ and $[0, \Lambda]$ are also roots of f, then all the roots of $\sum_{k=0}^{n} b_k x^k$ are real.* □

Proof of Theorem 2: For a 2×2 matrix A, the discriminant of $\text{per}(zJ_2 + A)$ is

$$(a_{11} + a_{12} + a_{21} + a_{22})^2 - 8(a_{11}a_{22} + a_{12}a_{21})$$

$$= a_{11}^2 + a_{12}^2 + a_{21}^2 + a_{22}^2 + 2a_{11}a_{12} + 2a_{11}a_{21} + 2a_{12}a_{22} + 2a_{21}a_{22} \\ - 6a_{11}a_{22} - 6a_{12}a_{21}.$$

The roots will all be real if and only if this expression is non-negative. We now describe a method by which this becomes visibly evident if A is Ferrers.

By replacing z by $z - \min_{ij} a_{ij}$ we can assume without loss of generality that $a_{ij} \geq 0$. Since A is Ferrers, either

$$0 \leq a_{11} \leq a_{12} \leq a_{21} \leq a_{22} \tag{4}$$

or

$$0 \leq a_{11} \leq a_{21} \leq a_{12} \leq a_{22}, \tag{5}$$

or both. If (4) holds, replace a_{11} by w_1, a_{12} by $w_1 + w_2$, a_{21} by $w_1 + w_2 + w_3$, and a_{22} by $w_1 + w_2 + w_3 + w_4$. The inequalities in (4) then imply that all the w_i are non-negative. The discriminant now becomes

$$w_4^2 + w_2^2 + 4w_3^2 + 6w_4w_2 + 4w_4w_3 + 4w_2w_3.$$

Note that this is w-monomial positive, i.e. the coefficient of every possible monomial in the w_i is non-negative. If (5) holds, replace a_{11} by w_1, a_{21} by $w_1 + w_2$, etc. By symmetry, the discriminant will again be w-monomial positive. Thus Conjecture 1 holds if $n = 2$.

To the set of inequalities in (4), we can associate the array

$$\begin{matrix} 1 & 2 \\ 3 & 4 \end{matrix}$$

and to (5) the array

$$\begin{matrix} 1 & 3 \\ 2 & 4 \end{matrix}.$$

These arrays are special cases of what are known as standard tableau of square $n \times n$ shape, that is $n \times n$ Ferrers matrices the set of whose entries equals the set $\{1, 2, \ldots, n^2\}$. In the second tableau the 2 in square $(2, 1)$ indicates a_{21} is the second-smallest a_{ij}. More generally, we can clearly break up the set of $n \times n$ Ferrers matrices into subsets, according to how the a_{ij} are related to one another, with a standard tableau of square $n \times n$ shape associated to each subset. The square containing k in such a tableau will indicate the kth-smallest a_{ij}, which we can then parameterize as $w_1+\ldots+w_k$, with w_i non-negative. Note that if some of the a_{ij} are equal, there is some ambiguity in selecting a representative tableau, but different tableau choices will have the effect of simply replacing certain w_i by other w_i in the discriminant, which will not affect monomial positivity.

Since a cubic polynomial has only real roots if and only if its discriminant is non-negative, to complete the proof of Theorem 2 it suffices to show that for all possible standard tableaux in the shape of a 3×3 square, the discriminant of $T(z; A)$, when expressed in terms of the w_i, is w-monomial positive. This has been verified via a long Maple calculation (here the discriminant of $T(z; A)$ typically consists of over 1500 monomials in the w_i). □

A similar argument proves the $n = 2$ case of Conjecture 2; by Theorem 2 we can assume that A is of the form

$$A := \begin{bmatrix} w_1 & w_1 + w_2 \\ w_1 + w_2 + w_3 + w_4 & w_1 + w_2 + w_3 \end{bmatrix},$$

in which case the discriminant of $\mathrm{per}(A + zJ_n)$ equals

$$(w_2 - w_4)^2 + 4w_2w_3 + 4w_3^2 + 4w_3w_4.$$

Although not monomial positive, it is still clearly positive.

The $n = 3$ case of Conjecture 2 is still open, as the discriminant of $T(z; A)$ for a typical matrix A (parameterized to reflect the column monotone condition) consists of over a thousand monomials, many of which have negative coefficients.

3. Interlacing Arguments

Before presenting the proof of Theorem 3, we consider three particularly interesting special cases.

$\underline{U}$ $a_{ij} \in \{0, 1\}$ for all $1 \leq i, j \leq n$.

$\underline{\Pi}$ There are positive real numbers p_i for $1 \leq i \leq n$ and q_j for $1 \leq j \leq n$ such that

$$a_{ij} = p_i q_j \text{ for all } 1 \leq i, j \leq n.$$

$\underline{\Sigma}$ There exist positive real numbers p_i for $1 \leq i \leq n$ and q_j for $1 \leq j \leq n$ such that
$a_{ij} = p_i + q_j$ for all $1 \leq i, j \leq n$.

Lemma 1. *If A is a Ferrers matrix satisfying $\underline{U}$, then A satisfies $\underline{C1}$, $\underline{C2}$, and $\underline{C3}$.*

Proof : Condition $\underline{C1}$ is a trivial consequence of $\underline{U}$. The hypothesis A being Ferrers and $\underline{U}$, and the conclusion $\underline{C2}$ and $\underline{C3}$ are symmetric under transposition $A \mapsto A^{\top}$; hence we need only prove $\underline{C2}$. Fix any $1 \leq r < s \leq n$ and $0 \leq t < u < v \leq n+1$; there are ten cases for the 2-by-3 submatrix

$$\begin{bmatrix} a_{rt} & a_{ru} & a_{rv} \\ a_{st} & a_{su} & a_{sv} \end{bmatrix},$$

leading to eight cases for the differences

$$\begin{bmatrix} a_{ru} - a_{rt} & a_{rv} - a_{ru} \\ a_{su} - a_{st} & a_{sv} - a_{su} \end{bmatrix}.$$

In each case the inequality C2 is easily seen to hold. □

Theorem 3 therefore implies Theorem 1.

Lemma 2. *If A is Ferrers and satisfies C1 and II, then A satisfies C2 and C3.*

Proof : Again by symmetry, we only need to prove C2, so fix any $1 \le r < s \le n$ and $0 \le t < u \le v \le n+1$; we separate four cases for

$$\begin{bmatrix} a_{rt} & a_{ru} & a_{rv} \\ a_{st} & a_{su} & a_{sv} \end{bmatrix}.$$

If $1 \le t$ and $v \le n$ then $a_{ij} = p_i q_j$ for all these entries of A, and C2 reduces to

$$p_r(q_u - q_t)p_s(q_v - q_u) \le p_r(q_v - q_u)p_s(q_u - q_t),$$

which is trivial. If $1 \le t$ and $v = n+1$ then C2 reduces to

$$p_r(q_u - q_t)(1 - p_s q_u) \le p_s(q_u - q_t)(1 - p_r q_u),$$

which (since A Ferrers implies $q_t \le q_u$) reduces to $p_r \le p_s$, which again follows if A is Ferrers. If $0 = t$ and $v \le n$ then C2 reduces to

$$p_r q_u p_s(q_v - q_u) \le p_s q_u p_r(q_v - q_u),$$

which is trivial. Finally, if $0 = t$ and $v = n+1$ then C2 reduces to

$$p_r q_u(1 - p_s q_u) \le p_s q_u(1 - p_r q_u),$$

which (since $q_u \ge 0$), by C1 reduces to $p_r \le p_s$ which follows when A is Ferrers. □

Corollary 2. *If A is a Ferrers matrix satisfying C1 and II, then $per(zJ_n + A)$ has all its zeros in the interval* $[-1, 0]$.

One of the strongest theorems involving polynomials having only real zeros is Corollary 3 below, which Szegö proved under the more general assumption that the roots of $f(z)$ are all real but need not all be non-positive [Sze], [PoSz,Problem 154]. Our proof of the weaker result is quite different from his proof, which uses Grace's Apolarity Theorem [Gra], [Vle].

Corollary 3. *(Szegö) If $f(z) := \sum_{k=0}^{n} b_k z^k$ and $g(z) := \sum_{k=0}^{n} d_k z^k$ are polynomials of degree n all of whose roots are non-positive real numbers, then the polynomial*

$$\sum_{k=0}^{n} b_k d_k k!(n-k)! z^k$$

has only real roots.

Proof : Let $f(z) := b_n(z+p_1)(z+p_2)\cdots(z+p_n)$ and $g(z) := d_n(z+q_1)(z+q_2)\cdots(z+q_n)$, where $0 \le p_1 \le \ldots \le p_n$ and $0 \le q_1 \le \ldots \le q_n$. Let $e_k(X)$ denote the kth elementary symmmetric function in the set of variables $X := \{x_1, x_2, \ldots, x_n\}$. Then if $a_{ij} := p_i q_j$, $\mathrm{per}(zJ_n + A)$ becomes

$$\sum_{k=0}^{n} z^k k!(n-k)! e_{n-k}(P) e_{n-k}(Q),$$

and has only real zeros by Corollary 2. Since $e_{n-k}(P) = b_k/b_n$ and $e_{n-k}(Q) = d_k/d_n$, the thesis follows. □

Remark 1 : The case where $a_{ij} := p_i q_j$ with $p_1 \le p_2 \le \ldots \le p_n$ and $0 \le q_1 \le q_2 \le \ldots \le q_n$ doesn't follow from Corollary 2 since if some of the p_i are negative A need need be Ferrers. However, this matrix will still be weakly increasing down columns, and so Szegö's Theorem is equivalent to this special case of Conjecture 2.

Lemma 3. *If A is a Ferrers matrix satisfying $\underline{\Sigma}$, then A satisfies $\underline{C2}$ and $\underline{C3}$.*

Proof : Yet again, by symmetry, we need only prove $\underline{C2}$, so fix any $1 \le r < s \le n$ and $0 \le t < u < v \le n+1$; we again separate four cases for

$$\begin{bmatrix} a_{rt} & a_{ru} & a_{rv} \\ a_{st} & a_{su} & a_{sv} \end{bmatrix}.$$

If $1 \le t$ and $v \le n$ then $a_{ij} = p_i + q_j$ for all these entries of A, and $\underline{C2}$ reduces to

$$(q_u - q_t)(q_v - q_u) \le (q_v - q_u)(q_u - q_t),$$

which is trivial. If $1 \le t$ and $v = n+1$ then $\underline{C2}$ reduces to

$$(q_u - q_t)(1 - p_s - q_u) \le (q_u - q_t)(1 - p_r - q_u),$$

which (given that A is Ferrers implies $q_t \le q_u$) reduces to $p_r \le p_s$, and this also follows when A is Ferrers. If $0 = t$ and $v \le n$ then $\underline{C2}$ reduces to

$$(p_r + q_u)(q_v - q_u) \le (p_s + q_u)(q_v - q_u),$$

which reduces to $p_r \le p_s$ and follows from A being Ferrers. Finally, if $0 = t$ and $v = n+1$ then $\underline{C2}$ reduces to

$$(p_r + q_u)(1 - p_s - q_u) \le (p_s + q_u)(1 - p_r - q_u),$$

which reduces to $p_r \le p_s$ and follows if A is Ferrers. □

Corollary 4. *If A is a Ferrers matrix satisfying $\underline{C1}$ and $\underline{\Sigma}$, then all the roots of $per(zJ_n + A)$ are in the interval $[-1, 0]$.*

To prove Theorem 3 we need some lemmas about polynomials with only real zeros. A polynomial $P \in \mathbb{R}[z]$ is *standard* if either $P \equiv 0$ or the leading coefficient of P is positive. Let $P, Q \in \mathbb{R}[z]$ be two such polynomials, of degrees p and q respectively. Let the zeros of P be

$$\xi_1 \leq \cdots \leq \xi_p$$

and let the zeros of Q be

$$\theta_1 \leq \cdots \leq \theta_q.$$

We say that P *alternates left of* Q if $p = q$ and

$$\xi_1 \leq \theta_1 \leq \cdots \leq \xi_p \leq \theta_p.$$

This is denoted $P \langle\!\langle Q$.

We say that P *interlaces* Q if $q = p + 1$ and

$$\theta_1 \leq \xi_1 \leq \cdots \leq \theta_p \leq \xi_p \leq \theta_{p+1}.$$

This is denoted $P \dagger Q$.

Furthermore if $P \langle\!\langle Q$ or $P \dagger Q$, then we use the notation $P \prec Q$. Moreover by convention if P has only real zeros, then $P \langle\!\langle 0$, $0 \langle\!\langle P$, $P \dagger 0$, and $0 \dagger P$.

Lemmas 4 and 5 can be proved using the techniques from section 3 of [Wag1].

Lemma 4. *Let $P, Q, S \in \mathbb{R}[z]$ be standard, with all zeros real and in the interval $[\alpha, \beta]$, and with $S \not\equiv 0$.*

(a) $P \langle\!\langle Q$ *if and only if* $(z - \alpha)Q \langle\!\langle (z - \beta)P$.
(b) *If* $S \prec P$ *and* $S \prec Q$ *then* $S \prec P + Q$.
(c) *If* $P \prec S$ *and* $Q \prec S$ *then* $P + Q \prec S$.
(d) *If* $P \prec Q$ *then* $P \prec P + Q \prec Q$.

Lemma 5. *Let $P_1, \ldots, P_m \in \mathbb{R}[z]$ be standard, with only real non-positive zeros, and such that $P_i \not\equiv 0$ for all $1 \leq i \leq m$. If $P_1 \prec \cdots \prec P_m$ and $P_1 \prec P_m$ then $P_h \prec P_i$ for all $1 \leq h < i \leq m$.*

Lemma 6. *Let $P_1, \ldots, P_m \in \mathbb{R}[z]$ be standard, with only real non-positive zeros, and such that $P_h \prec P_i$ for all $1 \leq h < i \leq m$. For any $c_1, \ldots, c_m \geq 0$, $P_1 \prec c_1P_1 + \cdots + c_mP_m \prec P_m$.*

Proof : We may reduce to the case that $c_i > 0$ and $P_i \not\equiv 0$ for all $1 \leq i \leq m$. Since $P_1 \prec c_iP_i$ for all $1 \leq i \leq m$, Lemma 4(b) (and induction on m) show that $P_1 \prec c_1P_1 + \cdots + c_mP_m$. Similar reasoning shows that $c_1P_1 + \cdots + c_mP_m \prec P_m$. □

Lemma 7. *Let $P_1, \ldots, P_m \in \mathbb{R}[z]$ be standard, with only real non-positive zeros, and such that $P_h \prec P_i$ for all $1 \le h < i \le m$. For $1 \le i \le m$ let $S'_i := P_1 + \cdots + P_i$ and $S_i := P_i + \cdots + P_m$. Then*

$$S'_1 \prec \cdots \prec S'_m = S_1 \prec \cdots \prec S_m \qquad \text{and} \qquad S'_1 \prec S_m.$$

Proof : Since $P_1 \prec P_2$, Lemma 4 (d) implies that $S'_1 \prec S'_2 \prec P_2$. Assume, inductively, that $S'_1 \prec \cdots \prec S'_i \prec P_i$ where $i < m$. By the hypothesis and Lemma 6, $S'_i \prec P_{i+1}$; thus Lemma 4 (d) implies that $S'_i \prec S'_{i+1} \prec P_{i+1}$. By induction we conclude that $S'_1 \prec \cdots \prec S'_m$. A similar argument shows that $S_1 \prec \cdots \prec S_m$. Finally, $S'_1 = P_1 \prec P_m = S_m$ is part of the hypothesis. □

Lemma 8. *Let $P_1, \ldots, P_m \in \mathbb{R}[z]$ be standard, with only real non-positive zeros, and such that $P_h \prec P_i$ for all $1 \le h < i \le m$. Let $b_1, \ldots, b_m \ge 0$ and $c_1, \ldots, c_m \ge 0$ be such that $b_i c_{i+1} \le c_i b_{i+1}$ for all $1 \le i \le m-1$. Then $c_1 P_1 + \cdots + c_m P_m \prec b_1 P_1 + \cdots + b_m P_m$.*

Proof : We may reduce to the case that $b_i > 0$ and $c_i > 0$ for all $1 \le i \le m$ by a limiting argument. The inequalities imply that $b_h c_i \le c_h b_i$ for all $1 \le h < i \le m$, and thus we may reduce to the case that $P_i \not\equiv 0$ for all $1 \le i \le m$. Replacing P_i by P_i/c_i for $1 \le i \le m$ we reduce further to the case $c_1 = \cdots c_m = 1$ and $b_1 \le \cdots \le b_m$. For $1 \le i \le m$ let $S_i := P_i + \cdots + P_m$. Lemmas 5 and 7 imply that $S_h \prec S_i$ for all $1 \le h < i \le m$. Now, with $b_0 := 0$, Lemma 6 implies that

$$S_1 \prec \sum_{i=1}^{m} (b_i - b_{i-1})\, S_i \prec S_m,$$

proving the result. □

Proof of Theorem 3: We proceed by induction on n, dividing the induction hypothesis into two parts, I(n) and II(n).

I(n): For a square Ferrers matrix A of side $m \le n$ satisfying C1, C2, and C3, $\mathrm{per}(zJ_m + A)$ has all zeros in the interval $[-1, 0]$.
For an m-by-m matrix A and $1 \le i, j \le m$ let $T_{ij}A$ denote the $(m-1)$-by-$(m-1)$ submatrix of A obtained by deleting the i-th row and j-th column of A, and let $Q_{ij} := \mathrm{per}(zJ_{m-1} + T_{ij}A)$.

II(n): For a square Ferrers matrix A of side $m \le n$ satisfying C1, C2, and C3, both of the following conditions hold.

C4 For any $1 \le r \le m$: $Q_{rj} \lll Q_{rk}$ for all $1 \le j < k \le m$.

C5 For any $1 \le c \le m$: $Q_{hc} \lll Q_{ic}$ for all $1 \le h < i \le m$.

The bases of induction, I(1) and II(1), are trivial. The next cases, I(2) and II(2), also follow easily. We divide the induction step into two parts: I(n) and II(n) imply II($n+1$), and I(n) and II($n+1$) imply I($n+1$).

For the first part of the induction step, assume I(n) and II(n), and let A be an $(n+1)\times(n+1)$ Ferrers matrix satisfying C1, C2, and C3. By the symmetry $A \mapsto A^\top$ it suffices to prove C4. Fix a row $1 \le r \le n+1$ and two columns $1 \le j < k \le n+1$. By I(n) we know that both Q_{rj} and Q_{rk} have all their zeros in $[-1,0]$; we must show that $Q_{rj} \langle\langle Q_{rk}$. Notice that for all $1 \le i \le n$, $T_{i,k-1}T_{rj}A = T_{ij}T_{rk}A$, and define $P_i := \operatorname{per}(zJ_{n-1} + T_{ij}T_{rk}A)$ for all $1 \le i \le n$. By II(n) we have $P_h \langle\langle P_i$ for all $1 \le h < i \le n$. Also, defining

$$b_i := \begin{cases} a_{ij} & \text{if } 1 \le i \le r-1, \\ a_{i+1,j} & \text{if } r \le i \le n, \end{cases}$$

and

$$c_i := \begin{cases} a_{ik} & \text{if } 1 \le i \le r-1, \\ a_{i+1,k} & \text{if } r \le i \le n, \end{cases}$$

we have

$$Q_{rj} = \sum_{i=1}^{n}(z+c_i)P_i \quad \text{and} \quad Q_{rk} = \sum_{i=1}^{n}(z+b_i)P_i.$$

For $1 \le i \le n$ let $F_i := P_i + \cdots + P_n$, and let $F_{n+1} :\equiv 0$.

For $1 \le i \le n$, the polynomial P_i is standard, not identically zero, and has all its zeros in the interval $[-1,0]$, by I(n). Thus we see that

$$(z+1)P_1 \langle\langle \cdots \langle\langle (z+1)P_n \langle\langle zP_1 \langle\langle \cdots \langle\langle zP_n.$$

Thus, for any $1 \le g \le n$ we have

$$(z+1)P_g \langle\langle \cdots \langle\langle (z+1)P_n \langle\langle zP_1 \langle\langle \cdots \langle\langle zP_g \quad \text{and} \quad (z+1)P_g \langle\langle zP_g. \tag{6}$$

Apply Lemmas 5 and 7 to (6) to see that

$$S_1' \prec S_2' \prec \cdots \prec S_{n+1}' \prec S_1 \prec \cdots \prec S_{n+1} \quad \text{and} \quad S_1' \prec S_{n+1},$$

in which case S_i' is the sum of the first i terms of (6), and S_i is the sum of the last $n+2-i$ terms of (6).

From Lemma 5 applied to this sequence, we see in particular that $S_n' \prec S_2$. But

$$S_n' = (z+1)P_g + \ldots + (z+1)P_n + zP_1 + \ldots + zP_{g-1} = zF_1 + F_g$$

and

$$S_2 = (z+1)P_{g+1} + \ldots + (z+1)P_n + zP_1 + \ldots + zP_g = zF_1 + F_{g+1}.$$

Also, one sees directly that

$$zF_1 + F_1 = (z+1)F_1 \langle\langle zF_1 = zF_1 + F_{n+1}.$$

From Lemma 5 we conclude that

$$zF_1 + F_g \langle\langle zF_1 + F_h \qquad \text{for all} \qquad 1 \le g < h \le n+1.$$

Putting $c_0 := 0$ and $c_{n+1} := 1$ and $b_0 := 0$ and $b_{n+1} := 1$ we have

$$Q_{rj} = \sum_{g=1}^{n+1}(c_g - c_{g-1})(zF_1 + F_g) \quad \text{and} \quad Q_{rk} = \sum_{g=1}^{n+1}(b_g - b_{g-1})(zF_1 + F_g).$$

Condition $\underline{C3}$ implies that for all $1 \leq g \leq n$,

$$(b_g - b_{g-1})(c_{g+1} - c_g) \leq (c_g - c_{g-1})(b_{g+1} - b_g).$$

Lemma 8 now implies that $Q_{rj} \lll Q_{rk}$, completing the first part of the induction step.

For the second part of the induction step assume I(n) and II($n+1$), and let A be an $(n+1) \times (n+1)$ Ferrers matrix satisfying $\underline{C1}$, $\underline{C2}$, and $\underline{C3}$. Let $P_i := \mathrm{per}(zJ_n + T_{i1}A)$ for $1 \leq i \leq n+1$; each P_i has all its zeros in the interval $[-1, 0]$, by I(n), and $P_h \lll P_i$ for all $1 \leq h < i \leq n+1$, by II($n+1$). Moreover,

$$\mathrm{per}(zJ_{n+1} + A) = \sum_{i=1}^{n+1}(z + a_{i1})P_i.$$

If A is Ferrers and $\underline{C1}$ holds we have $0 \leq a_{i1} \leq \cdots \leq a_{n+1,1} \leq 1$, so for all $1 \leq i \leq n+1$ we have $a_{i1}(1 - a_{i+1,1}) \leq a_{i+1,1}(1 - a_{i1})$, where $a_{01} := 0$ and $a_{n+2,1} := 1$. By Lemmas 6 and 8 we have

$$P_1 \lll \sum_{i=1}^{n+1}(1 - a_{i1})P_i \lll \sum_{i=1}^{n+1} a_{i1}P_i \lll P_{n+1}.$$

Since the zeros of these polynomials are all in the interval $[-1, 0]$, Lemma 4 (a) implies that

$$(z+1)\sum_{i=1}^{n+1} a_{i1}P_i \lll z\sum_{i=1}^{n+1}(1 - a_{i1})P_i.$$

Applying Lemma 4 (d) to this we obtain

$$(z+1)\sum_{i=1}^{n+1} a_{i1}P_i \lll \sum_{i=1}^{n+1}(z + a_{i1})P_i \lll z\sum_{i=1}^{n+1}(1 - a_{i1})P_i,$$

and so $\mathrm{per}(zJ_{n+1} + A)$ has all its zeros in the interval $[-1, 0]$. This completes the proof. □

We can also show Conjecture 1 is true if the underlying tableau is the trivial staircase

$$\begin{array}{cccc} 1 & 2 & \ldots & n \\ n+1 & n+2 & \ldots & 2n \\ \vdots & & & \vdots \\ n^2-n+1 & & \ldots & n^2 \end{array}$$

Theorem 4. *Conjecture 1 is true provided the tableau associated to A is the one above, i.e. if* $a_{1,n} \leq a_{2,1}, a_{2,n} \leq a_{3,1}, \ldots, a_{n-1,n} \leq a_{n,1}$. *Furthermore, under these assumptions, with* $P(z) := per(zJ_n + A)$,

$$\begin{aligned} &P(z) > 0 \text{ for } z > -a_{11}, \\ &P(z) < 0 \text{ for } -a_{21} < z < -a_{1n}, \\ &\ldots, (-1)^n P(z) < 0 \text{ for } -a_{n1} < z < -a_{n-1,n}, \\ &(-1)^n P(z) > 0 \text{ for } z < -a_{nn}. \end{aligned} \tag{7}$$

Proof : By induction on n. First assume that the inequalities on the a_{ij} are strict, and note that (7) then guarantees that $P(z)$ has n distinct roots in $[-a_{nn}, -a_{11}]$. Let $P_{ij}(z)$ denote the permanent of the matrix obtained by deleting row i and column j of $zJ_n + A$. By expanding in minors we have

$$P(z) = (a_{11} + z)P_{11}(z) + (a_{12} + z)P_{12}(z) + \ldots + (a_{1n} + z)P_{1n}(z). \tag{8}$$

Applying the induction hypothesis to the P_{1j} we get

$$\begin{aligned} &P_{11}(z) > 0 \text{ for } z > -a_{22}, \\ &P_{12}(z) > 0 \text{ for } z > -a_{21}, \ldots, \\ &P_{1n}(z) > 0 \text{ for } z > -a_{21}. \end{aligned}$$

By (8) this implies

$$P(z) > 0 \text{ for } z > -a_{11}, \text{ and } P(z) < 0 \text{ for } -a_{21} < z < -a_{1n}.$$

Similarly using

$$\begin{aligned} &P_{11}(z) < 0 \text{ for } -a_{32} < z < -a_{2n}, \ldots, \\ &P_{1n}(z) < 0 \text{ for } -a_{31} < z < -a_{2,n-1} \end{aligned}$$

we get

$$P(z) > 0 \text{ for } -a_{31} < z < -a_{2n}.$$

Continuing in this manner we easily verify the rest of (7). The case where some of the a_{ij} are equal follows from a simple continuity argument using the argument principle. □

ACKNOWLEDGMENTS

The authors would like to thank E. R. Canfield for suggesting the parameterization of the matrix A in terms of the w_i, H. Wilf for bringing Szegö's result to their attention, and R. Stanley for observing that Conjecture 1 could be formulated in terms of permanents.

REFERENCES

[Bre] F. Brenti, *Unimodal, log-concave and Pólya frequency series in combinatorics*, Mem. Amer. Math. Soc. **413** (1989).

[GaWa] J. Garloff and D.G. Wagner, *Hadamard products of stable polynomials are stable*, J. Math. Anal. Appl. **202** (1996), 797-809.

[GJW] J. R. Goldman, J. T. Joichi, and D. E. White, *Rook theory I: Rook equivalence of Ferrers boards*, Proc. Amer. Math. Soc. **52** (1975), 485–492.

[Gra] J. H. Grace, *The zeros of a polynomial*, Proc. Cambridge Phil. Soc. **11** (1900-1902), 352-357.

[Hag1] J. Haglund, *Compositions, Rook Placements, and Permutations of Vectors*, Ph.D. Thesis, University of Georgia, 1993.

[Hag2] J. Haglund, *Rook theory, compositions, and zeta functions*, Trans. Amer. Math. Soc. **348** (1996), 3799–3825.

[HeLi] O. Heilmann and E. Lieb, *Theory of monomer-dimer systems*, Comm. Math. Phys. **25** (1972), 190–232.

[HLP] G. Hardy, J. Littlewood and G. Pólya, *Inequalities*, 2nd ed., Cambridge Univ. Press, Cambridge, 1952.

[KaRi] I. Kaplansky and J. Riordan, *The problem of the rooks and its applications*, Duke Math. J. **13** (1946), 259–268.

[Lag] E. Laguerre, *Oeuvres*, vol. I, Gauther-Villars, Paris, 1898.

[Lev] B. Levin, *Distribution of Zeros of Entire Functions*, Transl. of Mathematical Monographs, vol. 5, Amer. Math. Soc., Providence, Rhode Island, 1964.

[Nij] A. Niejenhuis, *On permanents and the zeros of the rook polynomials*, J. Combin. Theory (A) **21** (1976), 240–244.

[PoSz] G. Polya and G. Szegö, *Problems and Theorems in Analysis, Volume II*, Springer-Verlag, New York, 1976.

[Rio] J. Riordan, *An Introduction to Combinatorial Analysis*, John Wiley, New York, NY, 1958.

[Sch] I. Schur, *Zwei Sätze über algebraische gleichungen mit laiter reellen wurzeln*, J. reine angew. Math. **144** (1914), 75–88.

[Sim] R. Simion, *A multi-indexed Sturm sequence of polynomials and unimodality of certain combinatorial sequences*, J. Combin. Theory (A) **36** (1984), 15-22.

[Sze] G. Szegö, *Bemerkungen zu einem Satz von J. H. Grace über die Wurzeln algebraischer Gleichungen*, Math. Z. **13** (1922), 28–55.

[Vle] E. B. Van Vleck, *On the location of roots of polynomials and entire functions*, Bull. Amer. Math. Soc. **35** (1929), 643–683.

[Wag1] D.G. Wagner, *Total positivity of Hadamard Products*, J. Math. Anal. Appl. **163** (1992), 459-483.

[Wag2] D.G. Wagner, *Enumeration of functions from posets to chains*, Europ. J. Combin. **13** (1992), 313-324.

[Wag3] D.G. Wagner, *Multipartition series*, S.I.A.M. J. Discrete Math. **9** (1996), 529-544.

DEPARTMENT OF MATHEMATICS, MASSACHUSETTS INSTITUTE OF TECHNOLOGY, CAMBRIDGE, MASSACHUSETTS 02139

E-mail address: `haglund@math.mit.edu`

DEPARTMENT OF MATHEMATICS, PENN STATE UNIVERSITY. UNIVERSITY PARK, PENNSYLVANIA 18602

E-mail address: `ono@math.psu.edu`

UNIVERSITY OF WATERLOO, WATERLOO, ONTARIO, CANADA N2L 3G1

E-mail address: `dwagner@uwaterloo.ca`

AN EXAMPLE OF AN ELLIPTIC CURVE WITH A POSITIVE DENSITY OF PRIME QUADRATIC TWISTS WHICH HAVE RANK ZERO

KEVIN JAMES

ABSTRACT. If p is prime, then let E_p denote the elliptic curve over the rationals given by

$$E_p : y^2 = x^3 - 32p^3.$$

We prove that E_p has only the trivial point (at infinity) for at least $\frac{1}{3}$ of the primes p. In fact, we will show that for 1/3 of the primes p, $L(E_p, 1) \neq 0$.

1. INTRODUCTION

Given an elliptic curve $E : y^2 = x^3 + Ax^2 + Bx + C \quad (A, B, C \in \mathbb{Z})$ defined over $\mathbb{Q}$ and having minimal discriminant Δ_E, we define $a_p = p + 1 - \#E(\mathbb{F}_p)$. Then we associate to E its L-series:

$$(1) \qquad L(E, s) = \prod_{p|\Delta_E} \frac{1}{1 - a_p p^{-s}} \prod_{p\nmid\Delta_E} \frac{1}{1 - a_p p^{-s} + p^{1-2s}} = \sum_{n\geq 1} \frac{a_n}{n^s}.$$

This product converges for $Re(s) > 3/2$. In the case that E has complex multiplication, it is known that $L(E, s)$ has analytic continuation to the whole complex plane. The Coates-Wiles theorem [3] then tells us that if $L(E, 1) \neq 0$ then E has only a finite number of rational points.

Given an elliptic curve E as above and an integer D, we define the Dth quadratic twist of E to be the curve $E_D : y^2 = x^3 + ADx^2 + BD^2x + CD^3$. If we let $L(E_D, s)$ denote the L-function associated to E_D, then we have

$$\begin{aligned} L(E_D, s) &= \prod_{p|\Delta_{E_D}} \frac{1}{1 - \chi_D(p) a_p p^{-s}} \prod_{p\nmid\Delta_{E_D}} \frac{1}{1 - \chi_D(p) a_p p^{-s} + p^{1-2s}} \\ (1) \qquad &= \sum_{n\geq 1} \chi_D(n) \frac{a_n}{n^s}. \end{aligned}$$

where $\chi_D(n)$ denotes the Kronecker symbol $(\frac{D}{n})$. We will sometimes refer to $L(E_D, s)$ as the Dth quadratic twist of $L(E, s)$. In this paper, we will be interested in determining how often we have $L(E_D, 1) \neq 0$ as D varies over all square-free numbers.

S.D. Ahlgren et al. (eds.), Topics in Number Theory, 223–227.

There have been many papers which have investigated this question. For a historical account, particularly of the works of Lucas and Sylvester, see chapter XXI of [4]. In the more recent papers of Bump, Friedberg and Hoffstein [1, 2], Murty and Murty [8], Iwaniec [7], Friedberg and Hoffstein [6] and Ono [9], one can find general theorems on the vanishing and nonvanishing of the quadratic twists of the L-function associated to a given elliptic curve. These theorems ensure that an infinite number of the quadratic twists of such an L-function will have nonzero central critical value.

In the case of an elliptic curve E having rational 2-torsion, one can often say much more. For example in the papers of Frey [5] and Tunnell [17] it is proved for certain elliptic curves E having rational 2-torsion that $L(E_p, 1) \neq 0$ for a positive proportion of primes p.

In the general case, the situation was much less satisfactory until recently. In [10], Ono has shown several examples of elliptic curves E such that $L(E_p, 1) \neq 0$ for a set of primes p of density $1/3$. Ono also proves a Theorem which gives sufficient conditions under which the L-function associated to an elliptic curve will have this property. In order to understand the proof of this theorem however, one must have some understanding of the Galois representations attached to modular forms. Using this theory of Galois representations, Ono and Skinner (see [11] and [12]) have recently extended this theorem.

In this paper, we will show:

Theorem 1. *Let $E : y^2 = x^3 - 32p^3$. Then $L(E_p, 1) \neq 0$ for at least $\frac{1}{3}$ of the primes p.*

Although this theorem follows from the more general theorems of Ono and Skinner mentioned above, it is not included in the specific examples worked out in [10]. We would like to discuss a simpler proof of this result that does not explicitly involve the theory of Galois representations.

Using the Coates-Wiles theorem, we can then deduce the following:

Corollary 2. *The curve*

$$y^2 = x^3 - 32p^3$$

has only the trivial point (at infinity) for at least $\frac{1}{3}$ of the primes p.

2. RESULTS

Denote by E_D the elliptic curve

$$E_D : y^2 = x^3 + 4D^3$$

where D is any 6th power free integer, and let

$$L(E_1, s) = \sum_{n \geq 1} \frac{a_n}{n^s}.$$

Now, we note that E_D has complex multiplication by $\mathbb{Z}[\omega]$, where ω is a cube root of unity. Thus,

$$f_D(z) = \sum_{n=1}^{\infty} a_n \chi_D(n) q^n \in S_2(N_D) \quad (q = e^{2\pi i z})$$

where N_D is the conductor of E_D. Also, f_D is an eigenform for all of the Hecke operators.

Let,

$$g(z) = \frac{1}{2}\left(\sum_{x,y,z\in\mathbb{Z}} q^{x^2+27y^2+6z^2} - \sum_{x,y,z\in\mathbb{Z}} q^{4x^2+2xy+7y^2+6z^2}\right) = \sum_{n=1}^{\infty} b_n q^n.$$

Then by theorems of Schoenberg [13] and Siegel [15], we have that $g(z) \in S_{\frac{3}{2}}(216, \chi_2)$. We can check computationally that $g(z)$ is an eigenform for all of the Hecke operators, and that g lifts through the Shimura correspondence [14] to f_1. Now we can apply a theorem of Waldspurger [18 Corollary 2] to gain information about the values $L(E_D, 1)$. In our case Waldspurger's theorem specializes to the following.

Theorem 2.1. *For $t \equiv 1$ modulo 6,*

$$L(E_{-2t}, 1) = \frac{b_t^2}{\sqrt{t}}\beta,$$

where $\beta \approx 1.363$ (the value of β was computed using the Apecs package with MAPLE).

Thus, $L(E_{-2t}, 1) = 0$ if and only if $b_t = 0$.

Now, by a theorem of Sturm [16] we know that if we have two cusp forms in $S_k(N, \chi)$ that are congruent modulo some prime q up to the first $\frac{k}{12}N\prod_{p|N}(1+\frac{1}{p})$ coefficients, then the two forms are all congruent modulo q. Now, $f_1 \in S_2(108) \subseteq S_2(216)$, and $g(z)\theta(2z) \in S_2(216)$. Thus by Sturm's theorem, we can check computationally that $g(z)\theta(2z) \equiv f_1(z)$ modulo 2. Now, we notice that $\theta(2z) \equiv 1$ modulo 2. Hence $g(z) \equiv f_1(z)$ modulo 2.

Now, we note that if $x^3 + 4$ has no root modulo p for p, a prime then $a_p \equiv 1$ modulo 2. This is because $E_1(\mathbb{F}_p)$ has no point of order 2 in this case. This implies that $b_p \equiv 1$ modulo 2. In particular $b_p \neq 0$. Thus from Theorem 2.1, $L(E_{-2p}, 1) \neq 0$ for such primes. So, Theorem 1 follows from the following lemma.

Lemma 2.2. *$x^3 + 4$ has no root modulo p for $\frac{1}{3}$ of the primes p.*

Proof. Note that for $p \equiv 2$ modulo 3, cubing is an automorphism of $\mathbb{F}_p$. So, $x^3 + 4$ always has a root modulo p when $p \equiv 2$ modulo 3. Thus we restrict our attention to $p \equiv 1(3)$ from now on. Hence, we have $\left(\frac{-3}{p}\right) = 1$. Now, we note that $\left(\frac{p}{3}\right) = \left(\frac{-3}{p}\right) = 1$, which implies that p splits in $\mathbb{Z}[\omega]$.

We have

$$\begin{array}{ccc} \mathbb{Q}(\omega) & & \mathfrak{p}_1\mathfrak{p}_2 \\ | & & | \\ \mathbb{Q} & & p \equiv 1 \pmod 3 \end{array}$$

Now, x^3+4 is irreducible over $\mathbb{Z}[\omega]$ (ω is a cube root of unity). Also, x^3+4 has a root in $\mathbb{Z}[\omega]/\mathfrak{p}_i$ ($i = 1, 2$) if and only if it has a root modulo p. (In fact $\mathbb{Z}[\omega]/\mathfrak{p}_i \cong \mathbb{F}_p$.) The splitting of $x^3 + 4$ modulo $\mathfrak{p}_i$ determines the splitting of $\mathfrak{p}_i$ in $\mathbb{Z}[\omega, 4^{\frac{1}{3}}]$. In particular, if $x^3 + 4$ has no root modulo p, then the $\mathfrak{p}_i$'s remain inert in $\mathcal{O}_{\mathbb{Q}(\omega,4^{\frac{1}{3}})}$. Thus p splits into exactly two primes in $\mathcal{O}_{\mathbb{Q}(\omega,4^{\frac{1}{3}})}$

$$\begin{array}{ccc} \mathbb{Q}(\omega, 4^{\frac{1}{3}}) & & \mathfrak{P}_1\mathfrak{P}_2 \\ | & & | \\ \mathbb{Q} & & p \equiv 1 \pmod 3 \end{array}$$

Now, $\mathbb{Q}(\omega, 4^{\frac{1}{3}})/\mathbb{Q}$ is a Galois extension with Galois group S_3. So, the residual degrees $f(\mathfrak{P}_1)$ and $f(\mathfrak{P}_2)$ are the same, and the ramification indices of $\mathfrak{P}_1$ and $\mathfrak{P}_2$ are the same namely 1. Thus $f(\mathfrak{P}_1) = f(\mathfrak{P}_2) = 3$. This tells us that the order of the Frobenius $\sigma_{\mathfrak{P}_i}$ is 3. So, the size of the conjugacy class of $\sigma_{\mathfrak{P}_i}$ on S_3 is 2. The Lemma now follows from the Chebetorev Density Theorem.

Since f is an eigenform for all of the Hecke operators, it follows that the a_n's are multiplicative, that is if $\gcd m, n = 1$ then $a_{mn} = a_n a_n$. Thus we can deduce the following corollary form Theorem 1:

Corollary 2.3. *If D is a square free natural number with only prime factors $p \equiv 1$ modulo 3 such that $x^3 + 4$ has no root modulo p, $L(E_D, 1) \neq 0$.*

ACKNOWLEDGEMENTS

The author would like to thank Alice Silverberg for her helpful comments during the preparation of this manuscript.

REFERENCES

1. D. Bump, S. Friedberg, and J. Hoffstein, *Eisenstein series on the metaplectic group and nonvanishing theorems for automorphic L-functions and their derivatives*, Ann. of Math. **131** (1990), 53–127.
2. D. Bump, S. Friedberg, and J. Hoffstein, *Nonvanishing theorems for L-functions of modular forms and their derivatives*, Inventiones Math. **102** (1990), 543–618.
3. J. Coates and A. Wiles, *On the conjecture of Berch and Swinnerton-Dyer*, Invent. Math. **39** (1977), no. 3, 223–251.
4. L. E. Dickson, *Theory of Numbers*, vol. 2, Chelsea Publishing Co., 1966.
5. G. Frey, *Der Rang der Lsungen von $Y^2 = X^3 \pm p^3$ uber $\mathbb{Q}$.*, Manuscripta Math. **48** (1984), no. 1–3, 71–101.
6. S. Friedberg and J. Hoffstein, *Nonvanishing theorems for automorphic L-functions on $GL(2)$*, Ann. of Math. **142** (1995), no. 2, 385–423.

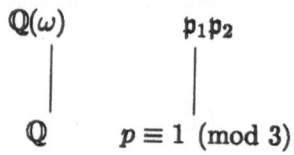

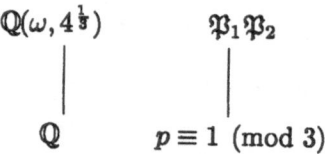

7. H. Iwaniec, *On the order of vanishing of modular L-series at the critical point*, Sém. de Théor. Nombres, Bordeaux **2** (1990), no. 2, 365–376.
8. M.R. Murty and V.K. Murty, *Mean values of derivatives of modular L-series*, Ann. of Math. **133** (1991), 447–475.
9. K. Ono, *Rank zero quadratic twists of modular elliptic curves*, Compositio Math. **104** (1996), 293-304.
10. K. Ono, *Twists of elliptic curves*, Compositio Math. **106** (1997), 349-360.
11. K. Ono and C. Skinner, *Fourier coefficients of half-integral weight modular forms modulo* ℓ, Annals of Math. **147** (1998), 453-470.
12. K. Ono and C. Skinner, *Non-vanishing of quadratic twists of modular L-functions*, Inventiones Math. (to appear).
13. B. Schoeneberg, *Das Verhalten von mehrfachen Thetareihen bei Modulsubstitutionen*, Math. Ann. **116**.
14. G. Shimura, *On modular forms of half integral wieght*, Ann. of Math. (2) **97** (1973), 440–481.
15. C. Siegel, *Gesammelte Abhandlungen Bd. 3*, Springer Verlag, 1966, pp. 326–405.
16. J. Sturm, *On the congruence of modular forms*, Springer Lect. Notes **1240** (1984), Springer-Verlag, 275–280.
17. J. Tunnell, *A classical Diophantine problem and modular forms of weight* 3/2, Invent. Math. **72** (1983), no. 2, 323–334.
18. J.L. Waldspurger, *Sur les coefficients de Fourier des formes modulaires de poids demi-entier*, J. Math. Pures. et Appl. **60** (1981), 375–484.

DEPARTMENT OF MATHEMATICS, PENN STATE UNIVERSITY, UNIVERSITY PARK, PENNSYLVANIA 16802

E-mail address: `klj@math.psu.edu`

EXPONENTS OF CLASS GROUPS OF QUADRATIC FIELDS

M. RAM MURTY

Dedicated to the memory of Sarvadaman Chowla

Given a positive integer $g \geq 2$, we would like to study the number of real and imaginary quadratic fields that have an element of order g in their ideal class group. Conjectures of Cohen and Lenstra predict a positive probability for such an event. Our goal here is to derive quantitative results in this direction. We establish for $g \geq 3$, the number of imaginary quadratic fields whose absolute discriminant is $\leq x$ and whose class group has an element of order g is $\gg x^{\frac{1}{2}+\frac{1}{g}}$. For g odd we show that the number of real quadratic fields whose discriminant is $\leq x$ and whose class group has an element of order g is $\gg x^{1/2g-\epsilon}$ for any $\epsilon > 0$. (The implied constant may depend on ϵ.)

Introduction. Given a positive integer $g \geq 2$, we would like to study the number of real and imaginary quadratic fields that have an element of order g in their ideal class group. Conjectures of Cohen and Lenstra [CL] predict a positive probability for such an event. Indeed, if $g = p$ is an odd prime, then in the case of imaginary quadratic fields, they forecast that

$$1 - \prod_{i=1}^{\infty}\left(1 - \frac{1}{p^i}\right)$$

as the probability that an imaginary quadratic field has an element of order p in the class group. In the real quadratic case, their prediction is

$$1 - \prod_{i=2}^{\infty}\left(1 - \frac{1}{p^i}\right)$$

as the corresponding probability.

Qualitative results in this line of thought have a long history. The case $g = 2$ is classical work of Gauss. The case $g = 3$ was studied by Davenport and Heilbronn [DH]. Earlier, Nagell [N], Honda [H], Ankeny and Chowla [AC], Hartung [Ha], Yamamoto [Y] and Weinberger [W] derived the infinitude of such fields, for any given g.

Research partially supported by NSERC.

S.D. Ahlgren et al. (eds.), Topics in Number Theory, 229–239.

Our goal here is to derive quantitative results in this direction. In an earlier paper [RM], we indicated how one can use the ABC conjecture to show that at least $\gg x^{1/g-\epsilon}$ imaginary quadratic fields with absolute discriminant $\leq x$ have an element of order g in their class group. In the real quadratic case, we obtained a lower bound of $\gg x^{1/2g-\epsilon}$ for the number of such fields.

In this paper, we will establish stronger results *without* the ABC conjecture. We will prove:

Theorem 1. *Let $g \geq 3$. The number of imaginary quadratic fields whose absolute discriminant is $\leq x$ and whose class group has an element of order g is $\gg x^{\frac{1}{2}+\frac{1}{g}}$.*

Theorem 2. *Let g be odd. The number of real quadratic fields whose discriminant is $\leq x$ and whose class group has an element of order g is $\gg x^{1/2g-\epsilon}$ for any $\epsilon > 0$. (The implied constant may depend on ϵ.)*

Theorem 2 is still valid if g is even and not divisible by 4 and we briefly indicate how to derive this at the end of the proof of Theorem 2. If g is divisible by 4, the argument of Theorem 2 gives us $\gg x^{1/4g-\epsilon}$ real quadratic fields with discriminant $< x$ and whose class group contains an element of exponent g.

A lot of work has centered around the complementary question of finding class groups not divisible by g. For example, the recent work of Kohnen and Ono [KO] shows that for odd primes ℓ that there are $\gg \sqrt{x}/\log x$ imaginary quadratic fields with discriminants $|D| \leq x$ with $\ell \nmid h(D)$ where $h(D)$ denotes the class number of the field. For real quadratic fields, if ℓ is an odd prime satisfying a very mild condition, then Ono [O] shows that there are $\gg \sqrt{x}/\log x$ discriminants $0 < D < x$ for which $\ell \nmid h(D)$ and $\ell \nmid R_p(D)$, where $R_p(D)$ denotes the p-adic regulator of $\mathbb{Q}(\sqrt{D})$.

Here is a brief outline of the method. In the imaginary quadratic case, our strategy is to consider numbers of the form

$$m^g - n^2$$

with $2X^{1/g} < m < 3X^{1/g}$ and $X^{1/2} < n < 2X^{1/2}$. Then,

$$3^g X > m^g - n^2 > 2^g X - 4X \geq 4X$$

since $g \geq 3$. Since the probablity that a random number is squarefree is $6/\pi^2$, one expects a positive proportion of such (m, n) with m and n in the stated range. This idea is made precise using the sieve of Eratosthenes in section 2. Hence, it is therefore reasonable to expect $\gg X^{1/2+1/g}$ numbers of the form

$$d = m^g - n^2$$

which are squarefree. It is then not difficult to show that in $\mathbb{Q}(\sqrt{-d})$, the ideal $(n + \sqrt{-d})$ is the g-th power of an ideal which has order g in the ideal class group of that field. Finally, we make a count of how many distinct numbers d arise in this fashion and show there are not too many repetitions. This is how Theorem 1 is derived.

For the real quadratic case, the strategy is different. We modify an argument of Weinberger to first establish that if $D = n^{2g} + 4$ and $n > g$ is prime, then with suitable restrictions, the ideal $(n^2, 2 + \sqrt{D})$ has order g or $g/2$ in the ideal class group of $\mathbb{Q}(\sqrt{D})$. One then counts the number of distinct real quadratic fields that arise in this way by making use of a classical result of Sprindzuk on the number of integral points on certain hyper-elliptic curves.

2. Squarefree values of quadratic polynomials.

Let $f(n) = n^2 + c$ with c an integer. We will need the following elementary lemma.

Lemma 1. *Let p be an odd prime. If $p \nmid c$, then the congruence $n^2 + c \equiv 0 \pmod{p^2}$ has at most 2 solutions. If $p|c$, and $p^2 \nmid c$, there are no solutions. If $p^2|c$, then $p|n$, so there are p solutions mod p^2. If $p = 2$, there are either 2 solutions or no solutions according as $c \equiv 0, -1 \pmod 4$ or not.*

Proof. The congruence $n^2 + c \equiv 0 \pmod p$ has at most two solutions mod p. If n_0 mod p is a solution, then we begin by counting how many lifts it has to mod p^2. Indeed, suppose

$$(n_0 + tp)^2 \equiv -c \pmod{p^2}$$

then,

$$\frac{n_0^2 + c}{p} + 2n_0 t \equiv 0 \pmod p.$$

This has a unique solution $t \pmod p$ provided $p \nmid 2c$. Therefore, if $p \nmid 2c$, then there are at most 2 solutions to the congruence $n_0^2 + c \equiv 0 \pmod{p^2}$. The last three assertions are obvious.

Lemma 2. *Let d be an odd squarefree number and denote by $\rho_c(d^2)$ the number of solutions of the congruence $n^2 + c \equiv 0 \pmod{d^2}$. Then*

$$\rho_c(d^2) \leq 2^{\nu(d)}\gamma(c)$$

where $\nu(d)$ denotes the number of distinct prime divisors of d and $\gamma(c)$ is the product of the distinct primes dividing c.

Proof. By the Chinese remainder theorem, $\rho_c(m)$ is a multiplicative function of m. By Lemma 1, $\rho_c(p^2) \leq 2$ if $p \nmid 2c$. If $p|c$, then $\rho_c(p^2) \leq p$ again by Lemma 1. Therefore,

$$\rho_c(d^2) \leq 2^{\nu(d)}\gamma(c),$$

as desired.

Lemma 3. *Let $\nu(m)$ denote the number of distinct prime divisors of m. Then,*

$$\sum_{m<T} \nu(m) = T \log\log T + O(T).$$

Proof. The sum is clearly bounded by

$$\sum_{p<T} \frac{T}{p}$$

and by the elementary formula

$$\sum_{p<T} \frac{1}{p} = \log\log T + O(1)$$

the result is now immediate.

Let $S(x, y; c)$ enumerate the number of $x < n < y$ such that $f(n) = n^2 + c$ is squarefree. For each squarefree d, let $\rho_c(d)$ be the number of solutions of the congruence

$$n^2 + c \equiv 0 \pmod{d}.$$

We will also introduce

$$P(z) = \prod_{p \le z} p,$$

where the product is over primes less than or equal to z. Observe that by elementary number theory, namely Chebychev's theorem, we have

$$P(z) < e^{\kappa z},$$

for some constant κ.

Theorem 3. *If $|c| < y$, then*

$$S(x, y; c) \ge (y - x)\frac{\phi(c)}{c} \prod_p \left(1 - \frac{2}{p^2}\right) + O\left(\frac{y\nu(c)}{\log y}\right),$$

where ϕ denotes Euler's function.

Proof. It is clear that

$$S(x, y; c) \le \sum_{x<n<y} \sum_{d^2 | (n^2+c, P(z)^2)} \mu(d).$$

On the other hand, let $N_p(x,y;c)$ to be the number of n satisfying $x < n < y$ and $p^2|n^2+c$. Then,

$$S(x,y;c) \geq \sum_{x<n<y} \sum_{d^2|(n^2+c,P(z)^2)} \mu(d) - \sum_{z<p<y+|c|} N_p(x,y;c).$$

By Lemma 1, we know that

$$N_p(x,y;c) \leq 2\Big(\frac{y}{p^2}+1\Big)$$

if $p \nmid 2c$. Again by Lemma 1, we have

$$N_p(x,y;c) \leq \frac{y}{p}$$

if $p|c$. Therefore (for $z \geq 2$),

$$\sum_{z<p<y+|c|} N_p(x,y;c) \ll \sum_{z<p<y+|c|,p\nmid c} \Big(\frac{y}{p^2}+1\Big) + \sum_{z<p<y+|c|,p|c} \frac{y}{p}.$$

Observe that

$$\sum_{z<p<y+|c|,p|c} \frac{y}{p} \leq \frac{y}{z}\nu(c)$$

and that

$$\sum_{z<p<y+|c|,p\nmid c} \Big(\frac{y}{p^2}+1\Big) \ll \frac{y}{z} + \frac{y}{\log y}.$$

Therefore, for $|c| < y$, we have

$$\begin{aligned} S(x,y;c) &\geq \sum_{x<n<y} \sum_{d^2|(n^2+c,P(z)^2)} \mu(d) - \sum_{z<p<y+|c|} N_p(x,y;c) \\ &\geq \sum_{d|P(z)} \mu(d) \sum_{\substack{x<n<y \\ n^2+c\equiv 0 \pmod{d^2}}} 1 + O(y\nu(c)/z) + O(y/\log y) \\ &= \sum_{d|P(z)} \mu(d)\Big\{\frac{y-x}{d^2}\rho_c(d^2) + O(\rho_c(d^2))\Big\} + O(y\nu(c)/z + y/\log y). \end{aligned}$$

We will choose $z = \kappa \log y$ for some appropriate constant $\kappa > 0$. Then the error term above is $O(y\nu(c)/\log y)$. Since $\rho_c(d^2) \leq \gamma(c)2^{\nu(d)}$ by Lemma 2, we find

$$\sum_{d|P(z)} \rho_c(d^2) \leq \gamma(c) \sum_{d\leq P(z)} 2^{\nu(d)}.$$

The elementary estimate

$$\sum_{d\leq w} 2^{\nu(d)} \ll w \log w$$

yields the result

$$S(x,y;c) \geq (y-x) \sum_{d|P(z)} \frac{\mu(d)}{d^2}\rho_c(d^2) + O(y^{\kappa(1+\epsilon)}) + O(y\nu(c)/\log y).$$

We will choose κ sufficiently small, so that this quantity is

$$\geq (y-x) \sum_{d|P(z)} \frac{\mu(d)}{d^2}\rho_c(d^2) + O(y\nu(c)/\log y).$$

The sum can be written as a product:

$$\sum_{d|P(z)} \frac{\mu(d}{d^2}\rho_c(d^2) = \prod_{p\leq z}\left(1 - \frac{\rho_c(p^2)}{p^2}\right) \geq \frac{\phi(c)}{c}\prod_{p}\left(1-\frac{2}{p^2}\right)$$

by an application of Lemma 2. This completes the proof of Theorem 3.

We now apply Theorem 3 to $c = -m^g$, m odd and consider those m, n satisfying

$$X^{1/2} < n < 2X^{1/2}, \quad 2X^{1/g} < m < 3X^{1/g}$$

so that the number of squarefree values in the sequence

$$m^g - n^2$$

is

$$\gg X^{1/2} \sum_{\substack{2X^{1/g}<m<3X^{1/g} \\ m \text{ odd}}} \frac{\phi(m)}{m} + O\left(\frac{X^{\frac{1}{g}+\frac{1}{2}} \log\log X}{\log X}\right),$$

by an application of Lemma 3 in estimating the error term. Since

$$\begin{aligned}\sum_{m\leq T, m \text{ odd}} \frac{\phi(m)}{m} &= \sum_{d\leq T, d \text{ odd}} \frac{\mu(d)}{d}\left[\frac{T}{d}\right] \\ &= \sum_{d\leq T, d \text{ odd}} \frac{\mu(d)}{d}\left(\frac{T}{d} + O(1)\right) = \frac{8}{\pi^2}T + O(\log T),\end{aligned}$$

we deduce that there are $\gg X^{1/2+1/g}$ squarefree values $\leq X$ that are of the form $m^g - n^2$ with m, n in the specified ranges and m odd. Let us signal for future use the important fact that for these m, n, we have $4X \leq (2^g - 4)X \leq m^g - n^2 \leq (3^g - 1)X$.

3. Proof of Theorem 1.

For each of the $\gg X^{1/2+1/g}$ squarefree values of $m^g - n^2$ produced in the previous section, we consider the factorization

$$m^g = (n + \sqrt{-d})(n - \sqrt{-d})$$

in $\mathbb{Q}(\sqrt{-d})$. Recall that in the construction of the previous section, m is odd. Since d is squarefree, m and n are coprime for otherwise the square of a prime will divide $m^g - n^2$ and hence d. Also, if the ideals $(n + \sqrt{-d})$ and $(n - \sqrt{-d})$ are not coprime, then any common factor must divide 2 which implies m is even, which is not the case, by construction. Therefore, each of the ideals $(n + \sqrt{-d})$ and $(n - \sqrt{-d})$ must be a perfect g-th power. Thus,

$$\mathfrak{a}^g = (n + \sqrt{-d})$$

for some ideal $\mathfrak{a}$ of norm m. If $\mathfrak{a}$ has order $r \leq g - 1$ (say), then

$$\mathfrak{a}^r = (u + v\sqrt{-d}) \quad \text{or} \quad (\frac{u + v\sqrt{-d}}{2}).$$

Observe that $v \neq 0$ for otherwise $\mathfrak{a}$ and $\mathfrak{a}'$, the conjugate ideal would have a common factor, which is not the case. Hence, taking norms of both sides of the above equation,

$$3^{g-1}X^{(g-1)/g} \geq m^{g-1} \gg d = m^g - n^2 \geq 4X$$

a contradiction for sufficiently large X. This produces an element of order g in the ideal class group.

We now count the number of distinct squarefree numbers in the enumeration above. Indeed, let S be the set of squarefree d's counted above which appear twice in the list. Then we have for each $d \in S$,

$$d = m_1^g - n_1^2 = m_2^g - n_2^2$$

for $(m_1, n_1) \neq (m_2, n_2)$. This means

$$m_1^g - m_2^g = n_1^2 - n_2^2 = (n_1 - n_2)(n_1 + n_2).$$

For fixed m_1, m_2, the choices for n_1 and n_2 are derived from the divisors of $m_1^g - m_2^g$. It is an elementary fact of number theory that the number of divisors of a number n is $O(n^\epsilon)$. Thus, the number of divisors of $m_1^g - m_2^g$ cannot exceed $O(X^\epsilon)$. The number of possible values of m_1 and m_2 are $O(X^{2/g})$ and therefore the total number of elements in S cannot exceed $O(X^{2/g+\epsilon})$. The final enumeration gives

$$\gg X^{1/2+1/g} - O(X^{2/g+\epsilon})$$

distinct squarefree values of $d < X$ such that the class group of $\mathbb{Q}(\sqrt{-d})$ has an element of order g. Since $g \geq 3$, this completes the proof of Theorem 1.

4. An effective version of Weinberger's theorem.

We will begin by making a result of Weinberger effective. In [W], it is proved that:

Proposition. Let $D = n^{2g} + 4$.

(i) If $n > g$ is prime and $\mathbb{Q}(\sqrt{D}) \neq \mathbb{Q}(\sqrt{5})$, then the fundamental unit of $\mathbb{Q}(\sqrt{D})$ is $(n^g + \sqrt{D})/2$.

(ii) Suppose that the polynomial $T^k - 4$ is irreducible mod n for every divisor k of g. Then, the ideal $\mathfrak{a} = (n^2, 2 + \sqrt{D})$ has order g if g is odd, or $g/2$ if g is even, in the ideal class of group of $\mathbb{Q}(\sqrt{D})$.

For the sake of completeness, we will indicate the proof of the Proposition. Let r and s be integers and ρ_1, ρ_2 the roots of

$$T^2 - rT - s = 0.$$

Define $c_j(r, s) = \rho_1^j + \rho_2^j$. Let

$$\alpha = \frac{n^g + \sqrt{D}}{2}.$$

Then, $N(\alpha) = -1$ and $\alpha > 1$. If $\epsilon > 1$ denotes the fundamental unit of $\mathbb{Q}(\sqrt{D})$, then $\alpha = \epsilon^j$ for some odd j. Since $\epsilon > 1$, we have $r = Tr(\epsilon) = \epsilon + \epsilon^{-1} > 0$ and the minimal polynomial of ϵ is

$$T^2 - rT - 1, \quad r > 0.$$

Then

$$n^g = Tr(\alpha) = Tr(\epsilon^j) = \rho_1^j + \rho_2^j = c_j(r, 1)$$

where ρ_1 and ρ_2 are zeros of $T^2 - rT - 1 = 0$. Thus,

$$c_j(r, 1) - n^g = 0.$$

It is not difficult to see that

$$c_j(r, s) = \sum_{\nu=0}^{j/2} f_\nu r^{j-2\nu} s^\nu$$

for integers f_ν. Moreover, when j is odd, $f_{(j-1)/2} = j$ and $f_0 = 1$. In addition, let us observe that if $r, s > 1$, then $c_j(r, s) > r^j$ as an easy computation shows. Thus, $r | c_j(r, 1)$ and so $r | n^g$. If $r = 1$, then $\epsilon = (1 + \sqrt{5})/2$ contrary to hypothesis. Since n is prime, $r = n^k$ for some $k \leq g$. By the observation above, $n^g = c_j(r, 1) \geq r^j = n^{jk}$ and so $jk \leq g$. Since $n > g$ is prime, it follows that $(j, n) = 1$ since $j \leq g$. As j is odd, $c_j(r, 1)/r \equiv j \pmod{n}$ by what we have said above. We conclude that $c_j(r, 1)/r$ is coprime to n so that $n^{g-k} = 1$. Thus, $g = k$ and $j = 1$ as desired. This proves (i).

To prove (ii), we consider the ideal

$$\mathfrak{a} = (n^2, 2 + \sqrt{D}).$$

Now,

$$\mathfrak{a}^g = (n^{2g}, n^{2g-2}(2 + \sqrt{D}), \cdots, (2 + \sqrt{D})^g)$$

which is easily seen to be

$$(\sqrt{D} + 2)(\sqrt{D} - 2, n^{2g-2}, \cdots, (\sqrt{D} + 2)^{g-1}).$$

This means that $\mathfrak{a}^g \subseteq (2 + \sqrt{D})$ while $N(\mathfrak{a}^g) = n^{2g} = N((2 + \sqrt{D})$. Hence, $\mathfrak{a}^g = (2 + \sqrt{D})$. It is then not difficult to show that in fact, the order of $\mathfrak{a}$ in the ideal class group is g when g is odd and $g/2$ when g is even whenever n is prime and the polynomial $T^k - 4$ is irreducible mod n for every divisor k of g.

5. Proof of Theorem 2.

We will now prove Theorem 2. We will need the following lemma.

Lemma. *The number of primes $p < x$ such that the polynomial $T^k - 4$ is irreducible mod p is $\gg x/\log x$.*

Proof. Let K be the splitting field of the polynomial $T^k - 4$. Let $L = \mathbb{Q}(\zeta_k)$ and consider the Galois extension K/L. By the Chebotarev density theorem, there are $\gg x/\log x$ prime ideals $\mathfrak{p}$ of first degree with absolute norm $\leq x$ in L such that $T^k - 4$ is irreducible mod $\mathfrak{p}$. This completes the proof.

Now suppose g is odd. To apply the effective Weinberger theorem, we must ensure that n is a prime such that $T^k - 4$ is irreducible mod n for every $k|g$. Using the Chebotarev density theorem, one can show that the number of $n < X^{1/2g}$ such that $T^{2g} - 4$ is irreducible mod n is $\gg X^{1/2g}/\log X$. This immediately implies that $T^k - 4$ is irreducible mod n for every $k|2g$. Thus, for each of these values of n, the class group of $\mathbb{Q}(\sqrt{n^{2g}+4})$ has an element of order g provided $\mathbb{Q}(\sqrt{n^{2g}+4}) \neq \mathbb{Q}(\sqrt{5})$. First, let us observe that the number of integral solutions of

$$5y^2 = x^{2g} + 4$$

is absolutely bounded (by a classical theorem of Siegel). This eliminates only a finite set of values of n from our consideration. We now need to count the number of distinct quadratic fields obtained in this way. This means, for a fixed d, we need to count the number of integral solutions of $dy^2 = n^{2g} + 4$ which is a hyper-elliptic curve. By a result of Sprindzuk [S] or Evertse - Silverman [ES], we find that the number of such n cannot exceed $C^{\nu(d)}$ for some absolute constant C, where $\nu(d)$ denotes the number of prime factors of d. Since the number of prime factors of d cannot exceed

$$\ll \frac{\log d}{\log \log d}$$

by a classical estimate of Ramanujan, we immediately deduce that the number of distinct quadratic fields among $\mathbb{Q}(\sqrt{n^{2g}+4})$ is $\gg X^{1/2g-\epsilon}$. This completes the proof of Theorem 2.

We remark that the above theorem is also valid if $g = 2r$ with r odd. Indeed, if $g/2$ is odd then, the argument of Weinberger used above gives us an element of order g or $g/2$. If a positive proportion of the fields we have above have an element of order g then we are done. Otherwise, let us suppose that we have $\gg X^{1/2g}/\log X$ fields with an element of order $g/2$. If the discriminant of a quadratic field is not prime, we have (by genus theory) an element of order 2 in the class group which we can multiply to an element of order $g/2$ which is odd, to get an element of order g. By an elementary sieve estimate, we know that the number of $n < X^{1/2g}$ with n prime and $n^{2g}+4$ a prime power is $\ll X^{1/2g}/\log^2 X$. Since this is negligible compared to the number of fields we have, namely $\gg X^{1/2g}/\log X$, we can now proceed as above and derive our estimate.

If g is divisible by 4, then it is clear that our argument produces at least $\gg X^{1/4g-\epsilon}$ real quadratic fields with an element of exponent g in the class group.

6. Concluding remarks.

All of the above results can be generalized to the function field case with better results in the analogue of the real quadratic case and this work will appear in the joint paper [CM]. Even sharper results can be obtained if one only considers quadratic extensions of $\mathbb{F}_p$ generated by the square root of either cubic or quartic polynomials. This is because of the obvious connection to elliptic curves. In fact, the connection beautifully intertwines the Lang-Trotter conjectures for primitive points of elliptic curves over function fields over finite fields with the Cohen-Lenstra conjectures. This work will appear jointly with Rajiv Gupta in [GM]. Finally we mention related work of Jiu-Kang Yu [Yu] which establishes the Cohen-Lenstra conjecture as $p \to \infty$ and fixed discriminantal degree. We are of course interested in the orthogonal direction, namely with p fixed and the discriminantal degree tending to ∞.

References

[AC] N. Ankeny and S. Chowla, On the divisibility of the class numbers of quadratic fields, *Pacific Journal of Math.*, **5** (1955) p. 321-324.

[CL] H. Cohen and H.W. Lenstra Jr., Heuristics on class groups of number fields, *Springer Lecture Notes*, **1068** in Number Theory Noordwijkerhout 1983 Proceedings.

[CM] D. Cardon and M. Ram Murty, Exponents of class groups of quadratic fields of function fields over finite fields, preprint.

[DH] H. Davenport and H. Heilbronn, On the density of discriminants of cubic fields II, *Proc. Royal Soc., A* **322** (1971) p. 405 - 420.

[ES] J.-H. Evertse and J.H. Silverman, Uniform bounds for the number of solutions to $Y^n = f(x)$, *Math. Proc. Camb. Phil. Soc.*, **100** (1986) p. 237-248.

[GM] R. Gupta and M. Ram Murty, Class groups of quadratic function fields, in preparation.

[Ha] P. Hartung, Proof of the existence of infinitely many imaginary quadratic fields whose class number is not divisible by 3, *J. Number Theory*, **6**(1974), 276-278.

[H] T. Honda, A few remarks on class numbers of imaginary quadratic fields, *Osaka J. Math.*, **12** (1975), 19-21.

[Ho] C. Hooley, Applications of sieve methods, *Cambridge Tracts in Mathematics*, 1976.

[KO] W. Kohnen and K. Ono, Indivisibility of class numbers of imaginary quadratic fields and orders of Tate-Shafarevich groups of elliptic curves with complex multiplication, to appear in *Inventiones Math.*

[MMS] M. Ram Murty, V. Kumar Murty and N. Saradha, Modular forms and the Chebotarev density theorem, *Amer. J. Math.*, **110** (1988), no. 2, p. 253-281.

[RM] M. Ram Murty, The ABC conjecture and exponents of quadratic fields, *Cont. Math.***210**, (1997) pp. 85-95, in Number Theory, edited by V. Kumar Murty and Michel Waldschmidt, Amer. Math. Soc., Providence.

[N] T. Nagell, Über die Klassenzahl imaginär quadratischer Zahlkörper, *Abh. Math. Seminar Univ. Hamburg*, **1** (1922) p. 140-150.

[O] K. Ono, Indivisibility of class numbers of real quadratic fields, to appear in *Compositio Math.*

[S] V.G. Sprindzuk, On the number of solutions of the Diophantine equation $x^3 = y^2 + A$ (in Russian), *Dokl. Akad. Nauk. BSSR* **7** (1963) p. 9-11.

[W] P. Weinberger, Real Quadratic Fields with Class Numbers Divisible by n, *Journal of Number Theory*, **5** (1973) p. 237-241.

[Y] Y. Yamamoto, On unramified Galois extensions of quadratic number fields, *Osaka J. Math.*, **7** (1970) 57-76.

[Yu] Jiu-Kang Yu, Toward the Cohen-Lenstra conjecture in the function field case, preprint.

Department of Mathematics, Queen's University, Kingston, Ontario, K7L 3N6, Canada

E-mail address: murty@mast.queensu.ca

A LOCAL-GLOBAL PRINCIPLE FOR DENSITIES

BJORN POONEN AND MICHAEL STOLL

ABSTRACT. This expository note describes a method for computing densities of subsets of $\mathbf{Z}^n$ described by infinitely many local conditions.

1. INTRODUCTION

The aim of this note is to present a general method for studying questions such as the following.

Fix $g \geq 1$. What is the 'probability' that a curve of the form

$$y^2 = f(x) = a_{2g+2}x^{2g+2} + a_{2g+1}x^{2g+1} + \cdots + a_2x^2 + a_1x + a_0 \tag{1}$$

with $a_j \in \mathbf{Z}$ has genus g and has $\mathbf{Q}_v$–rational points for all completions $\mathbf{Q}_v$ of $\mathbf{Q}$?

To make sense of this quesion, we have to make precise what we mean by 'probability.' We choose the coefficients a_j randomly from the integers of absolute value at most N and ask for the probability that the resulting curve has the property in question; then we take the limit of this as $N \to \infty$ and call the result a *density*.

Definition. For $v = (v_1, v_2, \ldots, v_d) \in \mathbf{Z}^d$, define $|v| := \max_i |v_i|$. If $S \subseteq \mathbf{Z}^d$, then the *density* of S is defined to be

$$\rho(S) := \lim_{N\to\infty} \frac{\#\{v \in S : |v| \leq N\}}{(2N+1)^d},$$

if the limit exists. Define the *upper density* $\overline{\rho}(S)$ and *lower density* $\underline{\rho}(S)$ similarly, except with the limit replaced by a lim sup or lim inf, respectively.

Note that in our question, the condition that (1) has genus g is equivalent to the non-vanishing of $\Delta(a_0, a_1, \ldots, a_{2g+2})$, the discriminant of f. We have chosen to exclude such curves for our density calculation, but this is of no consequence since the entire zero locus of Δ in $\mathbf{Z}^{2g+3}$ has density zero.

What makes our question difficult is that we are imposing conditions at infinitely many primes. If we wanted only an estimate for the density of curves (1) that had points over $\mathbf{R}$, $\mathbf{Q}_2$, and $\mathbf{Q}_{17}$, say, but required neither the existence nor the lack of points over the other $\mathbf{Q}_p$, then the question could easily be

Date: February 26, 1998.
Much of this research was done while the first author was supported by an NSF Mathematical Sciences Postdoctoral Research Fellowship at Princeton University.

S.D. Ahlgren et al. (eds.), Topics in Number Theory, 241-244

reduced to the computation of corresponding local probabilities, by invoking weak approximation to prove the 'independence' of the conditions being imposed.

We will show that our original question also can be reduced to the computation of local probabilities. Most of the proofs will be left out for reasons of space; they can be found in [PSt]. In that paper, the method is applied to prove results on the density of hyperelliptic curves whose Jacobians have a Shafarevich-Tate group of non-square order (if finite). But the method undoubtedly has many other interesting applications.

First we need some more notation. If S is a set, then 2^S denotes its power set. If K is a number field, let M_K denote the set of places of K. For example, $M_{\mathbf{Q}} = \{\infty\} \cup \{p : p \text{ prime}\}$. Finally, we let μ_∞ denote the standard Lebesgue measure on $\mathbf{R}^d$, and let μ_p denote the Haar measure on $\mathbf{Z}_p^d$ normalized to have total mass 1.

2. The results

We formalize the method for obtaining density results in the following lemma.

Lemma 1. *Suppose that U_∞ is a subset of $\mathbf{R}^d$ such that $\mathbf{R}^+ \cdot U_\infty = U_\infty$ and $\mu_\infty(\partial U_\infty) = 0$. Let $U_\infty^1 = U_\infty \cap [-1,1]^d$, and let $s_\infty = 2^{-d}\mu_\infty(U_\infty^1)$.*[1] *Suppose that for each finite prime p, U_p is a subset of $\mathbf{Z}_p^d$ such that $\mu_p(\partial U_p) = 0$. Let $s_p = \mu_p(U_p)$. Finally, suppose that*

$$\lim_{M\to\infty} \bar{\rho}\left(\{a \in \mathbf{Z}^d : a \in U_p \text{ for some finite prime } p > M\}\right) = 0. \tag{2}$$

Define a map $P : \mathbf{Z}^d \longrightarrow 2^{M_{\mathbf{Q}}}$ as follows: if $a \in \mathbf{Z}^d$, let $P(a)$ be the set of places v such that $a \in U_v$. Then

1. *$\sum_v s_v$ converges.*
2. *For $\mathfrak{S} \subseteq 2^{M_{\mathbf{Q}}}$, $\nu(\mathfrak{S}) := \rho(P^{-1}(\mathfrak{S}))$ exists, and ν defines a measure on $2^{M_{\mathbf{Q}}}$.*
3. *The measure ν is concentrated at the finite subsets of $M_{\mathbf{Q}}$: for each finite subset S of $M_{\mathbf{Q}}$,*

$$\nu(\{S\}) = \prod_{v\in S} s_v \prod_{v\notin S}(1 - s_v), \tag{3}$$

and if $\mathfrak{S} \subset 2^{M_{\mathbf{Q}}}$ consists of infinite subsets of $M_{\mathbf{Q}}$, then $\nu(\mathfrak{S}) = 0$.

Proof. See [PSt, Lemma 20]. □

For our original question, we will eventually take $d = 2g+3$ and let U_p (resp. U_∞) be the set of $(a_0, a_1, \ldots, a_{2g+2})$ with $a_i \in \mathbf{Z}_p$ (resp. $a_i \in \mathbf{R}$) such that the curve (1) has genus g and has *no* $\mathbf{Q}_p$-rational point (resp. no real point). Finally we will use (3) with $S = \emptyset$.

The main point of Lemma 1 is to isolate (2) as the non-trivial condition that must be checked in order to obtain density results with infinitely many local

[1] Since U_∞^1 is the union of the open set $(U_\infty^1)^0$ (its interior) and a subset of a measure zero set, U_∞^1 is automatically measurable.

conditions. The following result can be used to show that (2) is satisfied in many interesting cases.

Lemma 2. *Suppose f and g are relatively prime polynomials in $\mathbf{Z}[x_1, x_2, \dots, x_d]$. Let $S_M(f,g)$ be the set of $a \in \mathbf{Z}^d$ for which there exists a finite prime $p > M$ dividing both $f(a)$ and $g(a)$. Then* $\lim_{M\to\infty} \overline{\rho}(S_M(f,g)) = 0$.

Proof. See Section 3. □

Remark. Once one has Lemma 2, it is easy to apply Lemma 1 to obtain a formula for the density of $a \in \mathbf{Z}^d$ such that $f(a)$ and $g(a)$ are relatively prime, in terms of the number of solutions to $f(a) \equiv g(a) \equiv 0$ in $\mathbf{F}_p^d$ for each p. The same can of course be done for $\{a \in \mathbf{Z}^d : \gcd(f_1(a), \dots, f_n(a)) = 1\}$, provided that the polynomials $f_i \in \mathbf{Z}[x_1, \dots, x_d]$ define a subvariety of codimension at least 2 in $\mathbf{A}^d_{\mathbf{C}}$. This generalizes a result of Hafner, Sarnak, and McCurley [HSM].

For example, regarding the question we asked at the beginning, we obtain the following.

Lemma 3. *Fix $g \geq 1$. Let R_M be the set of $a = (a_0, a_1, \dots, a_{2g+2}) \in \mathbf{Z}^{2g+3}$ for which (1) is a curve X of genus g that fails to admit a $\mathbf{Q}_p$-rational point at some finite prime p greater than M. Then* $\lim_{M\to\infty} \overline{\rho}(R_M) = 0$.

Proof. An easy lemma [PSt, Lemma 15] shows that for p large compared to g, a necessary condition for the curve given by (1) to have no $\mathbf{Q}_p$-rational point is that the reduction of f mod p be a (non-square) constant times the square of some polynomial. If $g \geq 1$, then the Zariski closure[2] V of the image of the squaring map $\mathrm{Pol}_{g+1} \longrightarrow \mathrm{Pol}_{2g+2}$ (where Pol_n denotes the affine space of polynomials of degree $\leq n$) is of codimension at least 2 in $\mathrm{Pol}_{2g+2} = \mathbf{A}^{2g+3}$, so we can find two relatively prime polynomials $f, g \in \mathbf{Z}[a_0, \dots, a_{2g+2}]$ that vanish on V. For all but finitely many primes p, it is true that if $a \in \mathbf{Z}^{2g+3}$ and $f(x) \bmod p$ is a square in $\overline{\mathbf{F}}_p[x]$ then p divides $f(a)$ and $g(a)$. By Lemma 2, the claim follows. □

Lemma 3 supplies us with the condition (2) needed for the application of Lemma 1. We obtain the following answer to our question. Let ρ_g denote the density we asked for, and let $s_{g,v}$ be the s_v in Lemma 1 for the U_v (or U_∞) chosen in the paragraph after Lemma 1, so that $s_{g,v}$ is the 'probability' that the curve (1) with coefficients in $\mathbf{Z}_p$ (or $[-1,1]$ for ∞) has *no* $\mathbf{Q}_p$-rational point (resp. no real point). Then

$$\rho_g = \prod_{v \in M_{\mathbf{Q}}} (1 - s_{g,v}),$$

and the product converges. Furthermore it is easy to show that $0 < s_{g,v} < 1$ for all $g \geq 1$ and for all v, so $0 < \rho_g < 1$.

[2]In fact, it is easy to show that the image of the squaring map is already Zariski closed, but we do not need this.

Using Lemma 1, we could prove also that for any finite set S of places of $\mathbf{Q}$, and for any $g \geq 1$, there exists a genus g curve X over $\mathbf{Q}$ of the form (1) such that $\{v \in M_{\mathbf{Q}} : X(\mathbf{Q}_v) = \emptyset\} = S$ (and in fact, we would prove that such curves have positive density). A straightforward generalization of the method could be used to show that if K is any number field, S is any finite set of non-complex places of K, and $g \geq 1$, then there exists a genus g curve X over K of the form (1) such that $\{v \in M_K : X(K_v) = \emptyset\} = S$.

These last results fail for $g = 0$: it is well known that in addition $\#S$ must be even in order for there to exist X as above. The reason our method (luckily!) does not prove a false result for $g = 0$ is that (2) breaks down. More precisely, the proof of Lemma 3 fails for $g = 0$, since the image of the squaring map $\mathrm{Pol}_1 \to \mathrm{Pol}_2$ no longer has codimension at least 2.

3. Notes added in revision

We recently learned that T. Ekedahl developed in [Ek] very similar methods for computing densities when infinitely many local conditions are imposed. For instance, our Lemma 2 is a corollary of his Theorem 1.2, applied to the subscheme of $\mathbf{A}^d_{\mathbf{Z}}$ defined by the equations $f = g = 0$. Ekedahl gives applications of the method that are different from ours.

References

[Ek] Ekedahl, T., An infinite version of the Chinese remainder theorem, *Comment. Math. Univ. St. Paul.* **40** (1991), no. 1, 53–59.

[HSM] Hafner, J., Sarnak, P., and McCurley, K., Relatively prime values of polynomials, *A tribute to Emil Grosswald: number theory and related analysis*, 437–443, *Contemp. Math.* **143**, Amer. Math. Soc., Providence, RI, 1993.

[PSt] Poonen, B. and Stoll, M., The Cassels–Tate pairing on polarized abelian varieties, preprint, 1998.

Department of Mathematics, University of California, Berkeley, CA 94720-3840, USA
E-mail address: poonen@math.berkeley.edu

Mathematisches Institut, Universität Düsseldorf, Universitätsstr. 1, D–40225 Düsseldorf, Germany.
E-mail address: stoll@math.uni-duesseldorf.de

DIVISIBILITY OF THE SPECIALIZATION MAP FOR TWISTS OF ABELIAN VARIETIES

JOSEPH H. SILVERMAN

ABSTRACT. Let $A/\mathbb{Q}(T)$ be a non-trivial $\mathbb{Q}(T)/\mathbb{Q}$ cyclic twist of an abelian variety defined over $\mathbb{Q}$, and let $\Gamma \subset A(\mathbb{Q}(T))$ be an indivisible subgroup of $A(\mathbb{Q}(T))$. We show that for almost all (in the sense of density) $t \in \mathbf{Z}$, the specialization $\Gamma_t \subset A_t(\mathbb{Q})$ is indivisible in $A_t(\mathbb{Q})$, and similarly for $t \in \mathbb{Q}$.

§1. NOTATION AND STATEMENT OF THE MAIN RESULTS

Let $A/\mathbb{Q}(T)$ be an abelian variety, and let (B,τ) be a $\mathbb{Q}(T)/\mathbb{Q}$ trace of A. That is, B is the "largest" abelian variety defined over $\mathbb{Q}$ such that there is an inclusion $\tau : B \hookrightarrow A$ defined over $\mathbb{Q}(T)$. (For the precise definition of the trace in terms of universal mapping properties, see [6] or [7, chapter 6].) Let $\Gamma \subset A(\mathbb{Q}(T))$ be a (finitely generated) group which injects into $A(\mathbb{Q}(T))/\tau(B(\mathbb{Q}))$. Then for all but finitely many $t \in \mathbb{Q}$, we can specialize $T \mapsto t$ to obtain an abelian variety $A_t/\mathbb{Q}$ and a homomorphism

$$\sigma_t : \Gamma \longrightarrow \Gamma_t \subset A_t(\mathbb{Q}).$$

Néron [9] proved that σ_t is injective for almost all t in the sense of density, and the author [11] (when $B = 0$) and Lang [7, chapter 12, theorem 2.3] (in general) proved that in fact σ_t is injective for all but finitely many $t \in \mathbb{Q}$.

In this paper we study how the image Γ_t of Γ sits inside $A_t(\mathbb{Q})$. More precisely, we ask to what extent Γ_t is divisible in $A_t(\mathbb{Q})$. Let

$$\Gamma_t^{\text{div}} = \{z \in A_t(\mathbb{Q}) \ : \ mz \in \Gamma_t \text{ for some } m \geq 1\}.$$

We define the ***divisibility index of*** Γ ***at*** t to be

$$I_t(\Gamma) = [\Gamma_t^{\text{div}} : \Gamma_t].$$

The following theorem summarizes what is currently known about this index.

1991 *Mathematics Subject Classification*. 11G10 14K15.
Key words and phrases. abelian variety, specialization, divisibility.
Research partially supported by NSF DMS-9424642.

S.D. Ahlgren et al. (eds.), Topics in Number Theory, 245–258.

Theorem 1.1. *Suppose that A is an elliptic curve (i.e., $\dim(A) = 1$), and assume that one of the following two conditions is true:*
(i) $j(A) \notin \mathbb{Q}$.
(ii) $j(A) \in \mathbb{Q}$, $B = 0$ *(i.e., A does not split), and Γ is free of rank 1.*
Then there exists a bound $M = M(A, \Gamma)$ such that the set

$$\{t \in \mathbb{Z} : I_t(\Gamma) \geq M\}$$

has density zero in $\mathbb{Z}$.

Part (i) of Theorem 1.1 is due to the author [13], and part (ii) is due to Gupta and Ramsay [4]. A consequence of Theorem 1.1 is that if Γ is not divisible in $A(\mathbb{Q}(T))$, then Γ_t is not divisible in $A_t(\mathbb{Q})$ for almost all (in the sense of density) $t \in \mathbb{Z}$.

In this paper we prove a generalization of the result of Gupta and Ramsay. We set some notation.

- $\mathcal{A}/\mathbb{Q}$ a fixed abelian variety.
- μ_n $\subset \operatorname{Aut}(\mathcal{A})$, an inclusion, as Galois modules, of the nth-roots of unity into the automorphism group of $\mathcal{A}$. Note that at least for $n = 2$, every abelian variety admits such an inclusion.
- $f(T)$ $\in \mathbb{Q}(T)$, a non-constant rational function with the property that $f(T)$ is not in $\mathbb{Q}^*\left(\mathbb{Q}(T)^*\right)^n$. In other words, $f(T)$ cannot be written in the form $ag(T)^n$ with $a \in \mathbb{Q}^*$ and $g(T) \in \mathbb{Q}(T)$.
- $A/\mathbb{Q}(T)$ the n-twist of $\mathcal{A}$ by f. More precisely, let χ be the image of f under the composition of maps

$$\mathbb{Q}(T)^* \longrightarrow \mathbb{Q}(T)^*/\mathbb{Q}(T)^{*n} \cong H^1\left(\operatorname{Gal}\left(\overline{\mathbb{Q}(T)}/\mathbb{Q}(T)\right), \mu_n\right) \longrightarrow H^1\left(\operatorname{Gal}\left(\overline{\mathbb{Q}(T)}/\mathbb{Q}(T)\right), \operatorname{Aut}(\mathcal{A})\right).$$

  Then the cohomology class χ determines an abelian variety $A/\mathbb{Q}(T)$ which is isomorphic, over $\overline{\mathbb{Q}(T)}$, to $\mathcal{A}$. (See [14, X §2] for basic properties of twists.)
- $\sigma_0(n)$ the number of divisors of n. That is, $\sigma_0(n) = \sum_{d|n} 1$.

Example. Let $F(X) \in \mathbb{Q}[X]$, let C be a smooth projective model for the affine curve

$$Y^n = F(X),$$

and let J be the Jacobian variety of C. (We assume that C is geometrically irreducible.) The natural inclusion

$$\mu_n \hookrightarrow \operatorname{Aut}(C), \qquad \zeta \mapsto \left((X, Y) \mapsto (X, \zeta Y)\right)$$

induces an inclusion $\mu_n \hookrightarrow \operatorname{Aut}(J)$. Then the n-twist of J by $f(T) \in \mathbb{Q}(T)$ is the Jacobian of the twisted curve

$$f(T)Y^n = F(X).$$

As a special case, elliptic curves admit three essentially different sorts of twists:

Equation of E	j-invariant	n	Equation of Twist
$Y^2 = X^3 + AX + B$	arbitrary	2	$f(T)Y^2 = X^3 + AX + B$
$Y^2 = X^3 + AX$	$j = 1728$	4	$Y^2 = X^3 + f(T)AX$
$Y^2 = X^3 + B$	$j = 0$	6	$Y^2 = X^3 + f(T)B$

These elliptic curves are the twists studied by Gupta and Ramsay [4].

Remark. Suppose that $f(T)$ is in $\mathbb{Q}^* \left(\mathbb{Q}(T)^*\right)^m$ for some $m \geq 2$ dividing n. In other words, $f(T)$ can be written in the form $ag(T)^m$ with $a \in \mathbb{Q}^*$, $g(T) \in \mathbb{Q}(T)$, $m|n$, and $m \geq 2$. Let $\mathcal{A}^{(a)}/\mathbb{Q}$ be the n-twist of $\mathcal{A}$ by a. (N.B. $\mathcal{A}^{(a)}$ is defined over $\mathbb{Q}$.) Then A is the n/m-twist of $\mathcal{A}^{(a)}$ by $g(T)$. In other words, A is actually a twist by the lower order n/m. So we lose no generality by assuming that f cannot be written in the form $ag(T)^m$.

Theorem 1.2. *In the notation above, assume further that $f(T) \notin \mathbb{Q}^* \left(\mathbb{Q}(T)^*\right)^m$ for all $m \geq 2$ dividing n. Then for all $\varepsilon > 0$ there are constants $M = M(A, \Gamma, \varepsilon)$ and $c = c(A, \Gamma, \varepsilon)$ so that for all $B \geq 1$:*

(a) $$\#\{t \in \mathbb{Z} : |t| \leq B \text{ and } I_t(\Gamma) \geq M\} \leq cB^{1/\sigma_0(n)+\varepsilon}.$$

(b) $$\#\{t \in \mathbb{Q} : H(t) \leq B \text{ and } I_t(\Gamma) \geq M\} \leq cB^{2/\sigma_0(n)+\varepsilon}.$$

(Here $H(t)$ is the height of t, defined for a rational number a/b written in lowest terms as $H(a/b) = \max\{|a|, |b|\}$.)

Remark. Since

$$\#\{t \in \mathbb{Z} : |t| \leq B\} \gg\ll B \quad \text{and} \quad \#\{t \in \mathbb{Q} : H(t) \leq B\} \gg\ll B^2,$$

the estimates in Theorem 1.2 imply that $I_t(\Gamma) < M$ for almost all t in the sense of density, provided (of course) that $n \geq 2$. For composite values of n, the exponent $1/\sigma_0(n)$ strengthens the exponent $(1 + (n-1)/\phi(n))^{-1}$ obtained by Gupta and Ramsay [4].

Theorem 1.2 bounds the divisibility of Γ_t in $A_t(\mathbb{Q})$. Let

$$\Gamma^{\text{div}} = \{P \in A(\mathbb{Q}(T)) : mP \in \Gamma \text{ for some } m \geq 1\}.$$

Then the specialization map σ_t clearly sends Γ^{div} into Γ_t^{div}. As an immediate consequence of Theorem 1.2 and Néron's work, we can show that they are equal for almost all t. In particular, if Γ is indivisible in $A(\mathbb{Q}(T))$, then Γ_t is almost always indivisible in $A_t(\mathbb{Q})$. For example, if we take $\Gamma = A(\mathbb{Q}(T))$, then we see that $A(\mathbb{Q}(T)) \cong A_t(\mathbb{Q})$ for almost all t satisfying $\operatorname{rank} A(\mathbb{Q}(T)) = \operatorname{rank} A_t(\mathbb{Q})$.

Corollary 1.3. *With assumptions as in Theorem 1.2, the sets*

$$\{t \in \mathbb{Z} : \Gamma^{\text{div}} \ncong \Gamma_t^{\text{div}}\} \quad \text{and} \quad \{t \in \mathbb{Q} : \Gamma^{\text{div}} \ncong \Gamma_t^{\text{div}}\}$$

have density 0 in $\mathbb{Z}$ and $\mathbb{Q}$ respectively.

§2. Power-free part of polynomial values

The following version of the large sieve is due, in essentially this form, to Gallagher [2]. (See also [5, chap. 1, sec. 4].)

Proposition 2.1. (Gallagher) *Let $S \subset \mathbb{Q} \subset \mathbb{P}^1(\mathbb{Q})$ be a finite set of rational numbers, and let T be a set of primes. Choose M to satisfy*

$$M \geq \max\left\{ab' - a'b : \frac{a}{b}, \frac{a'}{b'} \in S\right\},$$

where we write all fractions in lowest terms. Further, for each $p \in T$, fix an r_p satisfying

$$r_p \geq \#(\tilde{S} \bmod p),$$

where $\tilde{S} \bmod p$ denotes the reduction of S in $\mathbb{P}^1(\mathbb{F}_p)$. Then

$$\#S \leq \frac{\sum_{p\in T} \log p - \log M}{\sum_{p\in T} \frac{\log p}{r_p} - \log M}$$

provided that the denominator is positive.

Proof. For any prime p and any $\beta \in \mathbb{P}^1(\mathbb{F}_p)$, let

$$S(\beta,p) = \{\alpha \in S : \alpha \equiv \beta \pmod{p}\}.$$

Then for any $p \in T$ we have

$$\begin{aligned} |S|^2 &= \Big(\sum_{\beta\in\mathbf{P}^1(\mathbf{F}_p)} |S(\beta,p)|\Big)^2 \\ &\leq r_p \sum_{\beta\in\mathbf{P}^1(\mathbf{F}_p)} |S(\beta,p)|^2 \quad \text{(Cauchy-Schwarz).} \end{aligned}$$

Hence

$$\frac{|S|^2}{r_p} \leq \sum_{\beta\in\mathbf{P}^1(\mathbf{F}_p)} |S(\beta,p)|^2 = \sum_{\substack{(\alpha,\alpha')\in S\times S \\ \alpha\equiv\alpha' \pmod{p}}} 1 = |S| + \sum_{\substack{(\alpha,\alpha')\in S\times S \\ \alpha\equiv\alpha' \pmod{p} \\ \alpha\neq\alpha'}} 1.$$

Now multiply both sides by $\log p$ and sum over $p \in T$. This gives

$$|S|^2 \sum_{p\in T} \frac{\log p}{r_p} \leq |S| \sum_{p\in T} \log p + \sum_{\substack{(\alpha,\alpha')\in S\times S \\ \alpha\neq\alpha'}} \sum_{\substack{p\in T \\ \alpha\equiv\alpha' \pmod{p}}} \log p.$$

Now we observe that if $\alpha = a/b$ and $\alpha' = a'/b'$ are distinct elements of S written in lowest terms, then the innermost sum satisfies

$$\sum_{\substack{p \in T \\ \alpha \equiv \alpha' \pmod{p}}} \log p \le \sum_{p \mid ab' - a'b} \log p \le \log |ab' - a'b| \le \log M.$$

Hence

$$|S|^2 \sum_{p \in T} \frac{\log p}{r_p} \le |S| \sum_{p \in T} \log p + \left(|S|^2 - |S|\right) \log M.$$

This is the desired inequality.

Using ideas of Hooley [5, pp. 66–68], Gupta and Ramsay [4] give an estimate for the number of integers x such that the value of a squarefree polynomial f has a sizable n^{th}-power free part. The next proposition improves their estimate and extends it to the case that f is not squarefree and x is allowed to assume rational values.

For any non-zero rational number α, we write

$$\nu_n(\alpha) = \min\{b \in \mathbb{Z}, b \ge 1 : \alpha = b\gamma^n \text{ for some } \gamma \in \mathbb{Q}\},$$

and we call $\nu_n(\alpha)$ the nth-power free part of α.

Proposition 2.2. *Let $f(X) \in \mathbb{Q}(X)$ be a non-constant rational function, and assume that $f(X)$ is not an mth-power in $\mathbb{Q}(X)$ for any $m > 1$ dividing n.*
(a) *For any $\varepsilon > 0$, we have*

$$\#\{a \in \mathbb{Z} : |a| \le B \text{ and } \nu_n(f(a)) \le B^{\varepsilon}\} \ll B^{1/\sigma_0(n)+\varepsilon} \log B.$$

(b) *For any $\varepsilon > 0$, we have*

$$\#\{\alpha \in \mathbb{Q} : H(\alpha) \le B \text{ and } \nu_n(f(\alpha)) \le B^{\varepsilon}\} \ll B^{2/\sigma_0(n)+\varepsilon} \log B.$$

(We recall that $\sigma_0(n) = \sum_{d \mid n} 1$ is the number of divisors of n.)

Remark. If $f(X)$ is an mth-power for some $m \mid n$, then one can use the trivial relation

$$\nu_n(A^m) = \nu_{n/m}(A)^m$$

to deduce analogous bounds. For example, (a) becomes

$$\#\{a \in \mathbb{Z} : |a| \le B \text{ and } \nu_n(f(a)) \le B^{\varepsilon}\} \ll B^{1/\sigma_0(n/m)+\varepsilon/m} (\log B)^{1/m},$$

and similarly for (b). Notice that the bounds are non-trivial if and only if $m < n$.

Proof. We are free to replace f by fh^n for any $h \in \mathbb{Q}(X)^*$, so we may assume that $f \in \mathbb{Z}[X]$ is a polynomial with integer coefficients.

For any non-zero $b \in \mathbb{Z}$, let $C_b \subset \mathbb{P}^2$ be the (possibly singular) projective curve given by the affine equation

$$C_b : bY^n = f(X).$$

Our assumption that $f(X)$ is not an mth-power for $m > 1$ dividing n implies that C_b is (geometrically) irreducible over $\mathbb{Q}$. Later we will need to consider the irreducibility of C_b over finite fields $\mathbb{F}_p$. For this purpose, we define a ***reduced discriminant*** $D(f)$ of $f(X)$ as follows. Factor $f(X)$ in $\bar{\mathbb{Q}}[X]$ as

$$f(X) = a \prod (X - \alpha_i)^{e_i},$$

where the α_i's are distinct. Then

$$D(f) = \prod_{i \neq j} (a\alpha_i - a\alpha_j).$$

Notice that $D(f) \in \mathbb{Z}$, and that if $p \nmid D(f)$, then the factorization of f over $\bar{\mathbb{F}}_p$ mirrors the factorization of f over $\bar{\mathbb{Q}}$. In particular, if $p \nmid D(f)$, then $\tilde{f} \bmod p$ is not an mth-power in $\mathbb{F}_p[X]$ for any $m > 1$ dividing n. Hence if $p \nmid bD(f)$, then $\tilde{C}_b \bmod p$ is geometrically irreducible over $\mathbb{F}_p$.

We let

$$\text{proj}_1 : C_b \longrightarrow \mathbb{P}^1, \qquad (X, Y) \longmapsto X,$$

be the projection onto the first factor. Our immediate goal is to bound the size of $\text{proj}_1(\tilde{C}_b(\mathbb{F}_p))$. We need to consider two cases.

First, if $p | bD(f)$, we use the trivial estimate

$$\# \text{proj}_1(\tilde{C}_b(\mathbb{F}_p)) \leq \# \mathbb{P}^1(\mathbb{F}_p) = p + 1.$$

Second, if $p \nmid bD(f)$, then as noted above, the curve $\tilde{C}_b \bmod p$ is geometrically irreducible. Weil's estimate tells us that

$$\# \tilde{C}_b(\mathbb{F}_p) \leq p + c_1 \sqrt{p},$$

where the constant c_1 is independent of p. Now consider a point $(x, y) \in \tilde{C}_b(\mathbb{F}_p)$. Let $d = \gcd(n, p-1)$. Then $\mathbb{F}_p^*$ contains a primitive d^{th} root of unity ξ, so $\tilde{C}_b(\mathbb{F}_p)$ contains the points

$$(x, y), (x, \xi y), (x, \xi^2 y), \ldots, (x, \xi^{d-1} y),$$

and these points are distinct unless $y = 0$. Since C_b contains at most $\deg(f) = O(1)$ points with $y = 0$, we conclude in this case that

$$\# \text{proj}_1(\tilde{C}_b(\mathbb{F}_p)) \leq \frac{p + c_2 \sqrt{p}}{\gcd(n, p-1)}.$$

We have now shown that

$$\#\operatorname{proj}_1(\tilde{C}_b(\mathbb{F}_p)) \le r_p \stackrel{\text{def}}{=} \begin{cases} p+1 & \text{if } p|bD(f), \\ (p+c_2\sqrt{p})/\gcd(n,p-1) & \text{otherwise.} \end{cases}$$

Let

$$T = T(A) = \{\text{primes } p \text{ such that } p \le A\}.$$

(We will choose A later.) We want to estimate the sum $\sum_{p\in T}(\log p)/r_p$. The fact that

$$\sum_p \left(\frac{\log p}{p+c_3\sqrt{p}} - \frac{\log p}{p}\right)$$

converges for any $c_3 \ge 0$ allows some simplification, as does the fact that we are allowing our constants to depend on f. Hence

$$\begin{aligned} \sum_{p\in T}\frac{\log p}{r_p} &\ge \sum_{\substack{p\le A \\ p\nmid bD(f)}} \frac{\log p}{p/\gcd(n,p-1)+c_2\sqrt{p}} + \sum_{\substack{p\le A \\ p|bD(f)}} \frac{\log p}{p+1} \\ &\ge \sum_{p\le A} \gcd(n,p-1)\cdot\frac{\log p}{p} - \sum_{p|b}\gcd(n,p-1)\frac{\log p}{p+1} + O(1) \\ &\ge \sigma_0(n)\log A - \sigma_0(n)\log\log b + O(1), \end{aligned}$$

where the last line is a consequence of the following lemma.

Lemma 2.3. *Let $n, A, b \ge 1$. Then*

(a) $$\sum_{p\le A}\gcd(n,p-1)\cdot\frac{\log p}{p} = \sigma_0(n)\bigl(\log A + O(1)\bigr).$$

(b) $$\sum_{p|b}\gcd(n,p-1)\cdot\frac{\log p}{p} \le \sigma_0(n)\bigl(\log\log b + O(1)\bigr).$$

Proof. We begin with an elementary calculation, where ϕ and μ are Euler's and Mobius' functions respectively:

$$\sum_{\substack{d,e \\ de|n}} \frac{d\mu(e)}{\phi(de)} = \sum_{k|n}\frac{1}{\phi(k)}\sum_{d|k} d\mu\left(\frac{k}{d}\right) = \sum_{k|n}\frac{1}{\phi(k)}\cdot\phi(k) = \sigma_0(n).$$

We also require the estimates

$$\sum_{\substack{p\le A \\ p\equiv 1 \pmod N}} \frac{\log p}{p} = \frac{\log A}{\phi(N)} + O(1) \quad\text{and}\quad \sum_{\substack{p|b \\ p\equiv 1 \pmod N}} \frac{\log p}{p} \le \frac{\log\log b}{\phi(N)} + O(1).$$

The first is a consquence of Dirichlet's theorem on primes in an arithmetic progression, and the second follows from the same theorem and a standard calculation (cf. [8]).

To prove part (a) of Lemma 2.3, we compute

$$\begin{aligned}
\sum_{p\le A} \gcd(n,p-1)\cdot\frac{\log p}{p} &= \sum_{d\mid n} d \sum_{\substack{p\le A\\ \gcd(n,p-1)=d}} \frac{\log p}{p}\\
&= \sum_{d\mid n} d \sum_{\substack{p\le A\\ p\equiv 1 \pmod d}} \frac{\log p}{p} \sum_{\substack{e\mid n/d\\ e\mid (p-1)/d}} \mu(e)\\
&= \sum_{d\mid n} d \sum_{e\mid n/d} \mu(e) \sum_{\substack{p\le A\\ p\equiv 1 \pmod{de}}} \frac{\log p}{p}\\
&= \sum_{\substack{d,e\\ de\mid n}} d\mu(e)\left\{\frac{\log A}{\phi(de)} + O(1)\right\}\\
&= \left(\sum_{\substack{d,e\\ de\mid n}} \frac{d\mu(e)}{\phi(de)}\right)(\log A + O(1))\\
&= \sigma_0(n)\bigl(\log A + O(1)\bigr).
\end{aligned}$$

Lemma 2.3(b) is proven similarly:

$$\begin{aligned}
\sum_{p\mid b} \gcd(n,p-1)\cdot\frac{\log p}{p} &= \sum_{d\mid n} d \sum_{\substack{p\mid b\\ \gcd(n,p-1)=d}} \frac{\log p}{p}\\
&= \sum_{\substack{d,e\\ de\mid n}} d\mu(e) \sum_{\substack{p\mid b\\ p\equiv 1 \pmod{de}}} \frac{\log p}{p}\\
&\le \sum_{\substack{d,e\\ de\mid n}} d\mu(e)\left\{\frac{\log\log b}{\phi(de)} + O(1)\right\}\\
&= \sigma_0(n)\bigl(\log\log b + O(1)\bigr).
\end{aligned}$$

This completes the proof of Lemma 2.3.

We now resume the proof of Proposition 2.2. We will prove (a) and (b) simultaneously, so we let R denote either $\mathbb{Z}$ or $\mathbb{Q}$. For any integer $b\ge 1$, let

$$S(b) = \{\alpha \in R : H(\alpha) \le B \text{ and } \nu_n(f(\alpha)) = b\}.$$

If $\alpha \in S(b)$, then there is a rational number γ such that $f(\alpha) = b\gamma^n$. In other words, $(\alpha, \gamma) \in C_b(\mathbb{Q})$, so

$$S(b) \subset \mathrm{proj}_1(C_b(\mathbb{Q})).$$

It follows that

$$\#\widetilde{S}(b) \bmod p \leq \#\,\mathrm{proj}_1(\tilde{C}_b(\mathbb{F}_p)) \leq r_p,$$

where r_p is as above.

Next we observe that

$$M \stackrel{\text{def}}{=} \begin{cases} 2B = \max\{a - a' \,:\, a, a' \in S(b)\} & \text{if } R = \mathbb{Z}, \\ B^2 \geq \max\{ad' - a'd \,:\, a/d, a'/d' \in S(b)\} & \text{if } R = \mathbb{Q}. \end{cases}$$

In order to continue treating the two cases $R = \mathbb{Z}$ and $R = \mathbb{Q}$ together, we set

$$k = 1 \text{ if } R = \mathbb{Z}, \quad \text{and} \quad k = 2 \text{ if } R = \mathbb{Q},$$

so in both cases we have $M \gg\ll B^k$.

We want to apply the large sieve (Proposition 2.1), so we need to choose A so that $\sum_{p \in T} (\log p)/r_p$ is greater than (say) $1 + \log M$. We can ensure that this is true by choosing

$$A = c_4 B^{k/\sigma_0(n)} \log b$$

for an appropriate constant c_4 (depending on f and n, but independent of b). For this choice of A, Proposition 2.1 yields the upper bound

$$\#S(b) \leq \sum_{p \leq A} \log p - \log M \leq c_5 A \leq c_6 B^{k/\sigma_0(n)} \log b.$$

Finally, we compute

$$\begin{aligned} \#\{\alpha \in R \,:\, H(\alpha) \leq B \text{ and } \nu_n(f(\alpha)) \leq B^\varepsilon\} &= \sum_{1 \leq b \leq B^\varepsilon} \#S(b) \\ &\leq \sum_{1 \leq b \leq B^\varepsilon} c_6 B^{k/\sigma_0(n)} \log b \\ &\leq c_8 B^{k/\sigma_0(n) + \varepsilon} \log B. \end{aligned}$$

§3. PROOF THE MAIN RESULTS

In this section we prove Theorem 1.2 and Corollary 1.3. The proof of Theorem 1.2 relies heavily on the theory of canonical heights. For basic properties concerning heights and canonical heights, see [7].

We fix an ample symmetric divisor class $D \in \mathrm{Pic}(A)$ and let

$$\langle \cdot, \cdot \rangle : A(\mathbb{Q}(T)) \times A(\mathbb{Q}(T)) \longrightarrow \mathbb{R}$$

be the associated canonical height pairing on $A/\mathbb{Q}(T)$. For each $t \in \mathbb{Q}$ such that A_t is smooth, we can specialize D to get an ample divisor class $D_t \in \mathrm{Pic}(A_t)$ and use it to define a canonical height pairing

$$\langle \cdot, \cdot \rangle_t : A_t(\mathbb{Q}) \times A_t(\mathbb{Q}) \longrightarrow \mathbb{R}.$$

These pairings induce Euclidean metrics on $A(\mathbb{Q}(T)) \otimes \mathbb{R}$, respectively on $A_t(\mathbb{Q}) \otimes \mathbb{R}$, see [7, chapter 5]. The following result relates them to one another.

Proposition 3.1. (Silverman [11]) *Let $P, Q \in A(\mathbb{Q}(T))$. Then with notation as above,*

$$\lim_{\substack{t\in\mathbb{Q}\\ h(t)\to\infty}} \frac{\langle\sigma_t(P),\sigma_t(Q)\rangle}{h(t)} = \langle P,Q\rangle,$$

where $h(t) = \log H(t)$ is the logarithmic height of t.

Proof. See [7, chapter 12, section 2] or [11, theorem B].

Remark. It is an immediate consequence of Proposition 3.1 and the positive definiteness of the canonical height pairings that the specialization map $\sigma_t : A(\mathbb{Q}(T)) \to A_t(\mathbb{Q})$ is injective for all but finitely many $t \in \mathbb{Q}$. See [11].

Remark. There are stronger versions of Proposition 3.1 of the form

$$\langle\sigma_t(P),\sigma_t(Q)\rangle = \langle P,Q\rangle h(t) + (\text{explicit error term})$$

due to Call [1], Green [3], and Tate [15], but Theorem 3.1 will suffice for our purposes.

The other main tool we need for the proof of Theorem 1.2 is a lower bound for the canonical height on $A_t(\mathbb{Q})$. We obtain the desired bound by combining the sieve estimate in Proposition 2.2 with the following result.

Proposition 3.2. (Silverman [12]) *With notation and assumptions as in Theorem 1.2, there are constants $c_9 > 0$ and c_{10} so that for all $t \in \mathbb{Q}$ with A_t smooth and all points $z \in A_t(\mathbb{Q})$, either*

(1) $nz = 0$, *or*

(2) $\langle z,z\rangle_t \ge c_9 \log\left|\operatorname{Disc}\left(\mathbb{Q}(\sqrt[n]{f(t)})/\mathbb{Q}\right)\right| - c_{10}$.

Proof. This is an immediate consequence of [12, theorem 6], once we observe that $\mathbb{Q}(\sqrt[n]{f(t)})$ is the splitting field of the cohomology class χ corresponding to f (see section 1 for the definition of χ).

We also use the following elementary lemma concerning Kummer extensions.

Lemma 3.3. *For all $\alpha \in \mathbb{Q}^*$,*

$$\log\left|\operatorname{Disc}\left(\mathbb{Q}(\sqrt[n]{\alpha})/\mathbb{Q}\right)\right| \ge \log \nu_n(\alpha),$$

where $\nu_n(\alpha)$ is the nth-power free part of α as defined in section 2.

Proof. If p is a prime with $\operatorname{ord}_p(\alpha) \not\equiv 0 \pmod{n}$, then $\mathbb{Q}(\sqrt[n]{\alpha})$ is ramified at p, and in fact

$$\operatorname{ord}_p\left(\operatorname{Disc}\left(\mathbb{Q}(\sqrt[n]{\alpha})/\mathbb{Q}\right)\right) \ge n-1 \tag{1}$$

for such primes. Let $b = \nu_n(\alpha)$. Thus $\alpha = b\gamma^n$ for some $\gamma \in \mathbb{Q}$, and $b \ge 1$ is the smallest integer with this property, so in particular, $0 \le \operatorname{ord}_p(b) < n$ for all primes p. This means that (1) is true for all $p|b$, so

$$\begin{aligned}\log\left|\operatorname{Disc}(\mathbb{Q}(\sqrt[n]{\alpha})/\mathbb{Q})\right| &= \sum_p \operatorname{ord}_p\left(\operatorname{Disc}(\mathbb{Q}(\sqrt[n]{\alpha})/\mathbb{Q})\right)\log p \\ &\geq \sum_{p|b}(n-1)\log p \\ &\geq \log b.\end{aligned}$$

We now have all of the tools needed to prove our main result.

Proof of Theorem 1.2. As in the proof of Proposition 2.2, we let R denote either $\mathbb{Z}$ or $\mathbb{Q}$. For any $B, \varepsilon > 0$, we let

$$V_\varepsilon(B) = \{t \in R : H(t) \leq B \text{ and } \nu_n(f(t)) \leq B^\varepsilon\},$$

and we let

$$U_\varepsilon(B) = \{t \in R : H(t) \leq B \text{ and } t \notin V_\varepsilon(B)\}$$

be the complement of $V_\varepsilon(B)$ in the set of $t \in R$ such that $H(t) \leq B$. Notice that Proposition 2.2 tells us that

$$\#V_\varepsilon(B) \ll B^{\kappa/\sigma_0(n)+\varepsilon}\log B,$$

where $\kappa = 1$ or 2 according to whether $R = \mathbb{Z}$ or $\mathbb{Q}$.

Now suppose that $t \in U_\varepsilon(B)$. Then $\nu_n(f(t)) > B^\varepsilon \geq H(t)^\varepsilon$, so Lemma 3.3 tells us that

$$\log\left|\operatorname{Disc}(\mathbb{Q}(\sqrt[n]{f(t)})/\mathbb{Q})\right| \geq \log \nu_n(f(t)) \geq \varepsilon h(t), \tag{2}$$

and then Proposition 3.2 says that for all but finitely many $t \in R$, every $z \in A_t(\mathbb{Q})$ satisfies either

$$nz = 0 \qquad \text{or} \qquad \langle z, z\rangle_t \geq c_{11}h(t),$$

where $c_{11} > 0$ is independent of t and z. In particular, $A_t(\mathbb{Q})_{\mathrm{tors}}$ is killed by n, so the torsion subgroup of Γ_t is bounded independently of t. Hence discarding finitely many values of t, we may assume without loss of generality that Γ is free and that $\Gamma \cong \Gamma_t$ for all t's under consideration.

Let $P_1, \ldots, P_r \in A(\mathbb{Q}(T))$ be a basis for Γ, let $t \in U_\varepsilon(B)$ have height $h(t) \geq c_{12}$, and let $z_1, \ldots, z_r \in A_t(\mathbb{Q})$ be a basis for Γ_t^{div}. Then the index $I_t(\Gamma)$ of Γ_t in Γ_t^{div} can be computed in terms of the covolumes of the lattices Γ_t and Γ_t^{div} in the Euclidean space $A_t(\mathbb{Q}) \otimes \mathbb{R}$ relative to the canonical height pairing. Thus

$$I_t(\Gamma)^2 = \frac{\det(\langle\sigma_t(P_i), \sigma_t(P_j)\rangle_t)_{1\leq i,j\leq r}}{\det(\langle z_i, z_j\rangle_t)_{1\leq i,j\leq r}}. \tag{3}$$

We can compute the numerator using Proposition 3.1, which gives

$$\det(\langle\sigma_t(P_i),\sigma_t(P_j)\rangle_t)_{1\le i,j\le r} = \det(\langle P_i,P_j\rangle)_{1\le i,j\le r}\cdot h(t)^r + o(h(t)^r),$$

where the regulator $\det(\langle P_i,P_j\rangle)$ is non-zero because the height pairing is positive definite on $A(\mathbb{Q}(T))$ and the P_i's are independent. In particular

$$\det(\langle\sigma_t(P_i),\sigma_t(P_j)\rangle_t)_{1\le i,j\le r} \le c_{13}h(t)^r \qquad \text{for all } t\in\mathbb{Q} \text{ with } h(t)\ge c_{14}. \quad (4)$$

As for the denominator, we use the elementary fact from the geometry of numbers that the covolume of any r-dimensional lattice can be bounded below in terms of the rth-power of the shortest lattice vector. See, e.g., [7, chapter 5, corollary 7.8]. In our situation we find that

$$\det(\langle z_i,z_j\rangle_t)_{1\le i,j\le r} \ge c_{18} \min_{\substack{z\in A_t(\mathbb{Q})\\ \langle z,z\rangle_t\ne 0}} \langle z,z\rangle_t^r. \quad (5)$$

(It is not hard to give an explicit value for c_{18} in terms of r, but it is a very hard problem to give sharp values.) We now combine (5), Proposition 3.2, and (2) to obtain

$$\det(\langle z_i,z_j\rangle_t)_{1\le i,j\le r} \ge c_{19}h(t)^r, \quad (6)$$

valid for all $t\in U_\varepsilon(B)$ with $h(t)\ge c_{20}$.

Substituting (4) and (6) into (3) yields

$$I_t(\Gamma)^2 \le c_{21} \qquad \text{for all } t\in U_t(B) \text{ with } h(t)\ge c_{22}.$$

So if we take $M = 1+\sqrt{c_{21}}$, then

$$\begin{aligned}\#\{t\in R : H(t)\le B \text{ and } I_t(\Gamma)\ge M\} &\le \#V_t(B) + \#\{t\in R : h(t)\le c_{22}\}\\ &\le c_{25}B^{k/\sigma_0(n)+\varepsilon}\log B + c_{26},\end{aligned}$$

where $k=1$ or 2 depending on whether $R=\mathbb{Z}$ or $R=\mathbb{Q}$. This concludes the proof of Theorem 1.2.

Remark. It is possible to find explicit constants, in terms of r, n, ε, $\#\Gamma_{\text{tors}}$, and the height regulator $\det\langle P_i,P_j\rangle$ of a basis for the free part of Γ, so that the estimates in Theorem 1.2 hold for all sufficiently large B.

Proof of Corollary 1.3. As usual, we let R denote either $\mathbb{Z}$ or $\mathbb{Q}$. Now suppose that $t\in R$ satisfies $\Gamma^{\text{div}} \not\cong \Gamma_t^{\text{div}}$. Of course, what we really mean by this is that $(\Gamma^{\text{div}})_t \ne \Gamma_t^{\text{div}}$. Thus there is some $z\in A_t(\mathbb{Q})$ and some $m\ge 2$ so that $mz\in\Gamma_t$ and $mz\notin(\Gamma^{\text{div}})_t$. Clearly the smallest such m satisfies

$$m\le(\Gamma_t^{\text{div}} : (\Gamma^{\text{div}})_t) \le (\Gamma_t^{\text{div}} : \Gamma_t) = I_t(\Gamma).$$

Let M be the constant described in Theorem 1.2. Then

$$\{t \in R(B) : \Gamma^{\mathrm{div}} \not\cong \Gamma_t^{\mathrm{div}}\}$$
$$\subset \bigcup_{1<m<M} \left\{ t \in R(B) : \begin{array}{c} \text{there exists a } z \in A_t(\mathbb{Q}) \\ \text{such that } mz \in \Gamma_t \\ \text{and } mz \notin (\Gamma^{\mathrm{div}})_t \end{array} \right\} \cup \{t \in R(B) : I_t(\Gamma) \geq M\},$$

and Theorem 1.2 tells us that the rightmost set has density 0 in R. It remains to deal individually with each of the sets in the finite union. The argument that we give, based on the Hilbert Irreducibility Theorem, is due to Néron [9].

Fix an integer $m \geq 2$. We need to show that the set

$$\{t \in R : \Gamma_t^{\mathrm{div}}/(\Gamma^{\mathrm{div}})_t \text{ has an element of exact order } m\} \tag{7}$$

has density 0. We take a model $\pi : A \to C$ for A as a family of abelian varieties over an open subset $C \subset \mathbb{P}^1$. (In other words, we take any model for A over $\mathbb{P}^1$ and discard a finite number of bad fibers.) A point $P \in A(\mathbb{Q}(T))$ corresponds to a section $P : C \to A$, and the specialization P_t is just the image of t under the map P. Further, the multiplication-by-m map $[m] : A \to A$ is a morphism of A over C.

Choose representatives $P_1, P_2, \ldots, P_w \in \Gamma$ for the finite group $\Gamma/m\Gamma$. For each P_i, write the irreducible components of $[m]^{-1}(P_i(C))$ as

$$[m]^{-1}(P_i(C)) = \bigcup_j Q_{ij}(C) \cup \bigcup_k C_{ik},$$

where the $Q_{ij}(C)$'s are the components of degree 1 over C (hence equal to the image of a section $Q_{ij} \in A(\mathbb{Q}(T))$) and the C_{ik}'s are components of degree at least 2 over C. Then the set (7) is contained in the finite union

$$\bigcup_{i,k} \mathrm{Image}\big(C_{ik}(\mathbb{Q}) \xrightarrow{\pi} C(\mathbb{Q})\big). \tag{8}$$

Intuitively, the fact that the map π has degree at least 2 on each component C_{ik} forces the image $\pi\big(C_{ik}(\mathbb{Q})\big)$ to be thinly spread in $C(\mathbb{Q})$. This intuition is made precise by the Hilbert Irreducibility Theorem. In particular, the version of the Hilbert Irreducibility Theorem proven in [7] and [10] implies that the set (8) has density 0 in both $\mathbb{Q}$ and $\mathbb{Z}$. This completes the proof of Corollary 1.3.

Acknowledgements. I would like the thank the referee for numerous helpful suggestions.

REFERENCES

1. G. Call, *Variation of the local heights on an algebraic family of abelian varieties*, Number Theory (J.-M. De Koninck & C. Levesque, eds.), Walter de Gruyter, Berlin-New York, 1989.
2. P.X. Gallagher, *A larger sieve*, Acta Arith. **18** (1971), 77–81.

3. W Green, *Heights in families of abelian varieties*, Duke Math. J. **58** (1989), 617–632.
4. R. Gupta, K. Ramsay, *Indivisible points on families of elliptic curves*, J. Number Theory **63** (1997), 357–372.
5. C. Hooley, *Applications of Sieve Methods to the Theory of Numbers*, Cambridge Univ. Press, Cambridge, 1976.
6. S. Lang, *Abelian Varieties*, Interscience, New York, 1959.
7. ———, *Fundamentals of Diophantine Geometry*, Springer-Verlag, New York, 1983.
8. M.R. Murty, M. Rosen, J.H. Silverman, *Variations on a theme of Romanoff*, Intern. J. Math. **7** (1996), 373–391.
9. A. Néron, *Propriétés arithmétiques et géométriques rattachés à la notion de rang d'une courbe algébrique dans un corps*, Bull. Soc. Math. France **80** (1952), 101–166.
10. J.-P. Serre, *Lectures on the Mordell-Weil Theorem*, Aspects of Mathematics E15, Friedr. Vieweg & Sohn, Braunschweig/Wiesbaden, 1989.
11. J.H. Silverman, *Heights and the specialization map for families of abelian varieties*, J. Reine Angew. Math. **342** (1983), 197–211.
12. ———, *Lower bounds for height functions*, Duke Math. J. **51** (1984), 395–403.
13. ———, *Divisibility of the specialization map for families of elliptic curves*, Amer. J. Math. **107** (1985), 555–565.
14. ———, *The Arithmetic of Elliptic Curves*, Graduate Texts in Math., vol. 106, Springer-Verlag, Berlin and New York, 1986.
15. J. Tate, *Variation of the canonical height of a point depending on a parameter*, Amer. J. Math. **105** (1983), 287–294.

MATHEMATICS DEPARTMENT BOX 1917 BROWN UNIVERSITY PROVIDENCE, RI 02912 USA

E-mail address: `jhs@math.brown.edu`

Other *Mathematics and Its Applications* titles of interest:

P.H. Sellers: *Combinatorial Complexes. A Mathematical Theory of Algorithms.* 1979, 200 pp. ISBN 90-277-1000-7

P.M. Cohn: *Universal Algebra.* 1981, 432 pp. ISBN 90-277-1213- 1 (hb), ISBN 90-277-1254-9 (pb)

J. Mockor: *Groups of Divisibility.* 1983, 192 pp. ISBN 90-277-1539-4

A. Wwarynczyk: *Group Representations and Special Functions.* 1986, 704 pp. ISBN 90-277-2294-3 (pb), ISBN 90-277-1269-7 (hb)

I. Bucur: *Selected Topics in Algebra and its Interrelations with Logic, Number Theory and Algebraic Geometry.* 1984, 416 pp. ISBN 90-277-1671-4

H. Walther: *Ten Applications of Graph Theory.* 1985, 264 pp. ISBN 90-277-1599-8

L. Beran: *Orthomodular Lattices. Algebraic Approach.* 1985, 416 pp. ISBN 90-277-1715-X

A. Pazman: *Foundations of Optimum Experimental Design.* 1986, 248 pp. ISBN 90-277-1865-2

K. Wagner and G. Wechsung: *Computational Complexity.* 1986, 552 pp. ISBN 90-277-2146-7

A.N. Philippou, G.E. Bergum and A.F. Horodam (eds.): *Fibonacci Numbers and Their Applications.* 1986, 328 pp. ISBN 90-277-2234-X

C. Nastasescu and F. van Oystaeyen: *Dimensions of Ring Theory.* 1987, 372 pp. ISBN 90-277-2461-X

Shang-Ching Chou: *Mechanical Geometry Theorem Proving.* 1987, 376 pp. ISBN 90-277-2650-7

D. Przeworska-Rolewicz: *Algebraic Analysis.* 1988, 640 pp. ISBN 90-277-2443-1

C.T.J. Dodson: *Categories, Bundles and Spacetime Topology.* 1988, 264 pp. ISBN 90-277-2771-6

V.D. Goppa: *Geometry and Codes.* 1988, 168 pp. ISBN 90-277-2776-7

A.A. Markov and N.M. Nagorny: *The Theory of Algorithms.* 1988, 396 pp. ISBN 90-277-2773-2

E. Kratzel: *Lattice Points.* 1989, 322 pp. ISBN 90-277-2733-3

A.M.W. Glass and W.Ch. Holland (eds.): *Lattice-Ordered Groups. Advances and Techniques.* 1989, 400 pp. ISBN 0-7923-0116-1

N.E. Hurt: *Phase Retrieval and Zero Crossings: Mathematical Methods in Image Reconstruction.* 1989, 320 pp. ISBN 0-7923-0210-9

Du Dingzhu and Hu Guoding (eds.): *Combinatorics, Computing and Complexity.* 1989, 248 pp. ISBN 0-7923-0308-3

Other *Mathematics and Its Applications* titles of interest:

A.Ya. Helemskii: *The Homology of Banach and Topological Algebras*. 1989, 356 pp. ISBN 0-7923-0217-6

J. Martinez (ed.): *Ordered Algebraic Structures*. 1989, 304 pp. ISBN 0-7923-0489-6

V.I. Varshavsky: *Self-Timed Control of Concurrent Processes. The Design of Aperiodic Logical Circuits in Computers and Discrete Systems*. 1989, 428 pp. ISBN 0-7923-0525-6

E. Goles and S. Martinez: *Neural and Automata Networks. Dynamical Behavior and Applications*. 1990, 264 pp. ISBN 0-7923-0632-5

A. Crumeyrolle: *Orthogonal and Symplectic Clifford Algebras. Spinor Structures*. 1990, 364 pp. ISBN 0-7923-0541-8

S. Albeverio, Ph. Blanchard and D. Testard (eds.): *Stochastics, Algebra and Analysis in Classical and Quantum Dynamics*. 1990, 264 pp. ISBN 0-7923-0637-6

G. Karpilovsky: *Symmetric and G-Algebras. With Applications to Group Representations*. 1990, 384 pp. ISBN 0-7923-0761-5

J. Bosak: *Decomposition of Graphs*. 1990, 268 pp. ISBN 0-7923-0747-X

J. Adamek and V. Trnkova: *Automata and Algebras in Categories*. 1990, 488 pp. ISBN 0-7923-0010-6

A.B. Venkov: *Spectral Theory of Automorphic Functions and Its Applications*. 1991, 280 pp. ISBN 0-7923-0487-X

M.A. Tsfasman and S.G. Vladuts: *Algebraic Geometric Codes*. 1991, 668 pp. ISBN 0-7923-0727-5

H.J. Voss: *Cycles and Bridges in Graphs*. 1991, 288 pp. ISBN 0-7923-0899-9

V.K. Kharchenko: *Automorphisms and Derivations of Associative Rings*. 1991, 386 pp. ISBN 0-7923-1382-8

A.Yu. Olshanskii: *Geometry of Defining Relations in Groups*. 1991, 513 pp. ISBN 0-7923-1394-1

F. Brackx and D. Constales: *Computer Algebra with LISP and REDUCE. An Introduction to Computer-Aided Pure Mathematics*. 1992, 286 pp. ISBN 0-7923-1441-7

N.M. Korobov: *Exponential Sums and their Applications*. 1992, 210 pp. ISBN 0-7923-1647-9

D.G. Skordev: *Computability in Combinatory Spaces. An Algebraic Generalization of Abstract First Order Computability*. 1992, 320 pp. ISBN 0-7923-1576-6

E. Goles and S. Martinez: *Statistical Physics, Automata Networks and Dynamical Systems*. 1992, 208 pp. ISBN 0-7923-1595-2

M.A. Frumkin: *Systolic Computations*. 1992, 320 pp. ISBN 0-7923-1708-4

J. Alajbegovic and J. Mockor: *Approximation Theorems in Commutative Algebra*. 1992, 330 pp. ISBN 0-7923-1948-6

Other *Mathematics and Its Applications* titles of interest:

I.A. Faradzev, A.A. Ivanov, M.M. Klin and A.J. Woldar: *Investigations in Algebraic Theory of Combinatorial Objects*. 1993, 516 pp. ISBN 0-7923-1927-3

I.E. Shparlinski: *Computational and Algorithmic Problems in Finite Fields*. 1992, 266 pp. ISBN 0-7923-2057-3

P. Feinsilver and R. Schott: *Algebraic Structures and Operator Calculus. Vol. I. Representations and Probability Theory*. 1993, 224 pp. ISBN 0-7923-2116-2

A.G. Pinus: *Boolean Constructions in Universal Algebras*. 1993, 350 pp. ISBN 0-7923-2117-0

V.V. Alexandrov and N.D. Gorsky: *Image Representation and Processing. A Recursive Approach*. 1993, 200 pp. ISBN 0-7923-2136-7

L.A. Bokut' and G.P. Kukin: *Algorithmic and Combinatorial Algebra*. 1994, 384 pp. ISBN 0-7923-2313-0

Y. Bahturin: *Basic Structures of Modern Algebra*. 1993, 419 pp. ISBN 0-7923-2459-5

R. Krichevsky: *Universal Compression and Retrieval*. 1994, 219 pp. ISBN 0-7923-2672-5

A. Elduque and H.C. Myung: *Mutations of Alternative Algebras*. 1994, 226 pp. ISBN 0-7923-2735-7

E. Goles and S. Martínez (eds.): *Cellular Automata, Dynamical Systems and Neural Networks*. 1994, 189 pp. ISBN 0-7923-2772-1

A.G. Kusraev and S.S. Kutateladze: *Nonstandard Methods of Analysis*. 1994, 444 pp. ISBN 0-7923-2892-2

P. Feinsilver and R. Schott: *Algebraic Structures and Operator Calculus. Vol. II. Special Functions and Computer Science*. 1994, 148 pp. ISBN 0-7923-2921-X

V.M. Kopytov and N. Ya. Medvedev: *The Theory of Lattice-Ordered Groups*. 1994, 400 pp. ISBN 0-7923-3169-9

H. Inassaridze: *Algebraic K-Theory*. 1995, 438 pp. ISBN 0-7923-3185-0

C. Mortensen: *Inconsistent Mathematics*. 1995, 155 pp. ISBN 0-7923-3186-9

R. Abłamowicz and P. Lounesto (eds.): *Clifford Algebras and Spinor Structures*. A Special Volume Dedicated to the Memory of Albert Crumeyrolle (1919–1992). 1995, 421 pp. ISBN 0-7923-3366-7

W. Bosma and A. van der Poorten (eds.), *Computational Algebra and Number Theory*. 1995, 336 pp. ISBN 0-7923-3501-5

A.L. Rosenberg: *Noncommutative Algebraic Geometry and Representations of Quantized Algebras*. 1995, 316 pp. ISBN 0-7923-3575-9

L. Yanpei: *Embeddability in Graphs*. 1995, 400 pp. ISBN 0-7923-3648-8

B.S. Stechkin and V.I. Baranov: *Extremal Combinatorial Problems and Their Applications*. 1995, 205 pp. ISBN 0-7923-3631-3

Other *Mathematics and Its Applications* titles of interest:

Y. Fong, H.E. Bell, W.-F. Ke, G. Mason and G. Pilz (eds.): *Near-Rings and Near-Fields*. 1995, 278 pp. ISBN 0-7923-3635-6

A. Facchini and C. Menini (eds.): *Abelian Groups and Modules*. (Proceedings of the Padova Conference, Padova, Italy, June 23–July 1, 1994). 1995, 537 pp. ISBN 0-7923-3756-5

D. Dikranjan and W. Tholen: *Categorical Structure of Closure Operators*. With Applications to Topology, Algebra and Discrete Mathematics. 1995, 376 pp. ISBN 0-7923-3772-7

A.D. Korshunov (ed.): *Discrete Analysis and Operations Research*. 1996, 351 pp. ISBN 0-7923-3866-9

P. Feinsilver and R. Schott: *Algebraic Structures and Operator Calculus*. Vol. III: Representations of Lie Groups. 1996, 238 pp. ISBN 0-7923-3834-0

M. Gasca and C.A. Micchelli (eds.): *Total Positivity and Its Applications*. 1996, 528 pp. ISBN 0-7923-3924-X

W.D. Wallis (ed.): *Computational and Constructive Design Theory*. 1996, 368 pp. ISBN 0-7923-4015-9

F. Cacace and G. Lamperti: *Advanced Relational Programming*. 1996, 410 pp. ISBN 0-7923-4081-7

N.M. Martin and S. Pollard: *Closure Spaces and Logic*. 1996, 248 pp. ISBN 0-7923-4110-4

A.D. Korshunov (ed.): *Operations Research and Discrete Analysis*. 1997, 340 pp. ISBN 0-7923-4334-4

W.D. Wallis: *One-Factorizations*. 1997, 256 pp. ISBN 0-7923-4323-9

G. Weaver: *Henkin–Keisler Models*. 1997, 266 pp. ISBN 0-7923-4366-2

V.N. Kolokoltsov and V.P. Maslov: *Idempotent Analysis and Its Applications*. 1997, 318 pp. ISBN 0-7923-4509-6

J.P. Ward: *Quaternions and Cayley Numbers*. Algebra and Applications. 1997, 250 pp. ISBN 0-7923-4513-4

E.S. Ljapin and A.E. Evseev: *The Theory of Partial Algebraic Operations*. 1997, 245 pp. ISBN 0-7923-4609-2

S. Ayupov, A. Rakhimov and S. Usmanov: *Jordan, Real and Lie Structures in Operator Algebras*. 1997, 235 pp. ISBN 0-7923-4684-X

A. Khrennikov: *Non-Archimedean Analysis: Quantum Paradoxes, Dynamical Systems and Biological Models*. 1997, 389 pp. ISBN 0-7923-4800-1

G. Saad and M.J. Thomsen (eds.): *Nearrings, Nearfields and K-Loops*. (Proceedings of the Conference on Nearrings and Nearfields, Hamburg, Germany. July 30–August 6, 1995). 1997, 458 pp. ISBN 0-7923-4799-4

Other *Mathematics and Its Applications* titles of interest:

L.A. Lambe and D.E. Radford: *Introduction to the Quantum Yang–Baxter Equation and Quantum Groups: An Algebraic Approach*. 1997, 314 pp. ISBN 0-7923-4721-8

H. Inassaridze: *Non-Abelian Homological Algebra and Its Applications*. 1997, 271 pp. ISBN 0-7923-4718-8

B.P. Komrakov, I.S. Krasil'shchik, G.L. Litvinov and A.B. Sossinsky (eds.): *Lie Groups and Lie Algebras*. Their Representations, Generalisations and Applications. 1998, 358 pp. ISBN 0-7923-4916-4

A.K. Prykarpatsky and I.V. Mykytiuk (eds.): *Algebraic Integrability of Nonlinear Dynamical Systems on Manifolds*. Classical and Quantum Aspects. 1998, 554 pp. ISBN 0-7923-5090-1

A.A. Tuganbaev: *Semidistributive Modules and Rings*. 1998, 362 pp. ISBN 0-7923-5209-2

M.V. Kondratieva, A.B. Levin, A.V. Mikhalev and E.V. Pankratiev: *Differential and Difference Dimension Polynomials*. 1999, 436 pp. ISBN 0-7923-5484-2

K. Yang: *Meromorphic Functions and Projective Curves*. 1999, 202 pp. ISBN 0-7923-5505-9

V. Kolmanovskii and A. Myshkis: *Introduction to the Theory and Applications of Functional Differential Equations*. 1999, 664 pp. ISBN 0-7923-5504-0

S.D. Ahlgren, G.E. Andrews and K. Ono (eds.): *Topics in Number Theory*. In Honor of B. Gordon and S. Chowla. 1999, 266 pp. ISBN 0-7923-5583-0